Estuaries of the World

Series editor

Jean-Paul Ducrotoy

More information about this series at http://www.springer.com/series/11705

Jing Zhang
Editor

Ecological Continuum from the Changjiang (Yangtze River) Watersheds to the East China Sea Continental Margin

Springer

Editor
Jing Zhang
State Key Laboratory of Estuarine and Coastal Research
East China Normal University
Shanghai
China

ISSN 2214-1553 ISSN 2214-1561 (electronic)
Estuaries of the World
ISBN 978-3-319-16338-3 ISBN 978-3-319-16339-0 (eBook)
DOI 10.1007/978-3-319-16339-0

Library of Congress Control Number: 2015934914

Springer Cham Heidelberg New York Dordrecht London

Printed on acid-free paper

Springer International Publishing AG Switzerland is part of Springer Science+Business Media
(www.springer.com)

Preface

The idea of publishing a monograph on the Changjiang (Yangtze River) Estuary came to us about 5 years ago, when a research team of State Key Laboratory of Estuarine and Coastal Research, East China Normal University (Shanghai) was awarded by the Natural Sciences Foundation of China (NSFC) with the title "Creative Research Group". This award offered an opportunity for creating a network for cross-disciplinary study from watersheds of Changjiang to the continental margin of East China Sea through joint efforts of East China Normal University with collaborations of other research institutions in China.

Among the top five largest rivers on this planet, Changjiang is unique in several aspects. For instance, the population density of Changjiang watersheds is on average ca. 250/km^2, which is high relative to other river systems like the Amazon, Congo, Orinoco, and Brahmaputra. Hydraulic engineering is also very extensive in the watersheds of Changjiang, more than 50×10^3 reservoirs have been created through dam constructions over the past 60 years; agricultural activities are intensive with application of huge amount of chemical fertilizers that cause continuous increase in the concentration of plant nutrients (e.g. dissolved inorganic nitrogen) in river water over the past five decades. Therefore, if someone would like to know how human activities affect the river drainage basin with consequences of change in seaward fluxes, the lessons we have learned through cross-disciplinary studies in the Changjiang would be an example to refer to.

The East China Sea has 71 % of shelf area, which is among the top 10 largest continental margins of the world. The western boundary current of the Pacific Ocean at sub-tropical and temperate region, i.e. Kuroshio, flows through the East China Sea with a water flux of 25–30 Sv (1 Sv = 10^6 m^3/s), similar to the Gulf Stream of the North Atlantic Ocean. So if a colleague would like to examine how the interaction between boundary current and large rivers over a wide shelf area can affect the coastal ecosystems through biogeochemical dynamics, the results from the East China Sea can be an illustration. Indeed, when examining the complex aspects of land-ocean interactions, watersheds of the river and the recipient coastal environment have to be taken into consideration together, both for the merit of scientific research and for the purpose of adaptive management of ecosystems.

People may believe that rivers have an important effect on the coastal environment, indeed standing by the bank and/or on the bridge, one can be deeply impressed by the large quantity of water and sediments carried by the river flow to the ocean; for example, the Changjiang has a water flux of ca. 30×10^3 m^3/s and a river mouth 90 km wide. This impression is, however, not necessarily true. In the case of East China Sea, the exchange of Kuroshio with shelf amounts to 3–4 Sv of water, including the water flow through Taiwan Strait, which is almost two orders of magnitude higher than the collective fresh water influx in which Changjiang accounts for 90 %. Thus, the exchange of energy and materials across the shelf break can have a dramatic impact not only on the circulation and biogeochemical cycles of coastal environment that drives the function of ecosystem through bottom-up mechanism, but also the fate of terrestrial pollutants over the shelf, and the export flux into the open ocean, with strong feedback to the atmosphere as well (e.g. uptake and/or emission of CO_2). In this aspect, the multidisciplinary studies of East China Sea have provided the audience with knowledge and experience.

The systematic research activities of Changjiang started in the 1950s, and the Institute of Estuarine and Coastal Research at East China Normal University was established in 1957, among the first groups of research institutions focusing on the estuarine and coastal sciences. In the 1990s, the Institute of Estuarine and Coastal Research at East China Normal University was promoted to be the State Kay Laboratory of Estuarine and Coastal Research with three cross-linked foci: geomorphology and climate change, hydrography and sedimentary dynamics, biogeochemistry and ecology, with broad collaborations with other research institutions in China. Thus, the results presented in this book show the recent research progress that is based on the synthesis of existing knowledge and digestion of data generated by a multidisciplinary team. It should be kept in mind, however, that this monograph provides by no means a historical overview of scientific approaches on the "Ecological Continuum from the Changjiang (Yangtze River) Watersheds to the East China Sea Continental Margin" nor a summary of previous research results. Rather, in this book we tried to examine the very dynamic system in the continuum from the watersheds of Changjiang to the continental margin of East China Sea through the critical point of view of ecosystem function, which needs absolutely an approach of cross-disciplinary studies. As the reader can see from this work, the individual chapters of this monograph cover the research disciplines of hydro-dynamics, sedimentary geology, geography and geo-morphology, ecology, fishery, as well as biogeochemistry and environmental sciences, etc.

We believe that knowledge generated from the watersheds of Changjiang across to the East China Sea is useful as reference to the scholar of land-ocean studies in the continuum of the river drainage basin to the continental margins. We hope that the lessons learned from the Changjiang watersheds and East China Sea will help colleagues to avoid having similar negative effects when promoting economic innovations in other places in the world.

Undertaking cross-disciplinary studies is a team work and needs continuous financial and logistic countenance. Although at the end of individual chapters my colleagues have expressed their gratitude to their colleagues, I would like to thank those people again, including our friends and students who have helped in the elaboration of field and sea-going observations as well as the generation of data in the laboratory. Also, I appreciate very much our funding agencies, particularly the Natural Sciences Foundation of China (NSFC), Ministry of Education (MOE), Ministry of Science and Technology (MOST) as well as the Municipality of Shanghai through its Commission of Science and Technology, for their continuous supports to the research efforts in this monograph.

Before ending this preface, we would like to take this opportunity to express our gratitude to Prof. Jean-Paul Ducrotoy who guided us in preparing for this monograph. Colleagues from Springer, including Alexandrine Cheronet, Judith Terpos, and Betty van Herk, etc., are acknowledged for their very kind support in publishing this work.

Jing Zhang

Prologue

The importance of understanding the estuary, its watershed and the adjacent coastal habitats as a continuum.

Coastal ecosystems, estuaries in particular, are under strong influence of their watershed as they consist of semi-enclosed basins receiving water directly from a riverine basin. They are permanently connected to the sea whose waters are diluted by freshwater drainage and run-off. Estuaries maintain exceptionally high levels of biological productivity and play important ecological roles, including water purification, 'exporting' nutrients and organic materials to outside waters through tidal circulation; providing habitats for a number of commercially or recreationally valuable fish species; and serving the needs of migratory bird species which require areas for breeding and/or sanctuary for their young. In fact, influences come both from the adjacent land and the ocean, particularly when the tide exerts a significant effect on currents dynamics.

The importance of hydrology to understand the functioning of estuaries has been widely recognized by the scientific community. In this book "Ecological Continuum from the Changjiang (Yangtze River) Watersheds to the East China Sea Continental Margin", in the series Estuaries of the World, the aim is to understand the relationship between hydrology and function of estuarine wetlands and peripheral estuarine areas, stressing their importance and the need for preserving and restoring them when necessary. In mega-tidal estuaries, differential degrees of sediment mobility show crucial effects on the zonation of flora and fauna in the tidal flat. Spatial and temporal variations of tide as well as its harmonic constituents, form and distortion in the Changjiang Estuary were obtained in order to better understand hydrological processes that are responsible for the evolution of the variability within and across the various habitats and their functional value. They govern the movement of materials and organisms between the estuary and adjacent marshes, the watershed, groundwater and the atmosphere. Wind, especially, can push forward intruding saltwater in the Changjiang Estuary.

Similarly, the estuary connects marine and continental influences by sedimentary processes that combine fluvial and marine processes. The physical forces involved in shaping the morphology of an estuary and especially sediment distribution vary along a gradient from marine sands in the mouth to muddy sand and mud at the head of the estuary. In the Changjiang, seasonal variations in water and sediment discharge are monsoon-governed. Particle fluxes to mid-water depths in the adjacent sea are mainly controlled by fluvial discharge and primary production. It is of note that fluvial discharge could be responsible for the higher lithogenic fluxes during autumn and winter, while high primary production could play a key role in generating biogenic particles during spring and summer. Siltation is related to sedimentation rates associating themselves with the highest tidal amplitudes (in relation to the lunar cycle), resulting from the increased availability of sediments associated with high concentration of suspended matter in the turbidity maximum. The evolution of such systems results in maturity stages more or less advanced, shaped by human intervention. The accelerated filling of estuaries develops from the movement of sedimentary deposits by reducing the available space in the upper estuary due to anthropic land claim. In the Changjiang, estuarine and coastal responses to the decline of river sediment discharge have had a substantial impact on the geography of the mouth of the estuary. In the future, the combined effects of the tide and the increase in average sea level will contribute to change the position of the coastline, with varying degrees of potential consequences from erosion. Monitoring its position requires the implementation of hydro-sedimentary modelling that takes into account both climate change, astronomical and average sea level using tide models, wave and sediment transport quantification. In this book, monitoring data and question-orientated research activities cover a period of ca. 20 years, with the erosion of mudflats being taken into account in the models.

Nutrient material recycled in the mudflats adopts a seasonal behaviour, which is mainly influenced by changes in hydrodynamics. The release, including ammonium, is mainly related to the mineralization of organic matter, while phosphate release is delayed, probably because of the combination of phosphate with organic matter and/or his co-precipitation with calcium. In the Changjiang, changes in nutrient concentrations have been recorded in relation to the construction of dams upstream, but in the lower estuary nitrification is rapid and complete. This behaviour is related to the different sedimentary hydro conditions of the systems because of nitrifying bacteria associated with suspended particles that depend on the estuarine dynamics. Here, multiple biomarkers have been used to elucidate the complex relationship between a large river and the estuarine system. Particulate phosphorus in the turbidity maximum area is a possible source of dissolved inorganic phosphate, available for algae growth. Instead of playing a nutrient conservation role, the turbidity maximum of the estuary area can promote coastal eutrophication as the biogeochemistry of nutrients and trace elements over the East China Sea Shelf is linked to river input.

Benthic organisms respond to changes in particulate fluxes of mineral and organic matter in relation to the depth of the photic layer. The succession of plant communities on salt marshes in the Changjiang Estuary is described in the book and the patterns of spatial and temporal variation on salt marshes are analysed on the basis of remote sensing mapping combined with field data from the survey of sampling plots. In particular, the expansion of the exotic species *Spartina alterniflora* after its introduction to the salt marshes in the Changjiang Estuary is described and analysed. The future development, management and preservation of salt marsh vegetation and biodiversity is discussed.

It is essential to understand how coastal discharge has a bearing on marine environment quality. However, despite organic and chemical contamination from human activities, estuaries, even industrialized, remain essential relay ecosystems. Estuaries constitute main transition zones or ecotones between land, the ocean and the atmosphere. As ecotones, they need to be preserved in their complexity. In order to define strategies compatible with conservation and sustainable development at the local, national and regional levels, environmental aspects must be integrated in the management of estuaries, which must rely on thorough collaboration between and mutual understanding of all actors and stakeholders. In Eastern China, domestic sewage discharge per capita has increased significantly in the past few decades, especially in metropolitan cities such as Shanghai. Illegal reclamation and dumping have also had adverse effects on the marine environment. Increased use of fertilizers to promote higher agricultural crop yields has directly impacted water quality in the East China Sea, leading to widespread eutrophication, followed by frequent harmful algal blooms. Incidents such as oil spills have had serious impacts on the marine ecosystem. River inputs represent the largest source of water pollution in the East China Sea, but sewage outfalls, mariculture, engineering activities, agriculture and non-point source pollutants are other key contributors. Resting on a rigorous scientific approach, restoring ecological functionalities in an estuary is dependent on efficient procedures of socio-ecological. For making interdisciplinarity work, socio-economics needs to be considered in the early stages of the elaboration of any restoration programme. Taking into account the ecological continuum between the estuary, its watershed and the adjacent coastal habitats is essential in providing robust scientific information to decision-makers and managers. However, the concept of continuum should also be applied to decision-making processes through the integration of science into decision-making. The prioritisation of human "well-being" might be a good way to focus on pressures and impacts inflicted to ecosystems, emphasising the ensuing suboptimal health of 'natural systems', including human and animal life in existence within these systems. Therefore, the eco-systemic approach requires understanding of, first, social, political, economic and, secondly, ecological structures and functions fundamental to healthcare provision for growing populations. Ecology stands out as a key component considering such a continuum and to understanding the instrumental factors to decision-making bottlenecks and to identify weaknesses in socio-economic systems.

Jing Zhang

Contents

Editor Biography and Contributors

Editor Biography

Jing Zhang is a full professor at the State Key Laboratory of Estuarine and Coastal Research, East China Normal University at Shanghai, China. While working as a mentor for postgraduate studies, he teaches chemical oceanography and biogeochemistry. He completed undergraduate studies in Nanjing University (China), received his M.Sc. from Ocean University of Qingdao (China) and Ph.D. from Universite Pierre et Marie Curie (Paris 6) in France.

The research experience of JZ has been in multidisciplinary studies from river watersheds to continental margins, where he and his colleagues worked on the transfer and transformation of chemical elements at land–ocean interfaces and the impacts on the food-web structure and ecosystem function.

Contributors

Hongyan Bao State Key Laboratory of Estuarine and Coastal Research, East China Normal University, Shanghai, China

Shuchai Gan State Key Laboratory of Estuarine and Coastal Research, East China Normal University, Shanghai, China

Qian Gao Chinese Academy of Fisheries Sciences, East China Sea Fisheries Research Institute, Shanghai, China

Leicheng Guo State Key Laboratory of Estuarine and Coastal Research, East China Normal University, Shanghai, China

Qing He State Key Laboratory of Estuarine and Coastal Research, East China Normal University, Shanghai, China

H.M. Huang South China Sea Marine Engineering and Environment Institute, State Oceanic Administration, Guangzhou, China

Lu Li State Key Laboratory of Estuarine and Coastal Research, East China Normal University, Shanghai, China

Hong Liu State Key Laboratory of Estuarine and Coastal Research, East China Normal University, Shanghai, China

Qian Long Central Library, East China Normal University, Shanghai, China

Xiao Min Shen Chinese Academy of Fisheries Sciences, East China Sea Fisheries Research Institute, Shanghai, China

Ya Wang State Key Laboratory of Estuarine and Coastal Research, East China Normal University, Shanghai, China

Hui Wu State Key Laboratory of Estuarine and Coastal Research, East China Normal University, Shanghai, China

Ying Wu State Key Laboratory of Estuarine and Coastal Research, East China Normal University, Shanghai, China

Zhao Li Xu Chinese Academy of Fisheries Sciences, East China Sea Fisheries Research Institute, Shanghai, China

H.F. Yang State Key Laboratory of Estuarine and Coastal Research, East China Normal University, Shanghai, China

S.L. Yang State Key Laboratory of Estuarine and Coastal Research, East China Normal University, Shanghai, China

L. Yuan State Key Laboratory of Estuarine and Coastal Research, East China Normal University, Shanghai, China

Jing Zhang State Key Laboratory of Estuarine and Coastal Research, East China Normal University, Putuo, Shanghai, China

L.Q. Zhang State Key Laboratory of Estuarine and Coastal Research, East China Normal University, Shanghai, China

Ying Ying Zhang School of Environmental Science and Engineering, Yancheng Institute of Technology, Yancheng, Jiangsu, China

Jianrong Zhu State Key Laboratory of Estuarine and Coastal Research, East China Normal University, Shanghai, China

Zhuoyi Zhu State Key Laboratory of Estuarine and Coastal Research, East China Normal University, Shanghai, China

1 Land–Ocean Interactions Between the Changjiang (Yangtze River) Watersheds to the East China Sea Continental Margin

Jing Zhang

Abstract

Energy and material fluxes from watersheds converge in coastal zones and at open ocean boundaries, and climate variability, biogeochemical processes, and anthropogenic perturbations in the coastal zone have strong feedbacks with other components of Earth's physical and social systems. Land–ocean interactions, forced by dynamic processes and natural and human activities, are concentrated in the zone between upstream watersheds, in this study the Changjiang (Yangtze River) Estuary, and the open boundary of the East China Sea. Studies of land–ocean interfaces in this region (e.g., Changjiang Estuary) have been carried out since the 1940s, and the knowledge based on previous research forms an important foundation for an integrated discussion of the hydrography, geochemistry, geography, ecology, and biology of the system. The sustainable development of coastal ecosystems depends on how well we understand the function and response of these systems under combined natural and human forcings, and adaptation and management policies depend on the understandings of the way that land–ocean interactions respond to these external forcings.

Keywords

Changjiang • Watersheds • Estuary • East China Sea • Multi-disciplinary studies

1.1 Geographic Setting

Estuaries and coasts are the major focus of land–sea interactions because it is in this narrow zone where land, atmosphere, and oceanic processes converge and have strong feedbacks with social systems. While dynamic ocean processes dominate in open waters, the coastal environment, including marginal seas and semi-enclosed basins, is also influenced by land-based forcings. Estuaries and coasts are linked to the early history of our civilizations, and more than 40 % of the human population is located within 100 km of the coast. Of the 20 largest cities in the world (with populations >10 million), 15 are located in coastal areas. The goods and services provided by the coastal areas accounts for 70 % to 80 % of the total economic value of the global ecosystem (Martínez et al. 2007; Moser et al. 2012). Moreover, coastal environments have played an important role in the evolution of human societies. For instance, around 1500 years ago, coastal areas began to be reclaimed in Asia for salt production and agriculture, and the early history of marine trade in the Middle East can be traced back to before the tenth century (Chen 2000).

There have also been conflicts over the past 200 years in many coastal areas between the limited spatial scale of coastal systems and resources, and rapid economic development and urbanization. In recent history, coastal systems have been stressed by a combination of physical processes (e.g., climate change and sea-level rise) and anthropogenic influences (e.g., land-use change, pollutant drainage, reclamation, and overfishing (Boesch 2002). The sustainable management of coastal environments therefore requires an

J. Zhang (✉)
State Key Laboratory of Estuarine and Coastal Research,
East China Normal University, 3663 Zhongshan Road North,
Putuo 200062, Shanghai, China
e-mail: jzhang@sklec.ecnu.edu.cn

J. Zhang (ed.), *Ecological Continuum from the Changjiang (Yangtze River) Watersheds to the East China Sea Continental Margin*, Estuaries of the World,
DOI 10.1007/978-3-319-16339-0_1, © Springer International Publishing Switzerland 2015

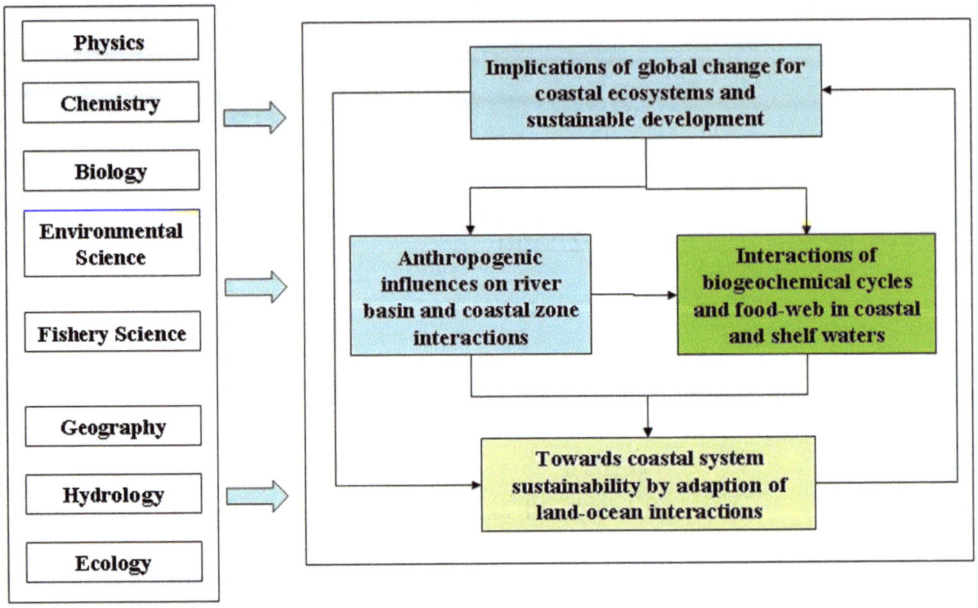

Fig. 1.1 Linkages between different research activities at the land–ocean interfaces to meet the requirements for ecosystem and human sustainability, modified from LOICZ (2005)

interdisciplinary approach from both the natural and social sciences (Fig. 1.1).

In the Asia–Pacific region, the Changjiang (Yangtze River) watersheds and East China Sea form a unique system. The area has experienced rapid economic development and strong population growth in countries such as China, Japan, and Korea. It is also impacted by a monsoon climate, strong western boundary circulation (i.e., the Kuroshio Current), as well as extreme weather events, such as typhoons in the northwest Pacific Ocean (Table 1.1). Thus, studies of land–sea interface processes in this area provide knowledge of how ecosystems evolve under rapid social and economic perturbations and how they respond to multiple sources of stress from both natural and anthropogenic forcings.

Table 1.1 Geographical and hydrological features of the study area, including the watersheds of Changjiang, Changjiang Estuary, and East China Sea

Area	Geography and hydrography
Watersheds of the Changjiang	Elevation difference (H) ≥ 5400 m and length (L) 6300 km, with an average H/L of ca. 0.86×10^{-3}. The watersheds is located at 90.5–122.5°E and 24.5–35.8°N with a surface area of 1.8×10^6 km^2
Changjiang Estuary	The tidal limit is the area up to Datong in Anhui Province, at ca. 650 km upstream from the river mouth. The estuary extends shelf-ward and is covered in an area by riverine effluent plumes, which have a strong seasonality depending on the freshwater discharge, tides, and the monsoon
East China Sea	An area of ca. 770×10^3 km^2 between the Chinese mainland and the Ryukyu Islands. There is a wide shelf in the west, and the deep Okinawa Trough (ca. 2720 m) in the east. The Changjiang, Taiwan Strait Warm Water, and Kuroshio currents are major water sources in this region

1.1.1 Watersheds of the Changjiang

The Changjiang (Note: here "jiang" means river in Chinese) originates in the Tanggulashan (mountainous region) at an elevation of higher than 5400 m above sea level (asl) and runs for more than 6300 km to the ocean (Fig. 1.2). The Changjiang watersheds covers an area of 1.8×10^6 km^2 and represents around 20 % of China's land surface. The long-term (1950–2000) average water discharge measured at Datong Hydrographic Station (i.e., 20 km upstream from Tongling in Fig. 1.1), approximately 650 km upstream from the river mouth, is 1×10^{12} m^3/yr. The interannual variation in water discharge from the Changjiang is also remarkably high as a result of the temperate climate and East Asian monsoon. For the period 1950–2000, discharge ranged from 0.7×10^{12} m^3/yr to 1.4×10^{12} m^3/yr. Annual rainfall in the watersheds ranges from 400 mm/yr to 2590 mm/yr with an average of 1100 mm/yr.

Historically, the seaward sediment flux of the Changjiang was about 0.5×10^9 tons/yr at Datong, with another 10 % carried as bed load. However, the sediment load of the Changjiang has fallen considerably over the last 10–20 years and was 0.185×10^9 tons/yr in 2010 and 0.0718×10^9 tons/yr in 2011 (CWRC 2012). The reduction in sediment transport in the river is most likely the result of reduced erosion rates in the catchment areas and tributaries, and the retention of river sediments in reservoirs is associated with

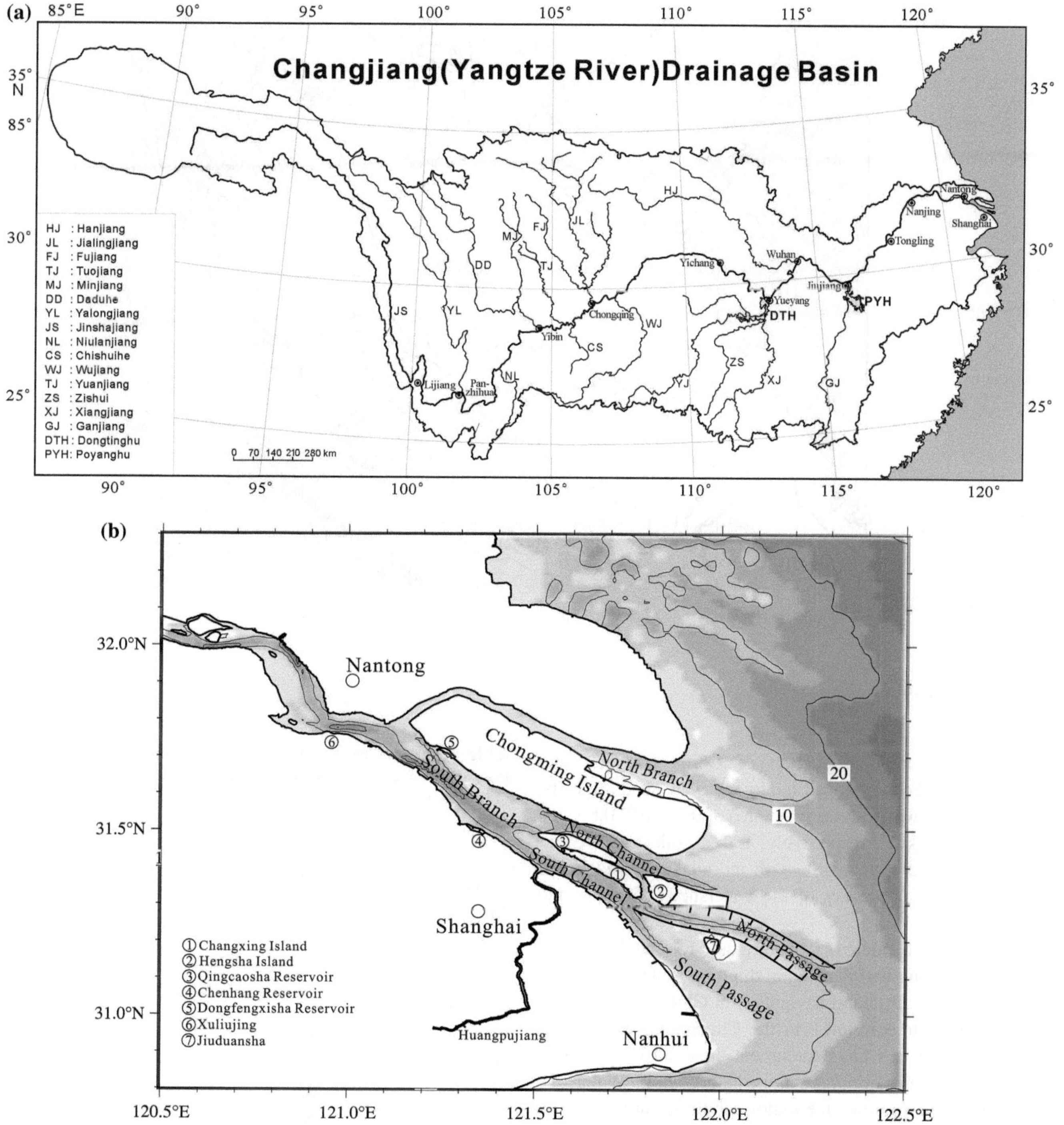

Fig. 1.2 Study area of field observations in this monograph, including, **a** the Changjiang watersheds, **b** Changjiang (Yangtze River) Estuary, and **c** the East China Sea, modified from Ouyang and Zhu (2011)

the construction of dams, notably the Three Gorges Dam (TGD) in the middle reaches of the river (Yang et al. 2007).

In 2007, the population of the Changjiang watersheds was 0.43×10^9, the population density was 240 individuals/km^2, and the GDP of the watersheds was 8.5×10^{12} Yuan RMB, or 35.5 % of the total for all of China (Data source: http://zh.wikipedia.org/wiki). The Changjiang watersheds is an important agricultural area, with cultivated fields covering an area of 0.31×10^6 km^2. In the upper part of the Changjiang watersheds, agriculture mainly consists of hillside dryland farming of maize and potatoes. In the middle and lower reaches of the watersheds, rice paddies are the dominant form of agriculture. In the Changjiang watersheds, rice production accounts for 70 %, cotton for 33 %, and freshwater fisheries

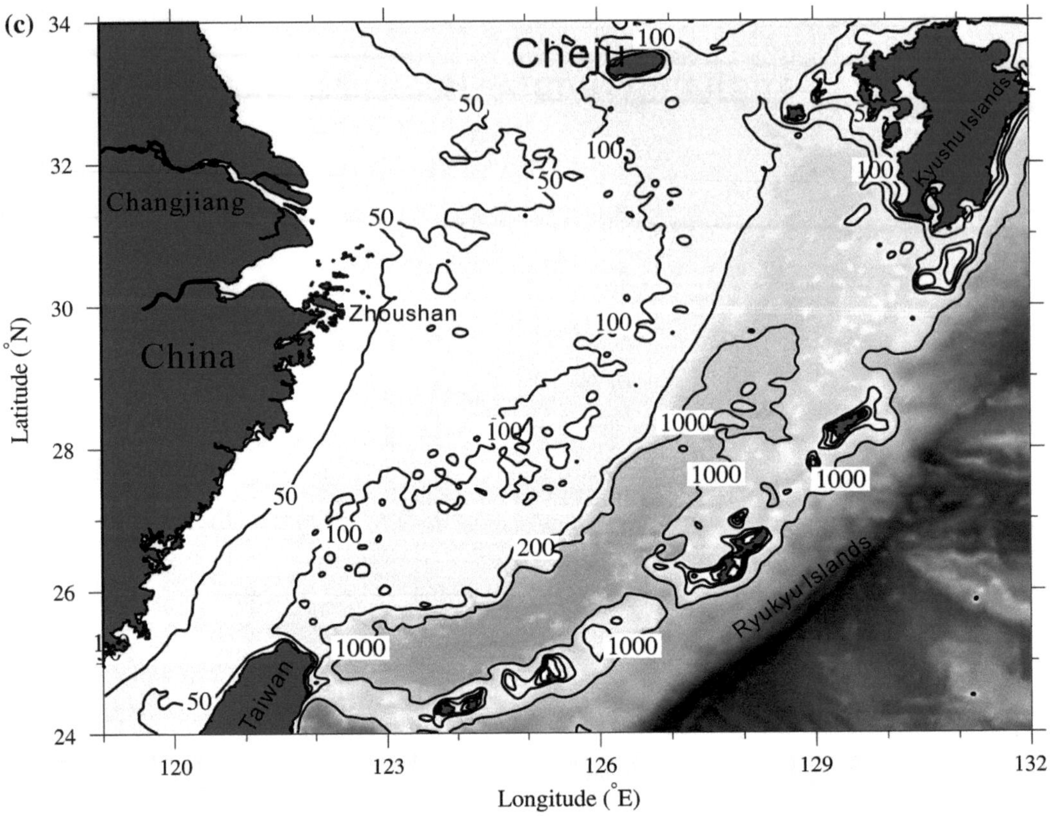

Fig. 1.2 (continued)

for 60 % of total production of China (Ouyang and Zhu 2011). Over the entire Changjiang watersheds, forests, grasslands, and wetlands cover an area of 0.7×10^6, 0.6×10^6 and 0.2×10^6 km^2, respectively. Lakes collectively cover an area of 0.015×10^6 km^2, with most lakes found in the middle and lower reaches of the watersheds (Ouyang and Zhu 2011).

The river channel of Changjiang for navigation is 70×10^3 km long and is a major shipping route, accounting for ca. 70 % of all inland navigation in China (Data source: http://baike.baidu.com/link). In the Changjiang watersheds, about 50 tributaries have catchment areas $>10 \times 10^3$ km^2 each (Data source: http://zh.wikipedia.org/wiki). Water discharge data for some of the major tributaries are shown in Table 1.2.

The upper reaches of the Changjiang, defined as the area from the water source to Yichang where TGD was constructed, has a total length of 4500 km (Photo 1.1). In this region, the river flows in deep valleys and has an elevation change (*H*) to water length (*L*) ratio of >1 ‰. The middle section of the river between Yichang and Hukou (i.e., Jiujiang in Fig. 1.2) has a total channel length of 1000 km. In this area, the river travels across alluvial plains and characterized by lakes and wetlands such as Poyanghu and Dongtinghu (Note: here "hu" means lake in Chinese). The middle reaches of the Changjiang are also an extensive agricultural area of China.

Table 1.2 Major tributaries of the Changjiang, including drainage areas and annual water discharge (*Data source* CWRC 2013 and http://zh.wikipedia.org/wiki)

Tributaries	Drainage area ($\times 10^3$ km^2)	Water discharge ($\times 10^9$ m^3/yr)
Yalongjiang	130	56.8
Minjiang	136	84.7
Jialingjiang	157	65.3
Tuojiang	27	15.8
Wujiang	83	48.8
Hanjiang	142	47.5
Xiangjiang	82	66.0
Yuanjiang	85.2	64.6
Zishui	26.7	23.0
Ganjiang	81	68.3
Qingjiang	17	14.3

The lower reaches of the river stretch 1000 km from Hukou to the river mouth, and the section of Changjiang from Nanjing to the river mouth is also called the "Yangtze River". The water level of the river in the lower reaches, downstream from Datong, is affected by tidal variations.

Photo 1.1 Upstream of Changjiang at Shigu in Yunnan Province, at ca. 4500 km from the river mouth. At this point, the river is known as the Jinshajiang and flows in deep valleys. The photograph was taken from a road on the left side of the river on September 10, 2009, during late-summer floods

This is also a densely populated area, which has experienced rapid urbanization over the last several decades (Photo 1.2).

Anthropogenic activities have had significant impacts on water resources in the Changjiang watersheds. By 2012, the total water impoundment from 217 large and 1259 midsize reservoirs was 153.8×10^9 m^3, which represents 15 % of the annual water flux from the Changjiang to the East China Sea (Wang 2012). In 2012, the total water consumption in the Changjiang watersheds was 203×10^9 m^3. Of this water use, agriculture accounted for 48.6 %, industry for 35.3 %, and domestic use for 15.3 %, leaving 0.8 % remaining for ecological use (Wang 2012). Wastewater drainage in the Changjiang watersheds was 34.8×10^9 tons in 2012, of which industrial sources accounted for 64.5 %, and another 35.5 % was generated in the domestic sector (Wang 2012). Sand extraction in the middle and lower reaches of the Changjiang is another important human perturbation; in 2011, the total sand production was ca. 44.1×10^6 tons, which is ca. 60 % of the annual sediment load at Datong (CWRC 2012).

1.1.2 The Changjiang (Yangtze River) Estuary

Although the tidal area of the Changjiang (i.e., the reach of the river that experiences water-level change between high and low tides) is 600–650 km from the river mouth, the area affected by the direct seawater intrusion (i.e., an increase in conductivity/salinity in water column) in the main channel of Changjiang Estuary is limited to 50 km upstream from the river mouth. The Changjiang Estuary system comprises the North Branch and the South Branch, which are separated by Chongming Island (Fig. 1.2). The South Branch splits into two channels: the North Channel and the South Channel, which are separated by Changxing–Hengsha Island. The South Channel

Photo 1.2 In the middle and lower reaches of the Changjiang, the river travels over fluvial plains with a width of several kilometers. Floods can cause the erosion of shoals in this region and damage to human property. The photograph was taken at Tongling in Anhui Province on October 8, 2009

bifurcates into the North Passage and the South Passage, which are separated by a shoal known as Jiuduansha (Photo 1.3).

In the Changjiang Estuary, the tidal range is upto 4–5 m with an average of 2.7 m at the river mouth. The estuary is therefore classified as a mesotidal system, dominated by the semidiurnal component (M_2). The tidal current can be as high as 1–2 m/s, and ebb currents are stronger than flood currents (Chen 1998). The turbidity maximum zone occurs in water channels, but its position shifts with the tidal phase by a distance of ca. 50 km. Salinity is usually <10 in the turbidity maximum zone, but the concentration of total suspended matter (TSM) can be as high as 5–10 g/L in near-bottom waters, or two orders of magnitude higher than riverine concentrations.

Seawater intrusion is an important feature of the Changjiang Estuary and can have a serious effect on water resources, particularly during spring tides in the dry (winter) season. Salty water can intrude upstream into the North Branch and then penetrate into the South Branch through the west tip of Chongming Island. The salinity of the North Branch can be as high as 10–20, and the intrusion of salt water into the South Branch can cause salinity to increase to between 1 and 5 (Li et al. 2014).

The pan–estuarine area of the Changjiang is ca. 210×10^3 km^2, including Shanghai, Jiangsu Province, and Zhejiang Province (Data source: http://baike.baidu.com/view/48994.htm?fr=aladdin). Delta and/or coastal wetlands occur in this area from water depths of 0 m to elevations of 5 m asl and cover an area of 2.6×10^3 km^2, including Chongming Island, Hengsha–Changxing Island, and Jiuduansha Shoal. (Fig. 1.2) (cf. Mao et al. 2014). The deltaic salt marsh ecosystem is relatively well studied, and key vegetation types from the water level upward include *Scirpus mariqueter*, *Phragmites australis* (reeds), and *Spartina alterniflora*. (Cui et al. 2014). *S. alterniflora* is an invasive species that is highly effective at trapping suspended

Photo 1.3 Mud flats are typical in the coastal area of the Changjiang. The photograph was taken on the east coast of Shanghai (i.e., Nanhui) on August 1, 2014. It shows a wetland with reeds and spartina, as well as the bridge–highway system connecting Shanghai with the Yangshan Deepwater Harbor. The groins used for reclamation are also visible (courtesy of L. Yuan)

sediments. It was introduced into China from the east coast of the USA in the 1980s to protect beaches from erosion and quickly became established along the coast of China. *S. alterniflora* eventually became the dominant wetland species in some parts of the Changjiang Estuary, invading the niche of other local plants, and changing the wetland ecosystem function (Photo 1.3).

In the Changjiang Estuary and adjacent coastal areas, reclamation is a traditional anthropogenic activity and can be traced back to the Han Dynasty (202–220 BC). Historically, reclamation in coastal areas was for aquaculture and salt extraction; however, since the 1950s, reclamation activities have occurred in response to rapid urbanization and industrial needs (CCICED 2011). Other human perturbations in the Changjiang Estuary include sediment dredging, training walls (i.e., jetty) for navigation, harbor construction, and construction of reservoirs in the delta region for freshwater extraction (e.g., Qingcaosha Reservoir) (Box 1.1).

Box 1.1 Qingcaosha Reservoir in the Changjiang Estuary

Historically, Shanghai was a metropolitan area that had only limited freshwater supplies available for human consumption. Traditionally, the main source of freshwater for Shanghai was the Huangpujiang, a tributary of the Changjiang. In the late 1990s, the city had a total population of more than 1×10^9 and was rapidly growing, but the total reservoir storage capacity was ca. 0.5×10^9 m^3, much less than for other large cities of the world (e.g., New York and Tokyo).

The Qingcaosha Reservoir was built between 2007 and 2010 at the bifurcation of the North and South passages of the Changjiang (cf. Fig. 1.2b). It has an area of 66.2 km^2 and a capacity of 0.44×10^9 m^3. Aside from providing freshwater, the Qingcaosha Reservoir will also help mitigate ground subsidence in Shanghai, which has been a serious problem since the 1970s because of the over extraction of groundwater for domestic and industrial use.

This area is affected by seawater intrusion into the Changjiang Estuary, particularly in periods of low freshwater discharge (i.e., winter) and during spring tides. Storm surges can also cause an enormous saltwater intrusion, which damages water quality (e.g., increases salinity inside the reservoir) (Le 2012; Lu et al. 2013). Hence, the sluices of the Qingcaosha Reservoir can only open to take waters from the Changjiang Estuary when in situ chlorinity is less than 250 mg/L. The relatively high turbidity of the Changjiang Estuary can also cause siltation inside the reservoir, which will reduce its life span.

Photo Box 1.1 Qingcaosha Reservoir in the Changjiang Estuary, the photo was taken from Changxing Island on 15 April 2013 (courtesy of C.F. Tong)

In the 1990s, two 50-km-long jetties were constructed in the North Passage to assist with cargo navigation in the Changjiang Estuary. Since that time, the North Passage has been continuously dredged, with an estimated 60×10^6 m^3/yr of sediment removed to maintain a water channel depth of 12.5 m (Jiang et al. 2012). The Qingcaosha Reservoir has a water storage capacity of ca. 0.4×10^9 m^3 and was built in the Changjiang Estuary between 2007 and 2010 to provide 7.2×10^6 m^3/day of freshwater to Shanghai to meet domestic and industrial requirements. This additional water supply is particularly important in winter when seawater intrusion affects freshwater availability (Le 2012).

The Yangshan Deepwater Harbor (YDH) was constructed on the Zhoushan Archipelagos (Dayangshan and Xiaoyangshan Islands) adjacent to the Changjiang Estuary. In 2012, the harbor had a surface area of 25 km^2, a water depth of ≥15 m, and a capacity of 14.1×10^6 TEU, ca. 50 % of the total cargo throughput of the Shanghai harbor system (Data Source: http://tieba.baidu.com/p/2121067985). The YDH is connected to the mainland and Shanghai via a 32-km-long bridge–highway system. Together with Shanghai harbor, this forms an important network of water- and land-based transportation systems (Box 1.2).

Box 1.2 Yangshan Deepwater Harbor

Historically, Shanghai was a commercial trade center for China and even in the East Asia. However, the old harbor system is inside the river mouth of Changjiang and along the Huangpujiang. Navigation is hampered by sandbars and shoals in the Changjiang Estuary, as well as tidal features because the water depth can be 5–10 m.

The YDH is located in the Zhoushan Archipelagos and is linked to Shanghai by a 32-km-long and 31.5-m-wide bridge–highway system (Photo Box 1.2). The construction of the YDH began in 2002 and will finish by 2020. Construction was planned in three phases.

- In 2002–2005, the construction included five berths with ability of operating cargos of 10×10^3 tons each in water depths of 15.5 m, and a total capacity was 2.2×10^6 TEU/yr.
- In 2006–2007, another four berths with ability of operating cargos of 10×10^3 tons each were constructed with a total capacity of 2.1×10^6 TEU/yr.
- In 2007–2008, seven berths with ability of operating 10×10^3 tons each were constructed in water depths of 17.5 m, and a total capacity of 5.0×10^6 TEU/yr was reached.

It is expected that by 2020, the total number of berths will be ≥50 with an annual cargo throughput capacity of $>15 \times 10^6$ TEU. In 2012, the throughput handled at YDH was 14.1×10^6 TEU, ca. 50 % of the total cargo throughput of the Shanghai harbor system.

It has been noted, however, that construction of the YDH has modified the tidal circulation at local scales, and this has resulted in morphological changes caused by erosion and accretion of seafloor sediments. Hence, regular dredging is required in some areas to maintain water depths of ≥15 m for navigation. (Data sources: http://baike.baidu.com/view/88787.htm; Ying et al. 2012)

Photo Box 1.2 YDH in the Zhoushan Archipelagos, the photo was taken from the hill top of Xiaoyangshan Island on 18 December 2014 (courtesy of F. Zuo)

Changjiang Estuary has also been affected by the rapid urbanization of Shanghai. This city had a population of 18.6×10^6 in 2007, an annual GDP of 2.0×10^{12} Yuan RMB in 2012 (Data source: http://zh.wikipedia.org/wiki/), and is located entirely on the Changjiang Estuary delta. The land surface area of Shanghai is ca. 6340 km^2, of which 13.2 % has been created through reclamation over the last five decades (Li et al. 2003; Mao et al. 2003), plus another 12.0×10^3 km^2 of sea area. Owing to the recent decline in terrestrial sediment supply to the estuary, the delta front of the Changjiang Estuary has been eroded and the submerged delta has degraded at water depths of ca. 10 m, creating challenges in terms of developing this area sustainably in the future (Yang et al. 2005 and 2007).

Moreover, in the Changjiang Estuary, there are two natural wetland reserves; the east coast of Chongming Island with a surface area of 242 km^2 and the Jiuduansha (shoal) wetland with surface area of 420 km^2. The Chongming Island reserve was recognized by the Ramsar Convention in 2002 (Data source: http://baike.baidu.com/view/322028.htm), and the Jiuduansha wetland was classified as a national reserve in China in 2005 (Data source: http://baike.baidu.com/view/2010947.htm).

1.1.3 East China Sea

The East China Sea is located in the midlatitudes in the northwest Pacific Ocean (117.2°–131.0°E and 23.0°–33.2°N), extends east to the Ryukyu Islands, and is separated from the Yellow Sea to the north by a line from the north tip of the Changjiang mouth (Qidong) to Jeju Island. It is connected to the Japan Sea/East Sea by the Tsushima Strait/Korean Strait in the northeast, and in the southwest, it is separated from the South China Sea by a line between Nanao Island and Eluanbi at the south tip of the Taiwan Strait.

The surface area of the East China Sea is 770×10^3 km^2, with an average water depth of 370 m (Zhang and Su 2006). The East China Sea is characterized by a wide shelf, of over 600 km, which extends from the Chinese mainland eastward. The Okinawa Trough is situated to the east, with water depths of up to 2719 m (Fig. 1.2).

The East China Sea is characterized by strong seasonal and interannual variability in circulation as a result of the monsoon climate and the western boundary current (i.e., Kuroshio) of the northwest Pacific Ocean. In summer, effluent plumes from the Changjiang, named Changjiang Diluted Waters (CDW), disperse across the surface of the East China Sea and are driven northeast by the monsoon. These waters cover an area of 10^4–10^5 km^2, following the floods of Changjiang with water discharge of 49.5×10^3 m^3/s in July, and terrestrial materials can be traced over a distance of 250–300 km off the Changjiang mouth. This material is an important source of plant nutrients for coastal ecosystems in the East China Sea (Wu et al. 2011). In winter, the Changjiang discharge is reduced considerably, for example, the average monthly water discharge in January (i.e., 11.3×10^3 m^3/s) was only 20 %–25 % of that in July in the period of 1952–2012 (Li et al. 2012). Also in winter, the northeast monsoon drives the riverine effluent plume southward along the coast of mainland China. This plume is visible in satellite images because of the high turbidity, namely Zhe-Min Coastal Current, which has a water flux of 0.2×10^6 m^3/s, or ten times higher than the winter discharge of the Changjiang (Wu et al. 2013).

Offshore from the Changjiang and in the East China Sea, a branch of the Kuroshio Current flows year-round along the Okinawa Trough with a water flux of ca. 20 Sv (1 Sv = 10^6 m/s). This current forms the eastern boundary of circulation in the East China Sea and leaves the area via the Tokara Strait (Guo et al. 2006). The Kuroshio impacts the shelf of the East China Sea mainly through mesoscale processes in the upper part of water column (i.e., the Kuroshio Surface Water, KSW), at shelf-break areas, and via the incursion of Kuroshio Subsurface Water (KSSW) on to the shelf east of Taiwan where water flow is estimated to be 1.5 Sv (Guo et al. 2006). The incursion of the Kuroshio across the shelf break has important seasonal, as well as interannual, variability; e.g., water flow onto the shelf of the East China Sea is usually stronger in autumn and winter than in spring and summer (Su 1998; Zhang and Su 2006).

Warm and salty water, known as the Taiwan Strait Warm Water (TSWW), moves into the East China Sea through the Taiwan Strait with an influx of 1.5–2 Sv (Guo et al. 2006). The shallow water depth of the Taiwan Strait (i.e., 60 m) causes the TSWW to affect the upper part of water column in the East China Sea, particularly the shelf area, and hence changes the vertical structure of hydrographic parameters such as salinity and temperature. The water column is highly stratified in this region in summer, with the CDW near the surface, the TSWW at mid-depths, and waters from the Kuroshio incursion (KSSW) occurring below the TSWW (Su 2005). In the northern East China Sea, water exchange with the Yellow Sea is mainly through the Yellow Sea Warm Current (YSWC) in the east and the Yellow Sea Coastal Current (YSCC) in the west (Su 1998). The shelf water leaves the East China Sea through circulation and water exchange, mainly via the Tsushima Strait/Korean Strait, as well as to the northwest Pacific Ocean through the Tokara Strait via mesoscale exchange processes.

Bottom sediments in the East China Sea have a zonal distribution (Qin et al. 1996). In water depths of less than 75 m, riverine sediments are distributed along the coast on the inner shelf, where the sediment composition is usually muddy and dominated by clay and silt size fractions. The sedimentation rate can be as high as 1–5 $g/cm^2/yr$ in areas such as the river mouth of Changjiang, illustrating the close link between riverine loads and sedimentation. In the mid- and open-shelf areas, with water depths of ca. 100–200 m, sedimentation rates are low and the sea floor is mainly composed of sandy sediments with carbonate debris (e.g., shell fragments), thought to be relics from the last glacial period when sea level was ca. 150 m lower than today (Deng et al. 2006). Sediment distribution in the Okinawa Trough, an expanding sub-abyssal back-arc basin, is dominated by fine materials (i.e., clay minerals), as well as the deposits of turbidities, volcanic material, and biological detritus (Qin et al. 1996).

Like the other coastal oceans in subtropical and temperate climate zones (e.g., the North Sea), phytoplankton blooms occur in spring and fall in the East China Sea. In this region, the spring bloom is stronger than the autumn bloom and plays a critical role in maintaining the structure and function of the food web. The fall bloom is important for biomass storage over winter.

1.2 Natural and Anthropogenic Forcings

The variability of climate and circulation over seasonal and interannual timescales is an important influence on dynamic processes in the East China Sea. This is particularly true when a range of natural stressors is at play, making it very difficult to predict ecosystem responses to these external forcings.

The northeasterly monsoon in winter can stimulate surface wind-driven circulation and induce deep convection in the East China Sea, particularly in the shelf area where water depths are shallow. This convection promotes the incursion of KSSW across the shelf region. For instance, the Zhe-Min Coastal Current is enhanced by northerly winds. This current can remobilize bottom sediments off the Changjiang Estuary and carry a huge amount of TSM southward along the east coast of China to the Taiwan Strait and even further to the South China Sea.

In winter, deep convection tends to homogenize the hydrographic properties in the water column and induce the upward flux of nutrient-rich deep waters to the surface through vertical mixing. In summer, high temperatures in the surface waters in combination with riverine plumes in the East China Sea induce strong stratification in the upper 10–20 m of the water column on the shelf. During this period, the flux of TSWW also increases. As the southerly monsoon and riverine influx both have interannual variations, the stratification of hydrographic properties also exhibits strong spatial and temporal changes. Moreover, summer typhoons and storms are quite common in this region, and these events temporarily disturb the vertical structure of the upper part of the water column and homogenize the profile of chemical parameters such as dissolved oxygen.

Anthropogenic forcings include both direct (e.g., overfishing and reclamation) and indirect components (e.g., coastal eutrophication and seasonal hypoxia). The East China Sea is one of the key traditional fishing grounds in East Asia, with a total catch, mainly from China, Japan, and Korea, of 7×10^6–8×10^6 tons/yr by the end of the last century (Guo et al. 2003). The major fishery products include yellow croakers, belt fish, horse-faced puffer, and squid. In 2000, the Chinese fish catch from the East China Sea reached 6.25×10^6 tons, which accounts for 42.3 % of the total fish catch of the country; the East China Sea, like other marginal areas of the northwest Pacific Ocean, is over fished system (Zheng et al. 2003). In the Zhoushan Archipelagos, for example, the number of fishing boats is 8000–8500, which have a total capacity of up to 800×10^3–850×10^3 tons (Shen 2009) (Photo 1.4). Overfishing is generally considered to be a top-down mechanism that has a negative effect on the ecosystem. By removing top predators, or even the omnivores, herbivores in the ocean can reduce the plankton biomass and thus affect primary productivity and biologically essential plant nutrients and trace elements (GLOBEC 1999).

The coastal zone of the East China Sea is also a region of significant reclamation, for example, the seawalls constructed in the Song Dynasty (960–1279 AD) along Hangzhou Bay are well documented in Chinese literature. Modern reclamation is mainly driven by the limited availability of land for economic activities, such as urbanization, marine aquaculture, and hydro-engineering construction. Reclamation occurs between the edge of the high watermark to around 2-m water depth, covering the most critical coastal wetland areas and tidal flats. As this coastal zone contains important spawning and hatching habitats for economically important fish species, the overall effect of reclamation is a net loss of habitat.

Pollutants and coastal eutrophication are other forcings related to anthropogenic activities (Muller et al. 2008). Through the over enrichment of plant nutrients, trace metals, and organic matter from terrestrial sources, episodic blooms of phytoplanktonic species occur in coastal areas. Blooms that occur adjacent to the river mouth are different to those regulated by the lunar calendar (i.e., spring vs. fall blooms) and result in the high accumulation of biomass in a short period of time and over a relatively large area (e.g., 100–200 km).

Coastal eutrophication can have a negative impact on ecosystems because it has a bottom-up effect on the marine food web (Zhou et al. 2001). In eutrophic waters, planktonic species, such as dinoflagellates and diatoms, profit through

Photo 1.4 In the East China Sea, overfishing impacts the health and sustainability of marine ecosystems. In the Zhoushan Archipelagos, thousands of fishing boats are berthed in the fishing harbors (e.g., Shenjiamen) in the winter. This photograph was taken at the Shenjiamen Fishing Harbor of Zhoushan Archipelagos in February 15, 2012 (courtesy of J.H. Gu)

competition and produce organic matter that is not taken up by function groups in higher niches in the food web (e.g., zooplankton) (GEOHAB 2001). Eutrophication can therefore cause a deterioration in water quality as well as a change in the vertical flux of organic matter to the sea floor (IMBER 2005). In the extreme cases, this can cause hypoxia and even anoxia in near-bottom waters off the Changjiang Estuary in summer (Li et al. 2002; Zhu et al. 2011). Along with the eutrophic character of the coastal environment, heterotrophic processes can also be stimulated, in which microorganisms (e.g., heterotrophic bacteria) use organic matter as an energy source. For this reason, the nearshore areas of the East China Sea can be a source of CO_2 and other greenhouse gases (Naqvi et al. 2010).

In summary, although the individual forcings discussed above have been well documented in previous studies, their combined effects and consequences remain poorly understood. From a management perspective, an understanding of the linkages between processes in the watersheds–ocean continuum will be one of the key sustainability challenges over future decades.

1.3 An Overview of Previous Studies in the Region

Early studies of land–ocean interactions in the coastal zone (LOICZ) adjacent to the Changjiang Estuary can be traced back to the 1930s–1940s in Chinese as well as in the international literature. Table 1.3 summarizes the major findings of previous studies in this region. For instance, in the 1940s, Chu and co-workers recognized the critical role of water discharge from the Changjiang in the delivery of plant nutrients to the East China Sea Shelf and its impacts on fish production in the area adjacent to the Zhoushan Archipelagos (Chu and Yang 1949). In Table 1.4 are summarized the major hydraulic engineerings in the Changjiang Estuary and adjacent coastal environment in last five decades.

Since the 1950s, Chen and co-workers have completed extensive studies of the development of delta areas in the Changjiang Estuary and particularly the variation of water channels under changing water and sediment loads in the river. They used historical data and field observations to develop a

Table 1.3 Major studies related to land–ocean interactions in the Changjiang watersheds to East China Sea continuum

Authors	Period	Major findings
Chu SP and co-workers	1940s–1950s	Impact of freshwater discharge from the Changjiang on the fishing grounds of the Zhoushan Archipelagos
Chen JY and co-workers	1950s–1990s	Evolution of delta morphology and sedimentary dynamics of the Changjiang Estuary with application to coastal engineering
Qin YS and co-workers	1960s–1990s	Sedimentary processes, morphological features, and geological evolution of the East China Sea
Gu HK and co-workers	1960s–1990s	Chemical aspects (e.g., plant nutrients and trace pollutants) of Changjiang Estuary and East China Sea
Hu DX and co-workers	1990s–2000s	China JGOFS and LOICZ studies in the Changjiang Estuary and East China Sea
Su JL and co-workers	1980s–2000s	Hydrographic and hydrodynamics of the Changjiang Estuary and East China Sea
Tang QS and co-workers	1990s–2010s	China GLOBEC and IMBER studies in the Changjiang Estuary and East China Sea

Table 1.4 Summary of major hydro-engineering activities in the Changjiang Estuary

Type of hydro-engineering work	Character and function
Jetty and navigation of Changjiang Estuary	Two jetties, each of ca. 50 km, were constructed in the north passage as part of the "deep waterway of navigation" project in the Changjiang Estuary over the period 1998–2002. This project plays an important role in maintaining a channel for cargo transportation
Yangshan Deepwater Harbor	The harbor is on Xiaoyangshan Island with a water depth of ≥15 m and a cargo throughput capacity of 14.1×10^6 TEU in 2012. The harbor is connected to the mainland through a highway–bridge system of ≥30 km
Qingcaosha Reservoir	The reservoir was constructed at the diversion between the north and south passages in 2007–2010 with a capacity of 0.44×10^9 m^3. The impoundment of the reservoir was stopped when in situ water chlorinity exceeded 250 ppm
Reclamation of delta	A total of 0.85×10^3 km^2 has been reclaimed in Shanghai over the last 5–6 decades, and >10 % of Chongming Island is made up of land reclaimed over the period of 1960–2010. Urbanization has been the driving force for recent reclamation activities
Marine protected area (MPA)	The wetlands of Jiuduansha (shoal) and the eastern part of Chongming Island are two marine protected areas in the Changjiang Estuary. The former was established in 2005 and has an area of 420 km^2, and the latter of 265 km^2, and was listed as part of the Ramsar Convention in 2002
Pudong International Airport (PVG)	PVG was constructed entirely on the delta of Changjiang Estuary between 1997 and 2008 and has an area of ca. 40 km^2. In 2011, the throughput was 3.1×10^6 tons of cargo and 41.4×10^6 of passengers

model of the evolution of the Changjiang Estuary, taking into consideration of ecosystem response to hydrodynamic processes (Chen 1957; Chen et al. 1959, 1979; Chen and Xu 1981; Chen et al. 1988). Their research was later applied to hydrodynamic and sedimentary studies in the Changjiang Estuary associated with hydraulic engineering projects (e.g., navigation, harbor construction, and reclamation (cf. Shen et al. 1980; Chen and Shen 1983; Chen 1995, 2000).

During the period 1979–1984, China and the USA conducted a joint study entitled "Sedimentary Dynamics of the Changjiang Estuary and Adjacent East China Sea Shelf," and results from this collaboration were published in a special issue of *Continental Shelf Research* (Milliman and Jin 1985). Later, the results of a Sino–French collaborative study, "Biogeochemical Dynamics of the Changjiang Estuary and Adjacent Marine Environment" in 1985–1990, were published as a monograph entitled *Biogeochemistry of the Changjiang Estuary* in China Ocean Press (Yu et al. 1990). In the late 1980s, China was involved in the international Joint Global Ocean Flux Study (JGOFS) program, which included the Changjiang Estuary and East China Sea as one of the key focus areas. Hu and co-workers synthesized the results of the Chinese JGOFS in a monograph entitled *Key Processes of Ocean Flux in the East China Sea* (Hu and Yang 2001). Gu and coworkers also summarized research activities in this region over the period from 1950–1990 and integrated the results into a monograph entitled *Marine Chemistry of Bohai, Yellow Sea, and East China Sea* (Gu 1991).

After the 1990, the scientific community in China actively participated in international initiatives such as the Global Ocean Ecosystem Dynamics (GLOBEC), LOICZ, and Integrated Marine Biogeochemistry and Ecosystem Research (IMBER) projects, in which the Changjiang Estuary and adjacent East China Sea were consistently one of the most prominent research areas. Contributions from the Chinese scientific community to the above-mentioned international initiatives can be found in Tang and Su (2000), Hu et al. (2001), Tang (2004), and Shen et al. (2009).

The major aspects of land–ocean interactions in the continuum from the Changjiang watersheds to the East China Sea can be summarized in but not limited to the following key points.

- Over the last 5–6 decades, although the annual water discharge of the Changjiang has remained relatively stable at ca. 1×10^{12} m^3/yr, the sediment load of the river has decreased considerably from 0.43×10^9 tons/yr over the period 1950–2000, to ca. 0.15×10^9 tons/yr for 2003–2010. Consequently, the submarine delta-front area switched from seaward progradation to retreat around the begining of this century. However, it seems that TSM is still high in the Changjiang Estuary turbidity maximum zone, presumably due to the strong tidal dynamics that are able to resuspend bottom sediments and transport them into the estuary from nearby deltaic and coastal areas.
- Over the same period, plant nutrient concentrations increased in the Changjiang watersheds, notably dissolved inorganic nitrogen (DIN = NO_3^- + NO_2^- + NH_4^+) and to a lesser extent dissolved phosphorus, particularly after the economic innovations of the late 1970s in China. In the 1950s–1970s, DIN concentrations in the Changjiang and its major tributaries were 40–50 μM. The DIN concentrations increased considerably after the 1980s, reaching ca. 100 μM at the beginning of the twenty-first century. The increase in the concentration of plant nutrients in the river after the 1980s was linked to changes in land use and the application of synthetic fertilizers and has also led to a high molar ratio of N/P.
- Three economic innovation centers in East China, namely Shanghai, as well as Jiangsu and Zhejiang provinces, are located in the Changjiang Delta and adjacent coastal areas. This is also an area of high population density (i.e., 500–1000 individuals/km^2). Shanghai has essentially developed on the delta of the Changjiang Estuary over a period of ca. 2000 years. The reclamation of coastal areas has occurred as a result of urbanization, the construction of harbors and jetties, dredging sediments for navigation, as well as aquaculture for shrimps and crabs. These are all land-based factors that can damage marine habitats. Moreover, the increase in invasive species, such as spartina and black-striped mussels (*Mytilopsis sallei*), has caused competition for niches, changes in the food web structure, and the deterioration of coastal ecosystems.
- In the East China Sea, the delivery of plant nutrients from land-based sources has induced eutrophication in shelf waters and frequent harmful algal blooms that can affect thousands of square kilometers. Furthermore, the accumulation of autochthonous and allochthonous organic matter in the water column and bottom sediments has stimulated heterotrophic processes that consume dissolved oxygen and result in seasonal hypoxia in near-bottom waters. China, Japan, and Korea all harvest the ecosystems of the East China Sea, and overfishing has exhausted natural and renewable resources and resulted in irrevocable damage to ecosystem.

1.4 Scientific Research and Relevance to Human Society

The coastal environment has sustained human economic activities in the past and will continue to be a major platform for the development in the future (Fig. 1.3). Consequently, ecosystems in coastal and adjacent marine waters are an over-stressed component of the Earth system, facing pressure from land, ocean, and atmospheric processes related to climate variability and anthropogenic perturbations. The changes that occur in this system have important feedbacks with other components of the Earth system through complex and rapid interactions. Coastal systems are also under pressure from the multiple, and sometimes conflicting, use of resources and ecosystem services, such as navigation and

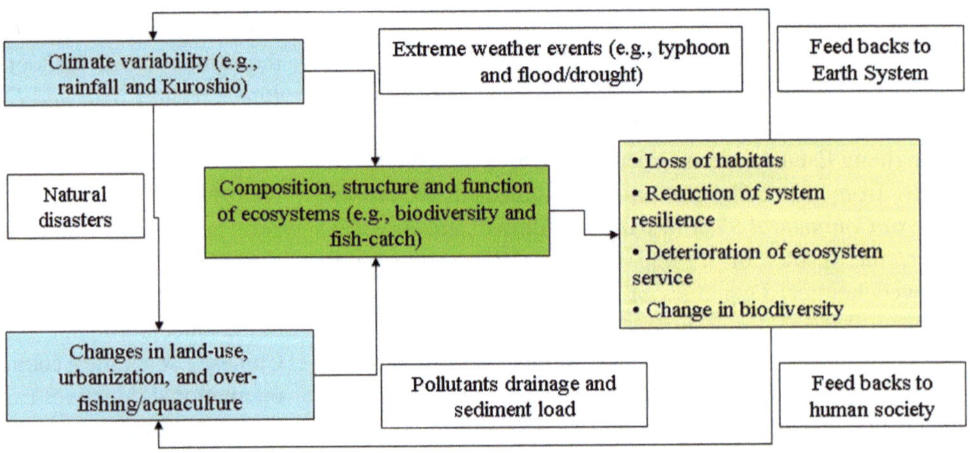

Fig. 1.3 Conceptual model showing the relationship between driving forces and ecosystem responses at the land–ocean interface using the Changjiang and the East China Sea as an example

cargo transportation, reclamation for land-use, national/marine protection areas, and reservoirs for freshwater storage, as well as fisheries and aquaculture.

Therefore, scientific research activities in the coastal zone that connects the Changjiang watersheds with the East China Sea are of direct relevance to human society and the extraction of natural resources and exploration of ecosystem services. The key issues are regional in nature, but have strong global significance due to the combined effects of climate and anthropogenic forcings.

The following key research topics have emerged in relation to the coastal environment in the Changjiang watersheds to East China Sea continuum.

- Circulation dynamics, the combined effects of tides and waves, their impacts on material fluxes in the estuary, and the dispersal of the riverine plumes over the broad shelf or even further offshore. Studies on these processes have an application in better understanding of saltwater intrusion and the variability of freshwater resources.
- Spatial and temporal dimensions of sedimentary processes in the estuary and adjacent coastal environment. This is of great concern to the evolution of the delta system and the protection of coastal areas and ecosystem services.
- Fate of plant nutrients and other pollutants in the land–ocean interface and the impact on biogeochemical cycles. This requires an examination of the role of the estuary and coastal zone as a "filter" and/or "amplifier" of terrestrial materials and has implications for the management of eutrophication and the bioaccumulation of pollutants in the food web.
- The coastal environment and estuary is a unique habitat (e.g., spawning and hatching grounds) for the economically valuable fish species, and the shelf region is a very important fishing ground. Thus, integrated studies of life history and changes in ecosystem function in this region will provide a sound scientific basis for, and contribute to, ecosystem-based management practices.

1.5 Scope and Structure of the Monograph

This monograph provides a synthesis of integrated studies of land–ocean interactions in the continuum from the Changjiang watersheds to the East China Sea. The term "integrated studies" has two interrelated meanings. The first refers to the integration of a study area that extends from the upper reaches of the Changjiang watersheds to the open boundary of the East China Sea. The second refers to multidisciplinary research activities, including basic disciplines such as physics, chemistry, biology, and geology through to the application of fisheries and social sciences. The overall structure of this monograph and linkages between individual chapters are shown in Fig. 1.4.

This chapter is an overview of the study area, including the geographic settings for the land–ocean interactions discussed here, and the major characteristics of the hydrological, environmental, social, and economic dimensions of the region. In Chap. 2, the dynamics of coastal circulation in the East China Sea are examined using cruise data and

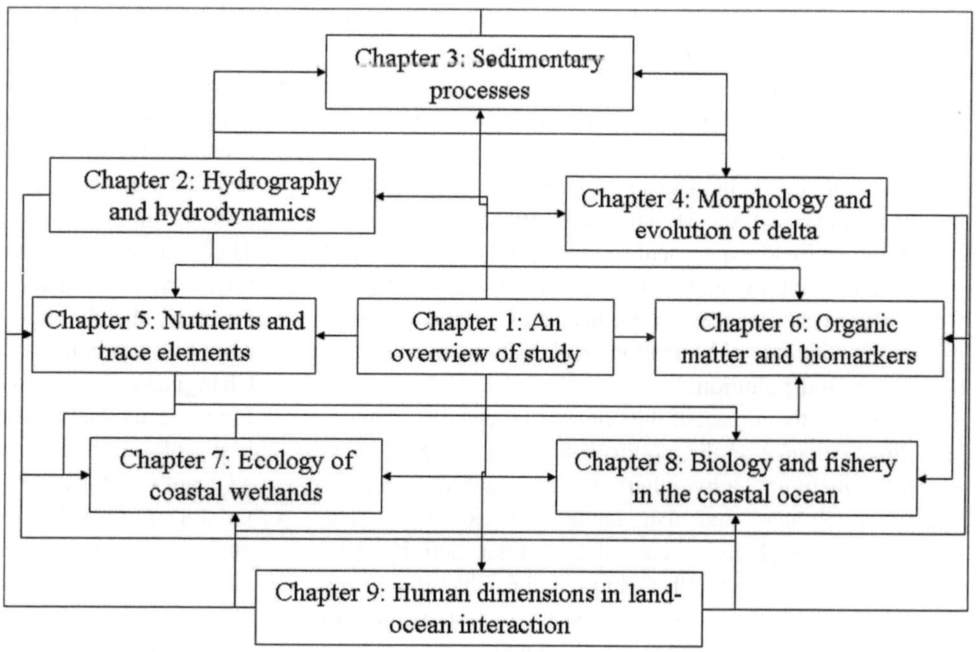

Fig. 1.4 Schematic representation of the linkages between the nine chapters in this monograph, which illustrates the necessity and advantages of integrated studies

numerical simulations and are used to develop a better understanding of the dispersal of the Changjiang effluent plume and seawater intrusion into the estuary. This chapter also discusses the effect of the monsoon climate (i.e., wind fields) and hydrodynamic processes (e.g., tides and waves) on coastal dynamics.

Chapter 3 uses field observations and laboratory experiments to discuss terrestrial sediment transport in the river and coastal waters. Suspended sediment tends to form flocks in the aquatic environment, which has an important impact on the properties, deposition, and resuspension of mineral particles. Chapter 4 examines the response of delta evolution to the changing sediment flux of the Changjiang. The underwater delta-front zone has recently become eroded, and the threshold from delta progradation to regression corresponds to the change in sediment load after the inundation of the TGR.

Chapter 5 uses literature data and observational results to provide an overview of the distribution of plant nutrients and trace elements over the continuum from the Changjiang watersheds to the East China Sea. A box model was applied to budgetary calculations for inorganic species in this area, and results show that although the Changjiang carries a tremendous amount of water and sediment, the incursion of the western boundary current (i.e., the Kuroshio) is of critical importance in modulating the supply and cycling of plant nutrients over the broad shelf of East China Sea. In Chap. 6, organic matter and biomarkers from the Changjiang watersheds to the East China Sea are examined. Hydrological sorting, primary production, and diagenetic processes regulate the distribution of biomarkers at the land–ocean interfaces. The TGD has had a profound effect on the supply and modification of terrigenous organic matter and pollutants and hence in the delivery of organic carbon to the open ocean.

Chapter 7 examines coastal wetlands in the estuarine region of Changjiang, which have been colonized by various types of salt marsh vegetation. It has been found that the expansion of an exotic species, *S. alterniflora*, has had a negative impact on the structure and function of benthic communities. Field experiments and a demonstration project were carried out to control the spread of *S. alterniflora*. Other negative influences on the health of coastal wetlands in this area include urbanization, reclamation, overfishing/harvesting, and pollution.

In Chap. 8, the impact of the human-induced disturbances and climate change on the food web structure and ecosystem function, including phytoplankton, zooplankton, benthic macroinvertebrates, and fish fauna, are examined. The accelerated eutrophication caused by human activity, particularly the increased influx through the Changjiang and skewed ratio of plant nutrient species, is responsible for some of the observed changes. Moreover, it has been observed that both the abundance and occurrence of subtropical and tropical zooplankton species has increased significantly. At higher trophic levels, commercially important demersal species (e.g., fish) have been replaced by smaller and pelagic species at lower trophic level, with less economic value.

Finally, Chap. 9 examines the social and economic factors in coastal areas that can have an adverse impact on the sustainability of ecosystems in the East China Sea. This analysis is based on a compilation of existing data from the literature and reviews the impacts on different parts of the coastal zone around the East China Sea. The study is mainly focused on changes in the social and economic situations over the last 50 years, and historical information and records are used to analyze trends over this time period.

1.6 Looking Forward to the Future: Challenges and Opportunities for Ongoing Research

The coastal provinces of China illustrate the convergence of various anthropogenic stressors, such as a high population that requires food and living space and rapid economic advance that requires resources and energy. As such, this area is a hot spot for studies that relate to conflicts in ecosystem services. Therefore, studies of land–ocean interactions in the area that links the Changjiang watersheds to the East China Sea must be in step with the rapid changes that are driven by human development. In combination with climate variability at a regional scale (e.g., monsoons), these factors result in complex forcing–response relationships across the region. Several key messages must be kept in mind as we look forward and face the challenges of the future.

- It can be expected that the sediment load of the Changjiang will continue to decrease in the near future, not only because of the full operation of the TGD in the middle reaches, but also as a result of other large hydropower stations under construction further upstream from the TGD that will further reduce the sediment supply to the lower reaches and river mouth.
- Faced by the shortage of terrestrial sediment supply and slow but continuous sea-level rise, the delta front of the Changjiang Estuary will continue to retreat in the near future. This makes the protection of coasts critical in terms of supporting socioeconomic development in the adjacent coastal environment.
- Urbanization in the delta area will be expanded further and requires more natural resources and energy. This means an increased requirement for food, water, and a clean environment.
- China has a long history of aquaculture. By 2010, fishery production in China exceeded 25×10^6 tons/yr with more

than 50 % from marine aquaculture. It can be expected that the combined effect of fish catch and aquaculture will further stress the marine ecosystem through the increased requirement for food production.

Given the examples above, it is clear that there will be challenges and opportunities for scientific research. This monograph is an overview of land–ocean interactions in the Changjiang watersheds to East China Sea continuum based on field observations, laboratory experiments, and numerical simulations since the 1990s. It is hoped that the knowledge and experience presented in this monograph will also be valuable and applicable to studies in other regions on the Earth.

Acknowledgments This monograph was developed within the framework of the Natural Science Foundation of China through the "Land–Sea Interactions in Highly Turbid Estuaries and Adjacent Coastal Environments" Project (Nos. 40721004 and 41021064). Observations in the East China Sea and data interpretation were also supported by the Ministry of Science and Technology (Nos. 2006CB400601 and 2011CB409801) as well as the Ministry of Education (No. ITR0427) and the Municipality of Shanghai through its Committee of Science and Technology (No. 088014192). On behalf of the authors of this collective work, I would like to take this opportunity to express our gratitude to colleagues and students for their continuous support in field campaigns and cruises, sample analysis, and simulation experiments since the 1990s.

References

Boesch DF (2002) Challenges and opportunities for science in reducing nutrients over-enrichment of coastal ecosystems. Estuaries 25 (4):886–900

CCICED (2011) Ecosystem issues and policy options addressing the sustainable development of China's Ocean and Coasts. China Council for International Cooperation on Environment and Development (CCICED), China Environmental Press, Beijing, pp 1–493

Chen JY (1957) Notes on the development of the Yangtze River Estuary. Acta Geogr Sin 23:241–253 (in Chinese)

Chen JY (ed) (1995) The deep waterway regulation and ports construction in the Changjiang Estuary. J East China Normal Univ 1–97 (in Chinese)

Chen JY (ed) (2000) Coastal reclamation in China. Press East China Normal University, Shanghai, pp 1–350 (in Chinese)

Chen JY, Shen HT (1983) Some key points on harnessing the Changjiang Estuary. Mar Sci 7(2):1–5 (in Chinese)

Chen JY, Xu HG (1981) Development processes of south branch channel of Yangtze River Estuary. J East China Normal Univ 2:93–108 (in Chinese)

Chen JY, Yu ZY, Yun CX (1959) Evolution of geomorphology of Yangtze River Delta. Acta Geogr Sin 25:201–220 (in Chinese)

Chen JY, Yun CX, Xu HG, Dong YF (1979) The development model of the Changjiang Estuary during last 2000 years. Acta Oceanol Sin 1:103–111 (in Chinese)

Chen JY, Shen HT, Yun CX (1988) Hydrodynamic processes and geomorphologic evolution of the Changjiang Estuary. Shanghai Science and Technology Press, Shanghai, pp 1–453 (in Chinese)

Chen ZS (ed) (1998) Annals of Coastal Areas of China, vol 14. China Ocean Press, Beijing, pp 1–799 (in Chinese)

Chu SP, Yang GX (1949) The variation with depth of certain nutrient salts for phytoplankton growth and some other properties of water in the fishing ground east of Chusan Islands in the East China Sea. Sciences 31:181–182

Cui L, Ge ZM, Yuan L, Zhang LQ (2014) Vulnerability assessment of the coastal wetlands in the Yangtze River Estuary, China to sea-level rise. Estuar Coast Shelf Sci. http://dx.doi.org/10.1016/j.ecss.2014.06.015

CWRC (2012) Changjiang Sediment Bulletin: 2011. Changjiang Water Resources Commission (CWRC), Changjiang Press, Wuhan, pp 1–64 (in Chinese)

CWRC (2013) Changjiang Sediment Bulletin: 2012. Changjiang Water Resources Commission (CWRC), Changjiang Press, Wuhan, pp 1–63 (in Chinese)

Deng B, Zhang J, Wu Y (2006) Recent sediment accumulation and carbon burial in the East China Sea. Glob Biogeochem Cycles 20: GB3014, doi:10.1029/2005GB002559

GEOHAB (2001) Global ecology and oceanography of harmful algal blooms. Science Plan SCOR IOC, Baltimore, Paris, pp 1–87

GLOBEC (1999) GLOBEC Implementation Plan. IGBP Report Series No. 47, IGPB Secretariat, Stockholm, pp 1–207

Gu HK (ed) (1991) Marine chemistry of Bohai, Yellow Sea and East China Sea. Science Press, Beijing, pp 1–500 (in Chinese)

Guo WL, Huang SL, Cao SJ (2003) Co-management and joint-conservation for the fishery resources in the East China Sea. J Nat Resour 18(4):394–401 (in Chinese)

Guo XY, Miyazawa Y, Yamagata T (2006) The Kuroshio onshore intrusion along the shelf break of the East China Sea: the origin of the Tsushima Warm Current. J Phys Oceanogr 36:2205–2231

Hu DX, Yang ZS (eds) (2001) Key processes of ocean flux in the East China Sea. China Ocean Press, Beijing, pp 1–204 (in Chinese)

Hu DX, Han WY, Zhang S (eds) (2001) Land–sea interactions in the Changjiang Estuary, Zhujiang Estuary and their adjacent Coastal Zones. China Ocean Press, Beijing, pp 1–218 (in Chinese)

IMBER (2005) Integrated marine biogeochemistry and ecosystem research, science plan and implementation strategy. IGBP Report No. 52, IGBP Secretariat, Stockholm, pp 1–76

Jiang CJ, Li JF, Swart HED (2012) Effects of navigational works on morphological changes in the bar area of the Yangtze River Estuary. Geomorphology 139–140:205–219

Le Q (2012) Research and analysis of the water supply potentiality of Qingcaosha reservoir during the low water period of Yangtze River. Water Wastewater Eng 38(9):123–127 (in Chinese)

Li DJ, Zhang J, Huang DJ, Wu Y, Liang J (2002) Oxygen depletion off the Changjiang (Yangtze River) Estuary. Sci China 45D:1137–1146

Li JF, Wan XN, Chen XH, Xu HG (2003) Coastal land resources in Shanghai with analysis on their sustainable exploitation. Resour Environ Yangtze Basin 12(1):17–22 (in Chinese)

Li L, Zhu JR, Wu H, Guo ZG (2014) Lateral saltwater intrusion in the North Channel of the Changjiang Estuary. Estuaries Coasts 37:36–55

Li P, Yang SL, Milliman JD, Xu KH, Qin WH, Wu CS, Chen YP (2012) Spatial, temporal, and human-induced variations in suspended sediment concentration in the surface waters of the Yangtze River Estuary and adjacent coastal areas. Estuaries Coasts 35:1316–1327

LOICZ (2005) Land–ocean interactions in the coastal zone, science plan and implementation strategy. IGBP Report No. 51/IHDP Report No. 18, Stockholm, pp 1–60

Lu YJ, Gu JS, Gu YL, Liu XC, Li R (2013) Study on technical problems of reservoir and dam construction for the Qingcaosha Reservoir. Adv Sci Technol Water Resour 33(3):45–49 (in Chinese)

Mao ZC, Li JF, Wu HL (2003) Study on siltation and reclamation of tidal flat in Shanghai. J Sediment Res 2:77–81 (in Chinese)

Mao ZC, Yu ZY, Xu HG (2014) Coastal wetlands of Shanghai. East China Normal University Press, Shanghai, pp 1–233 (in Chinese)

Martínez ML, Intralawan A, Vázquez G, Pérez-Maqueo O, Sutton P, Landgrave R (2007) The coasts of our world: ecological, economic and social importance. Ecol Econ 63:254–272

Milliman JD, Jin QM (eds) (1985) Sediment dynamics of the Changjiang Estuary and the adjacent East China Sea. Cont Shelf Res 4(1/2):1–251

Moser SC, Williams SJ, Boesch DF (2012) Wicked challenges at land's end: managing coastal vulnerability under climate change. Ann Rev Environ Resour 37:51–78

Müller B, Berg M, Yao ZP, Zhang XF, Wang D, Pfluger A (2008) How polluted is the Yangtze River? Water quality downstream from the Three Gorges Dam. Sci Total Environ 402:232–247

Naqvi SWA, Bange HW, Farias L, Monteiro PMS, Scranton MI, Zhang J (2010) Marine hypoxia/anoxia as a source of CH_4 and N_2O. Biogeosciences 7:2159–2190

Ouyang ZY, Zhu CQ (eds) (2011) Atlas of biodiversity and conservation in the Yangtze River Basin. Science Press, Beijing, pp 1–128

Qin YS, Zhao YY, Chen LR, Zhao SL (1996) Geology of the East China Sea. Science Press, Beijing, pp 1–357

Shen CH (2009) On the renewal and transformation of finishing boats of Zhoushan Islands. Zhoushan Fish 6:2–4 (in Chinese)

Shen HT, Mao ZC, Gu GC, Xu HG (1980) Seawater intrusion in the Changjiang Estuary—on the effect of water deviation from Changjiang to the North China. Yangtze River 3:20–26 (in Chinese)

Shen HT, Zhu JR, Wu HL (2009) Interfaces of ocean–land interaction in the Changjiang Estuary. China Ocean Press, Beijing, pp 1–210 (in Chinese)

Su JL (1998) Circulation dynamics of the China Seas north of 18°N. In: Robinson AR, Brink KH (eds) The Sea, vol 11. Wiley, New York, 483–505

Su JL (ed) (2005) The hydrography of China Seas. China Ocean Press, Beijing, pp 1–367 (in Chinese)

Tang QS (ed) (2004) Study on ecosystem dynamics in Coastal Ocean III: Atlas of resources and environment of East China Sea and Yellow Sea. Science Press, Beijing, pp 1–398 (in Chinese)

Tang QS, Su JL (eds) (2000) Study on ecosystem dynamics in Coastal Ocean I: key scientific questions and research strategy. Science Press, Beijing, pp 1–252 (in Chinese)

Wang XC (ed) (2012) Changjiang and Southwest Rivers Water Resources Bulletin. Changjiang Water Resources Commission (CWRC), Changjiang Press, Wuhan, pp 1–42 (in Chinese)

Wu H, Zhu JR, Shen J, Wang H (2011) Tidal modulation on the Changjiang river plume in summer. J Geophys Res 116:C08017. doi:10.1029/2011JC007209

Wu H, Deng B, Yuan R, Hu J, Gu JH, Shen F, Zhu JR, Zhang J (2013) De-tiding measurement on transport of the Changjiang derived buoyant coastal current. J Phys Oceanogr 43:2388–2399

Yang SL, Zhang J, Zhu J, Smith JP, Dai SB, Gao A (2005) Impact of dams on Yangtze River sediment supply to the sea and delta wetland response. J Geophys Res 110:F03006. doi:10.1029/2004JF000271

Yang SL, Zhang J, Xu XJ (2007) Influence of the Three Gorges Dam on downstream delivery of sediment and its environmental implications, Yangtze River. Geophys Res Lett 34:L10401. doi:10.1029/2007GL029472

Ying XM, Ding PX, Wang ZB, Van Maren DS (2012) Morphological impact of the construction of an offshore Yangshan Deepwater Harbor in the Port of Shanghai, China. J Coastal Res 28:163–173

Yu GH, Martin JM, Zhou JY (1990) Biogeochemistry of the Changjiang Estuary. China Ocean Press, Beijing, pp 1–898

Zhang J, Su JL (2006) Nutrient dynamics of the Chinese Seas: The Bohai, Yellow Sea, East China Sea and South China Sea. In: Robinson AR, Brink KH (eds) The Sea, vol 14. Harvard University Press, Cambridge MA, pp 637–671

Zheng YJ, Chen XZ, Cheng JH, Wang YL, Shen XQ, Chen WZ, Li CS (2003) Biological resources and environment of the East China Sea Shelf. Shanghai Science and Technology Press, Shanghai, pp 1–835 (in Chinese)

Zhou MJ, Zhu MY, Zhang J (2001) Status of harmful algal blooms and related research activities in China. Chinese Bull Life Sci 13:54–60 (in Chinese)

Zhu ZY, Zhang J, Wu Y, Zhang YY, Lin J, Liu SM (2011) Hypoxia off the Changjiang (Yangtze River) Estuary: oxygen depletion and organic matter decomposition. Mar Chem 125:108–116

Hydrodynamics of the Changjiang Estuary and Adjacent Seas

2

Jianrong Zhu, Hui Wu, and Lu Li

Abstract

In this chapter, we use our previous observational and modeling studies to analyze some basic dynamic features of the Changjiang (Yangtze River) Estuary, including the tidal and subtidal motion, saltwater intrusion, and the extension of a river plume into the ocean. The dynamics of this estuary are strongly regulated by tides on both intertidal and subtidal scales. Harmonic analysis shows that the tidal regime has one dominant semidiurnal constituent. The tidal range has significant seasonal variations, but is typically between 4 and 5 m during the spring tide and ~2 m during the neap tide. Tidal forcing induces significant subtidal circulation, resulting in a net landward flow into the North Branch of the Changjiang Estuary when river discharge is low. This residual transport forms a type of saltwater intrusion known as the saltwater-spilling-over (SSO). This causes high-salinity water in the North Branch to intrude upstream and finally spill over into the South Branch, which is the major channel of the estuary. The SSO has a major impact on the Changjiang Estuary and on the availability of freshwater for nearby cities. In areas near the river mouth, the tide-induced residual transport also results in significant inter-channel water mass exchange. Freshwater output from the Changjiang forms a large-scale plume, which plays a key role in physical processes in the adjacent East China Sea and Yellow Sea. We found that, in the region close to the river mouth, tidal forcing has a marked effect on the extension of the plume into the open ocean. Wind forcing also plays an important role in the dynamics of the Changjiang Estuary. Northerly winds can enhance the SSO and saltwater intrusion around the river mouth. In the plume-affected region, the wind significantly modulates the plume shape, especially farther from the river mouth.

Keywords

Tide • Subtidal circulation • Saltwater intrusion • River plume • Wind forcing

J. Zhu (✉) · H. Wu · L. Li
State Key Laboratory of Estuarine and Coastal Research,
East China Normal University, 3663 Zhongshan Road North,
Shanghai 200062, China
e-mail: jrzhu@sklec.ecnu.edu.cn

H. Wu
e-mail: hwu@sklec.ecnu.edu.cn

L. Li
e-mail: lilusklec@gmail.com

J. Zhang (ed.), *Ecological Continuum from the Changjiang (Yangtze River) Watersheds to the East China Sea Continental Margin*, Estuaries of the World,
DOI 10.1007/978-3-319-16339-0_2,

2.1 Introduction

The Changjiang, also known as the Yangtze River in the region between Nanjing and the river mouth, originates in the Tibetan Plateau and flows more than 6000 km through 10 provinces of China, before entering the East China Sea near Shanghai (Fig. 3.1; see Chap. 3). The Changjiang has the fourth largest mean sediment load and fifth largest discharge of rivers worldwide, delivering $\sim 9.24 \times 10^{11}$ m^3/year of freshwater to the Yellow Sea and East China Sea (Shen et al. 2003). The Changjiang is the most important source of terrestrial materials deposited into the East China Sea and the Yellow Sea (see Chaps. 3, 5 and 6).

The Changjiang Estuary has a tidal limit of nearly 640 km, a 90-km-wide river mouth, and is characterized by multiple bifurcations due to sand islands (Fig. 2.1). Downstream from Xuliujing, the estuary is divided by Chongming Island into the South Branch (SB) and North Branch (NB). The SB and its lower reaches form the main channel of the Changjiang and contain most of the river discharge, while the NB is heavily silted. The waterways in the Changjiang Estuary are far wider than most other estuaries, with widths of between ~ 10 and ~ 20 km. The subaqueous delta, adjacent to the Changjiang mouth, occurs at water depths that gradually increase to between 50 and 60 m. The lower SB is bifurcated into the South Channel (SC) and the North Channel (NC) by Changxing Island and Hengsha Island. Finally, the SC is bifurcated into the South Passage (SP) and the North Passage (NP) by Jiuduan **Sandbank**. Thus, four outlets, separated by extensive intertidal flats, discharge from the Changjiang Estuary into the sea.

The natural evolution and artificial reclamation of the intertidal zone from the 1950s to the 1970s have severely narrowed the upper reaches of the NB. As a consequence, the upper reaches have become almost orthogonal to the SB, while the lower reaches have become funnel-shaped (Shen et al. 2003). These modifications not only help to prevent runoff from entering the NB, especially during the dry season, but also make the tidal range larger than in the SB. This high tidal range results in a process known as the saltwater-spilling-over (SSO), whereby saltwater from the NB spills into the SB during dry-season floods.

The Changjiang Estuary is unique among well-studied estuaries, due to its large size, high river discharge, considerable tidal range, and relatively high suspended-sediment concentration (see Chap. 3). It also has an extremely

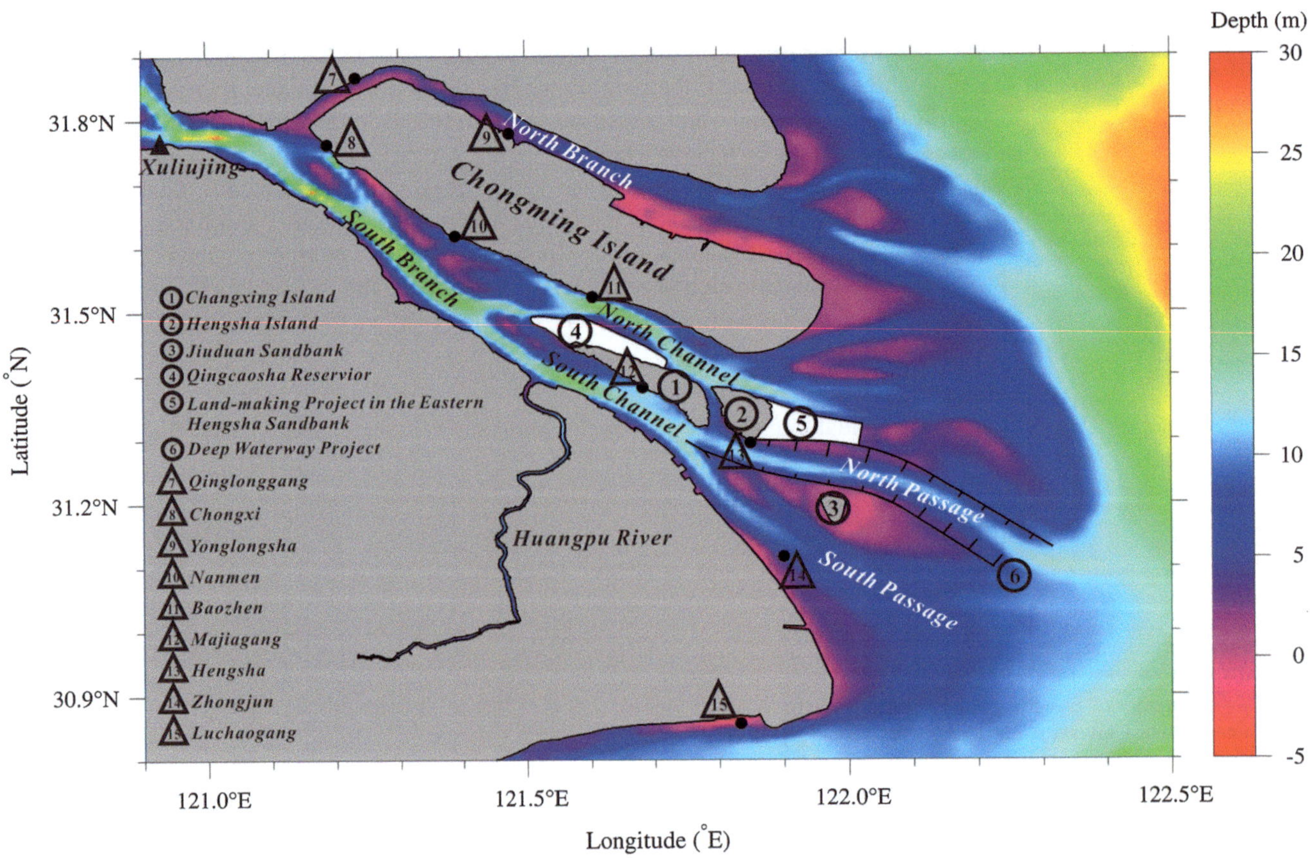

Fig. 2.1 Map of the Changjiang Estuary. Key geographic locations and tide gauges are *numbered* and marked by *circles* and *triangles*, respectively

dynamic hydrological environment due to significant shelf currents off the river mouth and the East Asian monsoon. These dynamics have a strong influence on tides, circulation, saltwater intrusion, plume dispersal, sedimentary erosion/deposition (see Chap. 4), and other biogeochemical processes (see Chaps. 5 and 6).

Runoff, tides, winds, the continental-shelf current, mixing, and topography are the main controls on hydrodynamic processes in the Changjiang Estuary (Wu et al. 2006; Xiang et al. 2009; Li et al. 2012). The river discharge and winds have a strong seasonality. Winds are controlled primarily by the monsoon, which brings weak southerly winds in summer and strong northerly winds in winter. Tides exhibit a medium tidal range and are the most energetic source of water movement in the Changjiang Estuary. The tidal range introduces a significant tidal current, which oscillates and mixes dissolved materials. Saltwater intrusion into the river mouth can also produce estuarine circulation (Pritchard 1956) and change the stratification (Simpson et al. 1990), thereby influencing the movement of sediments and producing maximum estuarine turbidities (Geyer 1993; see Chap. 3).

The East China Sea and the Yellow Sea are adjacent to the Changjiang mouth. The water depths and summertime circulation patterns of these seas are shown in Fig. 2.2. The East China Sea has the flattest and widest continental shelf in the world, which is bounded to the north by a line running northeast from the northern edge of the Changjiang mouth to the southwestern tip of Korea, to the east by the Ryukyu islands chain and Kyushu, to the south by Taiwan, and to the west by the east coast of the Chinese mainland (Fig. 2.2). The Yellow Sea is a shallow, marginal sea bounded to the north by the Bohai Sea, to east by Korea, to south by the East China Sea, and to the west by the east coast of China (Fig. 2.2).

Intense land–ocean interactions occur between the Changjiang Estuary and the adjacent seas. The coastal current brings sediment and other materials into the estuary, and the ocean produces tides and tidal currents that cause

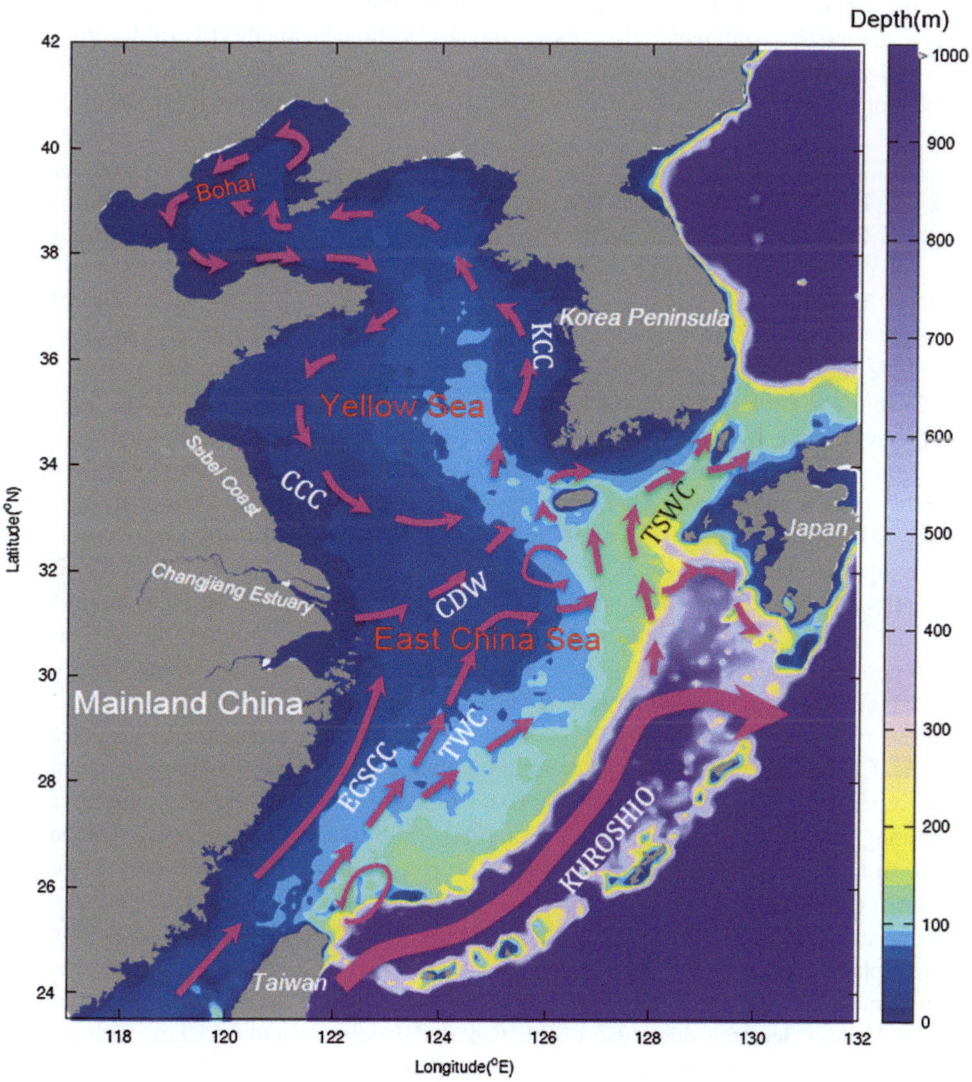

Fig. 2.2 Water depth and summertime circulation patterns in the East China Sea, Yellow Sea, and Bohai Sea. Circulation patterns in the Bohai Sea are adapted from Su and Yuan (2005), and circulation patterns in the East China Sea and Yellow Sea are adapted from Lü et al. (2006). TWC is the Taiwan Warm Current; ECSCC is the East China Sea Coastal Current; CDW is the Changjiang Diluted Water; CCC is the Chinese Coastal Current; KCC is the Korean Coastal Current; and TSWC is the Tsushima Warm Current

saltwater intrusion and that influence circulation, sediment flocculation, the ecological environment, and freshwater availability.

Huge quantities of freshwater from the Changjiang enter the East China Sea and Yellow Sea. This water flux, known as Changjiang diluted water (CDW), is a prominent hydrographic feature in the East China Sea and the Yellow Sea. In summer, it extends northeast and forms a surface plume (Fig. 2.2), and in winter it flows south in a narrow band along the East China coast. The Changjiang carries an enormous volume of water, sediments, and nutrients to the East China Sea and Yellow Sea (see Chaps. 3, 5, and 6), which significantly influences the water properties, circulation structure, sediment deposition, and ecosystems of the adjacent seas (Beardsley et al. 1985; Zhu and Shen 1997).

This chapter describes the hydrodynamics of the Changjiang mouth and the land–ocean interactions in this region. Section 2.2 describes the basic features of the tides and residual water transport, and Sect. 2.3 provides an overview of the impact of wind on saltwater intrusion. Section 2.4 describes the dispersal of the CDW riverine plume across the continental shelf based on observational data and numerical simulations of tidal modulation of plume dispersion.

2.2 Tides and Residual Water Transport

2.2.1 Tides

Tides are the most energetic source of water movement in the Changjiang Estuary (Shen et al. 2003). According to data from Luchaogang hydrographic station (Fig. 2.1) collected in 2006 and 2007, the tidal range exhibits significant monthly, semimonthly, and semidiurnal signals. During the spring tide, the amplitude is generally greater than 450 cm and is approximately twice that of the neap tide, which is typically less than 250 cm. Around the Changjiang Estuary, the tidal amplitude is increased due to shoaling effects. The mean annual tidal range at Zhongjun station, for example, is 267 cm and the maximum tidal range is 462 cm (Shen et al. 2003).

The Changjiang Estuary is influenced by two distinct tidal systems: a progressive tide from the East China Sea and a rotating tide from the Yellow Sea (Choi 1980). The tidal wave from the northwest Pacific propagates nearly parallel to the Ryukyu Islands, entering the East China Sea, and moving northwest toward the Changjiang Estuary. Shallow topography adjacent to the river mouth causes denser co-phase lines and a reduction in the velocity of the tidal wave. In the Yellow Sea, the tide rotates with two nodal points (Choi 1980), with the southern node having a dominant influence in the northern Changjiang Estuary. Convergence between the progressive tide and the rotating tide produces a special region along a line from the Subei Coast (Fig. 2.2) to the southwest tip of the Korean Peninsula, where the tide becomes standing.

Because of the complicated river regime, the tide acts in distinct ways in different outlets. For example, as the NB is funnel-shaped, there is significant reflection of the tidal wave, and a standing wave is formed. In the SB, on the other hand, the reflection is relatively weak and the tide propagates evenly, with only minor distortion of the tidal wave.

To conduct a detailed investigation of the tidal characteristics inside the Changjiang mouth, we used tidal data from six hydrological stations: Hengsha, Majiagang, Baozhen, Yonglongsha, Chongxi, and Nanmen (Fig. 2.1). Data were sampled hourly throughout 2009, providing sufficient resolution to reveal the temporal variability of tides in this region.

Although the in situ tidal elevation is a straightforward way to describe water-level fluctuations, as well as the associated fortnightly and seasonal variations, a better method is to decompose the observed data into a series of constituents with different periods using harmonic analysis. Although infinite numbers of harmonic constituents can be generated, it is generally only practical to consider several of the major constituents. In this study, we used harmonic analysis to characterize the key tidal constituents, along with the tidal form and distortion.

2.2.1.1 Tidal Elevation and Tidal Range

The tidal elevation not only exhibited semidiurnal variability, but also had a significant fortnightly variability (Fig. 2.3). Water levels in December were generally lower than in September due to the reduction in river discharge. Mean monthly water levels at the six stations all show significant seasonal variability, reflecting the variability in river discharge, with a peak values around August, and low values in January (Table 2.1). Mean monthly water levels at Yonglongsha station were lower than the other five stations (174 cm in January and 219 cm in August), because the tidal flat at the upper reaches of the NB restrains the runoff into the NB.

Figure 2.3 also clearly illustrates the temporal variation of the tidal range at the six stations in September and December 2009. For example, at Hengsha station, the maximum tidal range of 360 cm occurred during a spring tide on September 20 and was only 130 cm on September 27 during a neap tide. The difference between the spring and neap tidal ranges was primarily due to differences in the high-tide water level, which was more variable during the neap tide. Two high-tide water levels on September 14 during a neap tide were 235 and 305 cm, with a difference exceeding half of the tidal range at that time.

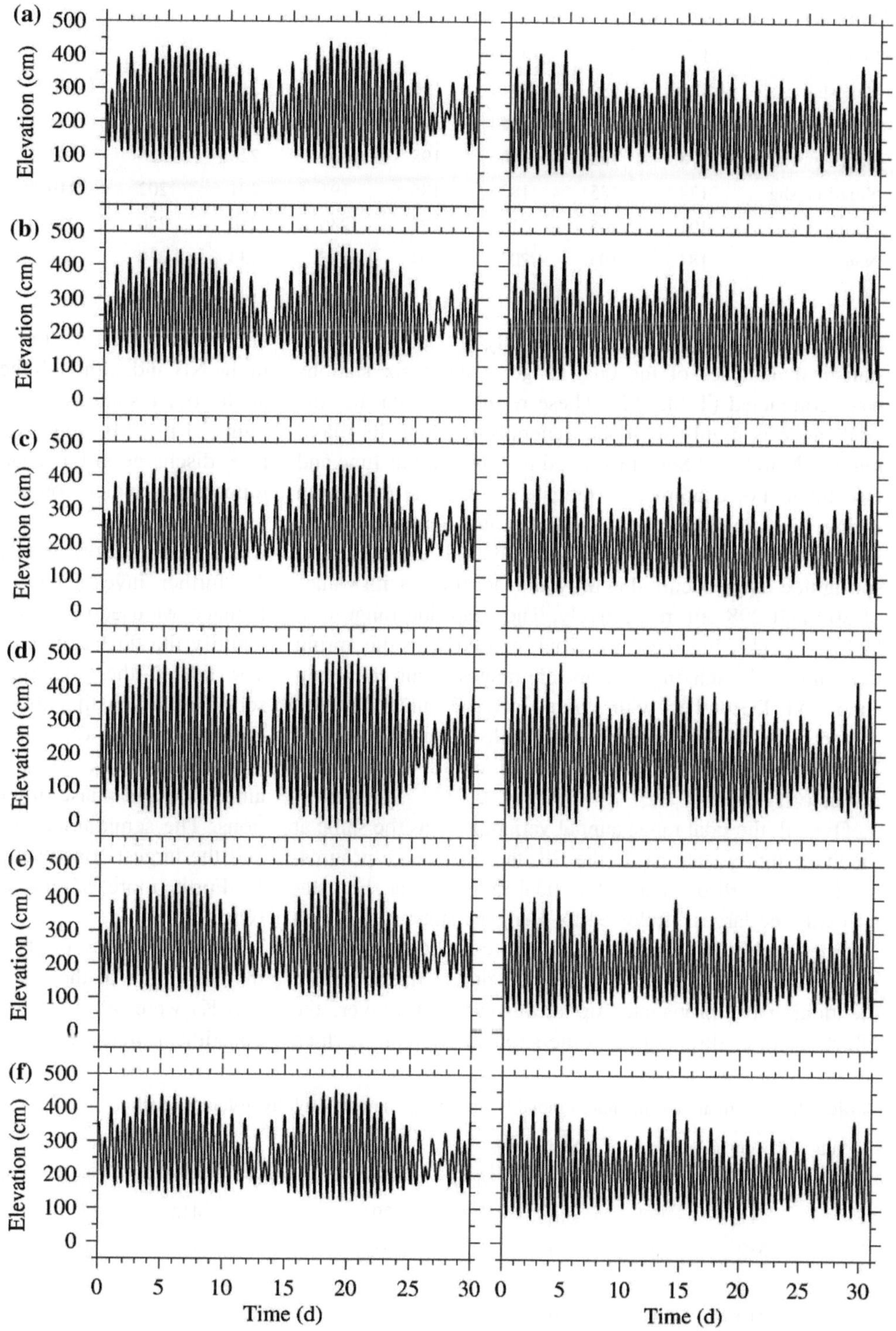

Fig. 2.3 Temporal variation of the measured water level (tidal elevation) in 2009 at: **a** Hengsha station; **b** Majiagang station; **c** Baozhen station; and **d** Yonglongsha station, during September (*left panels*) and December (*right panels*)

Such diurnal inequality is mainly caused by the gravitational pull of the earth, moon, and sun. The maximum tidal range in December was also smaller than in September when the earth, moon, and sun are aligned and the gravitational pull combines to produce a spring tide. The fortnightly variability of the tidal range was weaker in December than in September. The seasonal and fortnightly variations in September and December at the other five stations are comparable to those at Hengsha station (Fig. 2.3b–d). Of note, the tidal range at Yonglongsha station exceeded 500 cm and reached 400 cm in December. This range is considerably larger than at other stations due to the location of Yonglongsha station in the NB, where the funnel-shaped topography concentrates flow and the tidal forcing is stronger than river discharge.

Table 2.1 Mean monthly water level (cm) at each hydrological station

Month	1	2	3	4	5	6	7	8	9	10	11	12
Hengsha	178	189	191	196	206	215	217	236	234	215	206	195
Majiagang	184	198	204	207	217	225	230	251	249	228	220	210
Baozhen	178	190	196	198	208	220	226	243	238	213	205	192
Yonglongsha	174	185	182	185	191	201	202	219	216	201	201	188
Chongxi	191	205	218	221	236	250	256	277	267	234	219	206
Nanmen	181	195	203	204	215	234	237	257	250	221	206	193

To further investigate the variability in tidal intensity, statistical analyses of the tidal range at these six stations were conducted (Table 2.2). These results indicate that the tidal range is highly seasonal, generally reaching its maximum in March and September, and its minimum in June and December. For example, at Hengsha station, the maximum spring tide range occurred in March and September with values of 352 and 364 cm, respectively, and the minimum spring tide range occurred in May and December with values of 309 and 298 cm, respectively. The neap tide range also had two peaks, but these were anti-phased with the spring tide range. The maximum neap tide ranges occurred during June and December with values of 167 and 137 cm, respectively, and the minimum neap tide ranges occurred in February and September with values of 78 and 87 cm, respectively.

Overall, the tidal range annual variability was the same at the six gauging stations. In the SB, the tidal range progressively decreased upstream. The tidal range at Baozhen station was the largest, followed by Hengsha station, and then Nanmen and Chongxi stations, which are inside the SB. This pattern is caused by energy dispersion due to friction when the tidal wave propagates up the estuary. Moreover, the effects of river discharge become stronger upstream, which also reduces the tidal range. Yonglongsha station is located in the NB and, although the seasonal variation was similar to at the other stations, the tidal range was larger. The tidal range of the NB is amplified due to the comparatively low river discharge in this channel, as well as the funnel-shaped topography, which concentrates the tidal energy.

2.2.1.2 Harmonic Analysis

To further investigate the tidal form in the Changjiang Estuary, we used the harmonic analysis program T_TIDE to identify the tidal constituents from the tidal elevation data. The tide in the Changjiang Estuary is regulated by four semidiurnal constituents (M_2, S_2, N_2, and K_2); four diurnal constituents (K_1, O_1, P_1, and Q_1); and three shallow-water constituents (M_4, MS_4, and M_6). Table 2.3 gives the amplitude and phase of each constituent at all gauging stations. The semidiurnal constituents were dominant overall, and the largest constituents within this group were M_2 and S_2. Furthermore, M_2 has an amplitude that is approximately twice that of S_2.

At Hengsha station, the amplitude of M_2 was 114 cm and the amplitude of the other semidiurnal constituents S_2, N_2, and K_2 were 49, 19, and 14 cm, respectively. The diurnal constituents had considerably smaller amplitudes. Of this

Table 2.2 Maximum and minimum monthly tidal range (cm) at each hydrological station

Month	Hengsha		Majiagang		Baozhen		Yonglongsha		Chongxi		Nanmen	
	Spring	Neap	Spring	Neap	Spring	Neap	Spring	Neap	Spring	Neap	Spring	Neap
1	311	122	305	121	294	106	410	135	272	112	300	112
2	332	78	330	77	317	66	442	107	295	68	320	105
3	352	80	336	75	321	64	455	50	292	67	327	68
4	319	86	320	83	303	74	431	112	281	75	313	74
5	309	137	312	130	293	117	433	184	277	114	302	120
6	317	167	322	156	306	144	443	221	283	137	310	144
7	350	122	349	114	324	102	475	140	301	97	333	101
8	356	97	358	92	333	85	517	64	309	82	344	84
9	364	87	368	83	338	79	500	62	322	81	360	82
10	339	89	350	90	331	81	459	125	312	86	340	83
11	311	125	318	123	306	115	419	156	285	117	314	118
12	298	136	303	128	293	118	382	165	275	117	298	121

Table 2.3 Tidal harmonic constants (amplitude and phase) of the major tidal constituents at each tide gauge

Constituents	Hengsha		Majiagang		Baozhen		Yonglongsha		Chongxi		Nanmen	
	amp	phs	amp	phs	amp	phs	amp	phs	amp	phs	amp	phs
M_2	114	112	109	128	101	1401	137	143	93	185	100	161
S_2	49	153	47	168	44	179	61	182	39	226	43	201
N_2	19	102	19	117	18	130	25	133	16	172	18	149
K_2	14	147	13	164	13	173	18	174	11	222	12	196
K_1	25	83	25	90	23	90	23	71	22	115	23	102
O_1	16	43	16	49	15	48	13	30	15	72	15	60
P_1	6	93	6	98	6	99	6	93	6	121	6	110
Q_1	3	30	3	34	2	34	2	21	2	54	2	44
M_4	16	151	16	187	16	209	24	219	19	286	20	245
MS_4	13	192	14	229	14	251	22	265	16	328	17	288
M_6	4	149	4	184	2	238	3	213	4	149	4	301

group, K_1 was the strongest with an amplitude of 25 cm, followed by O_1 with an amplitude of 16 cm. Notably, the shallow-water tidal constituents around the Changjiang Estuary were relatively high, and the amplitudes of M_4 and MS_4 were both around 15 cm. M_6 was a very minor constituent with an amplitude of less than 5 cm. The phases of the semidiurnal tides were between 102° and 153°, and the phases of the diurnal tides were between 43° and 93°. The phases of shallow-water tides were larger overall; e.g., MS_4 had a phase of 192°.

The amplitudes and phases of each tidal constituent at Majiagang and Baozhen stations were similar to those at Hengsha station. Semidiurnal tides were dominant, but shallow-water constituents were also significant. The semidiurnal and diurnal tides at Nanmen and Chongxi stations were slightly smaller, and the shallow-water constituents were larger compared with downstream stations. However, due to bottom friction, their amplitudes were suppressed, particularly for the semidiurnal tidal constituents. Moreover, tidal distortion became more severe farther up the estuary, which is particularly evident in the increased amplitudes and phase of the M_4 and MS_4 constituents.

The tidal phase at Yonglongsha station was similar to the other stations. Amplitudes of the semidiurnal tidal constituents at Yonglongsha station were larger than at other stations, particularly for M_2, which had an amplitude of 137 cm. However, the amplitudes of the diurnal tides were comparatively small at this station, with an amplitude of only 22 cm for K_1. This indicates that the tidal form is more semidiurnal in the NB than in the SB. Moreover, the shallow-water tides at Yonglongsha station were amplified, with the amplitudes for M_4 and MS_4 being 50 % greater than those at Hengsha station. This indicates that the tidal distortion in the NB is also more pronounced. In extreme tides, a tidal bore occurs in the NB.

Based on the results of the harmonic analysis, the tidal form can be classified using the ratio of semidiurnal to diurnal tidal amplitudes:

$$F = (H_{O1} + H_{K1})/(H_{M2} + H_{S2}) \tag{2.1}$$

where F is the tidal form number, and H_{o1}, H_{k1}, H_{m2}, and H_{s2} are the amplitudes of the tidal constituents of O_1, K_1, M_2, and S_2, respectively.

Tidal form is classified as regular semidiurnal when $0.0 < F \le 0.25$; irregular semidiurnal when $0.25 < F \le 1.5$; irregular diurnal when $1.5 < F \le 3.0$; and regular diurnal when $F > 3.0$ (Chen 1980). F was greater than 0.25 at all stations aside from Yonglongsha station, where it was less than 0.25 (Table 2.4), indicating that the tidal form is regular semidiurnal in the NB and irregular semidiurnal in the SB, NC, SC, and NP.

Because of the influences of the friction, runoff, and topography, the tidal wave is often distorted in the shallow estuary region. The distortion can be quantified using the ratio of M_4 to M_2 amplitudes; i.e., $A = H_{M4}/H_{M2}$. The flood/ebb asymmetry can be quantified using phase difference; i.e., $G = 2\varphi_{M2} - \varphi_{M4}$. When $0° < G \le 180°$, the ebb duration is longer than the flood duration, and the flood current is stronger than the ebb current. In contrast, when $180° < G \le 360°$, the flood duration is longer than the ebb duration, and the ebb current is stronger than the flood current.

The value of A was greater than 0.1 at all stations (Table 2.4), indicating that tide distortion due to shallow water is significant in the Changjiang Estuary. The tide distortion was comparatively small at Hengsha station ($A = 0.137$), but stronger at Yonglongsha station ($A = 0.173$). The G values at all six stations were between 68.10 and 84.84, indicating that the ebb duration was longer than the

Table 2.4 The tidal form number, F, and the tidal distortion coefficients, A and G, at each tide gauge

	Hengsha	Majiagang	Baozhen	Yonglongsha	Chongxi	Nanmen
F	0.252	0.259	0.258	0.178	0.283	0.268
A	0.137	0.150	0.159	0.173	0.206	0.204
G	73.66	68.10	72.9	68.17	84.84	77.13

flood duration, and the flood current was stronger than the ebb current.

Although G values reveal the asymmetry of the ebb/flood duration in the Changjiang Estuary, they fail to represent spatial variability of ebb/flood currents. In the SB, SC, and NC, for example, the ebb current is stronger, while in the NB the flood current is stronger. Due to the very high river discharge, the ebb current is generally more dominant in the Changjiang Estuary (Qiu and Zhu 2013), which is in contrast to the results produced using G values. Because G considers only the tidal constituents and neglects the effects of river discharge, it should be used with caution in the areas with very high river discharge, such as the Changjiang Estuary.

2.2.2 Residual Water Transport

Subtidal circulation, also known as residual water transport, governs the long-term transport of salt and other soluble and conservation materials. Subtidal circulation in estuaries is mainly induced by low-frequency forces, such as runoff and wind, baroclinic gradient forces, and the tide itself (Pritchard 1956).

The most commonly used method to calculate residual movement is the Euler residual current, which uses arithmetic averages or low-pass filters to cut off the periodically oscillating component of the tidal current. However, in regions where the tidal range is large relative to the water depth, the Euler residual current might be misleading. To solve this problem, Ianniello (1977) introduced the nonlinear factor $\varepsilon = \zeta/D$, where ζ is the tidal range and D is the total depth. The Euler residual is reasonable only if ε is <0.3. In the extensive shallow intertidal zone of the Changjiang Estuary, ε is undoubtedly >0.3, so the Euler residual current is not a suitable method for describing mass transport in this region. Another method that can be used to calculate the residual is the Lagrangian residual, but this method requires a tracer study, is strongly related to the phase at which the tracking particle is released, and is computationally expensive.

We used the residual water transport method to describe the net water mass flux. The water mass flux is defined as the amount of water mass that passes through an area in a unit of time. In this method, the unit width flux is averaged over one or several tidal cycles to calculate the residual (net) unit width water mass flux (NUWF). The NUWF represents the tidally averaged transport rate and direction of water mass at a specified location and is calculated as follows:

$$\vec{R}(x,y) = \frac{1}{T}\int_0^T \int_{-H(x,y)}^{\zeta} \vec{V}(x,y,z,t)\,dz\,dt \qquad (2.2)$$

where $\vec{R}(x,y)$ is the NUWF, T is one or several tidal cycles, $H(x,y)$ is the steady depth water, ζ is the surface elevation, and $\vec{V}(x,y,z,t)$ is the horizontal current vector.

2.2.2.1 Numerical Model

ECOM-si was developed based on the POM (Princeton Ocean Model) (Blumberg and Mellor 1987; Blumberg 1994) with several improvements (Chen et al. 2001) to address the demand for numerical simulations of water bounded by complicated coastlines. This model incorporates the Mellor and Yamada level 2.5 turbulent closure scheme to provide a time-and-space-dependent parameterization of vertical turbulent mixing (Mellor and Yamada 1974, 1982; Galperin et al. 1988).

Under the assumption of incompressibility, using the Boussinesq and hydrostatic approximations, and introducing the horizontal non-orthogonal curvilinear and vertical-stretched sigma coordinate system, the governing equations of ocean circulation and water mass (consisting of momentum, continuity, temperature, salinity, and density equations) are as follows:

$$\begin{aligned}
&\frac{\partial DJu_1}{\partial t} + \frac{\partial DJ\hat{U}u_1}{\partial \xi} + \frac{\partial DJ\hat{V}u_1}{\partial \eta} + \frac{\partial J\omega u_1}{\partial \sigma} \\
&- Dh_2\hat{V}\left[v_1\frac{\partial}{\partial \xi}\left(\frac{J}{h_1}\right) - u_1\frac{\partial}{\partial \eta}\left(\frac{J}{h_2}\right) + Jf\right] \\
&- Dh_2u_1v_1\frac{\partial}{\partial \xi}\left(\frac{h_3}{h_1h_2}\right) \\
&= -h_2gD\frac{\partial \zeta}{\partial \xi} + \frac{gh_2D}{\rho_o}\frac{\partial D}{\partial \xi}\int_{\sigma}^{0}\sigma\frac{\partial \rho}{\partial \sigma}d\sigma \\
&- \frac{gh_2D^2}{\rho_o}\frac{\partial}{\partial \xi}\int_{\sigma}^{0}\rho d\sigma + \frac{1}{D}\frac{\partial}{\partial \sigma}\left(K_m\frac{\partial Ju_1}{\partial \sigma}\right) + DJF_x, \qquad (2.3)
\end{aligned}$$

$$\frac{\partial DJv_1}{\partial t}+\frac{\partial DJ\hat{U}v_1}{\partial \xi}+\frac{\partial DJ\hat{V}v_1}{\partial \eta}+\frac{\partial J\omega v_1}{\partial \sigma} + Dh_1\hat{U}\left[v_1\frac{\partial}{\partial \xi}\left(\frac{J}{h_1}\right)-u_1\frac{\partial}{\partial \eta}\left(\frac{J}{h_2}\right)+Jf\right] - Dh_1u_1v_1\frac{\partial}{\partial \eta}\left(\frac{h_3}{h_1h_2}\right)$$
$$= -h_1gD\frac{\partial \zeta}{\partial \eta}+\frac{gh_1D}{\rho_o}\frac{\partial D}{\partial \eta}\int_\sigma^0 \sigma\frac{\partial \rho}{\partial \sigma}d\sigma - \frac{gh_1D^2}{\rho_o}\frac{\partial}{\partial \eta}\int_\sigma^0 \rho d\sigma+\frac{1}{D}\frac{\partial}{\partial \sigma}\left(K_m\frac{\partial Jv_1}{\partial \sigma}\right)+DJF_y, \quad (2.4)$$

$$\frac{\partial \zeta}{\partial t}+\frac{1}{J}\left[\frac{\partial}{\partial \xi}(DJ\hat{U})+\frac{\partial}{\partial \eta}(DJ\hat{V})\right]+\frac{\partial \omega}{\partial \sigma}=0 \quad (2.5)$$

$$\frac{\partial JD\theta}{\partial t}+\frac{\partial JD\hat{U}\theta}{\partial \xi}+\frac{\partial JD\hat{V}\theta}{\partial \eta}+\frac{\partial J\omega\theta}{\partial \sigma} = \frac{1}{D}\frac{\partial}{\partial \sigma}\left(K_h\frac{\partial J\theta}{\partial \sigma}\right)+DJF_\theta \quad (2.6)$$

$$\frac{\partial JDs}{\partial t}+\frac{\partial JD\hat{U}s}{\partial \xi}+\frac{\partial JD\hat{V}s}{\partial \eta}+\frac{\partial J\omega s}{\partial \sigma} = \frac{1}{D}\frac{\partial}{\partial \sigma}\left(K_h\frac{\partial Js}{\partial \sigma}\right)+DJF_s \quad (2.7)$$

$$\rho_{total} = \rho_{total}(\theta, s) \quad (2.8)$$

where,

$$\omega = w-\sigma\left(\hat{U}\frac{\partial D}{\partial \xi}+\hat{V}\frac{\partial D}{\partial \eta}\right) - \left[(1+\sigma)\frac{\partial \zeta}{\partial t}+\hat{U}\frac{\partial \zeta}{\partial \xi}+\hat{V}\frac{\partial \zeta}{\partial \eta}\right] \quad (2.9)$$

In the above equations, the new coordinate (ξ, η, σ) is defined as: $\xi = \xi(x,y), \quad \eta = \eta(x,y), \quad \sigma = \frac{z-\zeta}{H+\zeta}$

The vertical coordinate σ varies from −1 at $z = -H$ to 0 at $z = \zeta$, where x, y, and z are the east, north, and vertical axes of the Cartesian coordinate, respectively; ζ is the sea surface elevation; and H the total water depth. The ξ and η components of velocity (defined as u_1, v_1) can be expressed in the form $u_1 = \frac{h_2}{J}(x_\xi u + y_\xi v), \quad v_1 = \frac{h_1}{J}(x_\eta u + y_\eta v)$, in which the u, v and x, y velocity components are $\xi_x = \frac{y_\eta}{J}, \quad \xi_y = -\frac{x_\eta}{J}, \quad \eta_x = -\frac{y_\xi}{J}, \quad \eta_y = \frac{x_\xi}{J}$, where J is the Jacobin function in the form of $J = x_\xi y_\eta - x_\eta y_\xi$, and the subscript symbols (ξ and η) indicate derivatives. The metric factors h_1 and h_2 of the coordinate transformation are defined as $h_1 = \sqrt{x_\xi^2+y_\xi^2}$, $h_2 = \sqrt{x_\eta^2+y_\eta^2}$, $\hat{U} = \frac{1}{J}\left(h_2u_1-\frac{h_3}{h_1}v_1\right), \quad \hat{V} = \frac{1}{J}\left(h_1v_1-\frac{h_3}{h_2}u_1\right)$, in which $h_3 = y_\xi y_\eta + x_\xi x_\eta$, where θ is the water temperature, s is the salinity, f is the Coriolis parameter, g is the gravitational acceleration, K_m is the vertical eddy viscosity coefficient, and K_h is the thermal vertical eddy friction coefficient. F_u, F_v, F_θ and F_s represent the horizontal momentum, thermal, and salt diffusion terms, respectively. ρ and ρ_o are the perturbation and reference density, which satisfy $\rho_{total} = \rho+\rho_o$. F_u, F_v, F_θ and F_s are calculated by Smagorinsky's (1963) formula in which horizontal diffusion is directly proportional to the product of horizontal grid sizes. K_m and K_h are calculated using the modified Mellor and Yamada (1974, 1982) level 2.5 turbulent closure scheme.

The surface and bottom boundary conditions for the momentum and heat equations are given by:

$$\frac{\rho_0K_m}{D}\left(\frac{\partial u_1}{\partial \sigma},\frac{\partial v_1}{\partial \sigma}\right) = (\tau_{0\xi},\tau_{0\eta});\ \frac{\rho_0K_H}{D}\left(\frac{\partial \theta}{\partial \sigma}\right) = Q_{net};\ \frac{\rho_0K_H}{D}\left(\frac{\partial s}{\partial \sigma}\right) = s(\hat{P}-\hat{E});\ \omega = 0,\ \text{at } \sigma = 0,$$
$$\frac{\rho_0K_m}{D}\left(\frac{\partial u_1}{\partial \theta},\frac{\partial v_1}{\partial \theta}\right) = (\tau_{b\xi},\tau_{b\eta});\ \frac{\partial \theta}{\partial \sigma} = 0;\ \frac{\partial s}{\partial \sigma} = 0;\ \omega = 0,\ \text{at } \sigma = -1$$

where ($\tau_{o\xi}$, $\tau_{o\eta}$) and $(\tau_{b\xi},\tau_{b\eta}) = C_d\sqrt{U^2+V^2}(U^2+V^2)$ are the ξ and η components of surface wind and bottom stresses; Q_{net} is the net surface heat flux; $\hat{P}$ is the precipitation flux; and $\hat{E}$ is the evaporation flux. The surface wind stress was calculated based on the neutral steady-state drag coefficient developed by Large and Pond (1981). The drag coefficient C_d at the bottom is determined by matching a logarithmic bottom layer to the model at a height z_{ab} above the bottom; i.e.,

$$C_d = \max\left[k^2/\ln\left(\frac{z_{ab}}{z_0}\right)^2, 0.0025\right]$$

where k = 0.4 is Karman's constant, and z_0 is the bottom roughness parameter.

Equations 2.3–2.9 are solved prognostically as initial value-problems of oceanic motion. The initial velocity takes the form $u_1 = v_1 = 0$, and the water elevation is also set to be $\zeta = 0$. The initial temperature and salinity are specified using observational data.

This model was developed by Wu and Zhu (2010) with the third HSIMT-TVD scheme to solve the advection term in the mass transport equation. This scheme is flux-based with third-order accuracy in space, second-order accuracy in time, and no numerical oscillation. A wet/dry scheme was included to characterize the intertidal zone due to tidal excursion, and the critical depth was set to 0.2 m.

The model domain covered the whole Changjiang Estuary, Hangzhou Bay, and adjacent seas from 117.5 to 125 °E in

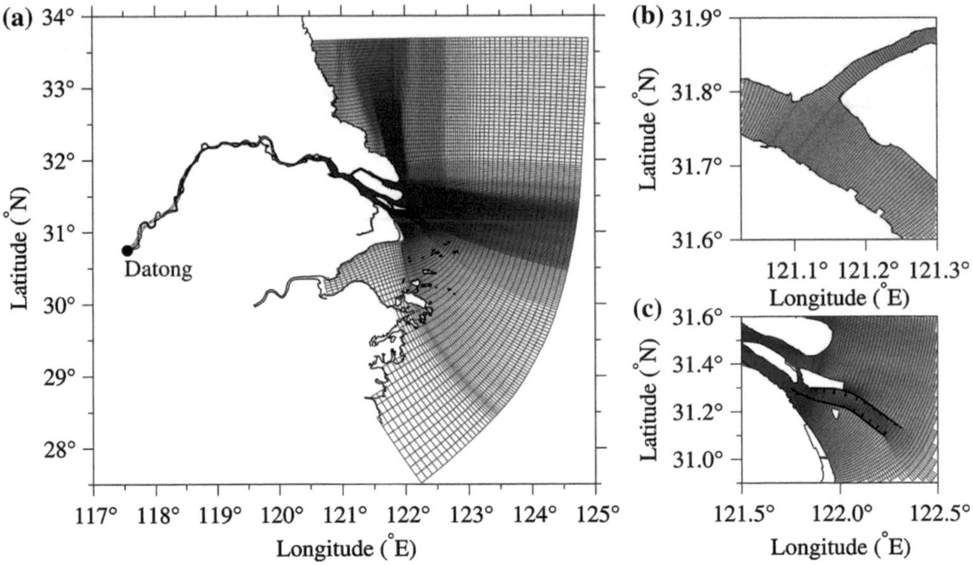

Fig. 2.4 **a** The numerical model mesh; **b** an enlarged view of the model mesh around the bifurcation of the NB and the SB; and **c** an *enlarged view* of the model mesh in the NP, showing the location of the Deep Waterway Project

longitude and 27.5 to 33.7 °N in latitude (Fig. 2.4a). The model grid mesh was composed of 307 × 224 cells horizontally and 10 uniform σ levels vertically. The mesh was designed to fit the coastline, with high-resolution grids near the Changjiang mouth, especially near the bifurcation of the SB and NB (Fig. 2.4b), and the NP where a deep waterway is maintained for navigation (Fig. 2.4c). A lower-resolution grid was used in open water. The grid resolution ranged from 300 to 600 m in proximity to the river mouth and was 15 km near open water. To simulate the SSO in our study, we reduced the grid near the upper NB to a resolution of about 100 m.

According to the *NaoTide* dataset (http://www.miz.nao.ac.jp/), the open sea boundary is driven by 16 astronomical constituents: M_2, S_2, N_2, K_2, K_1, O_1, P_1, Q_1, MU_2, NU_2, T_2, L_2, $2N_2$, J_1, M_1, and OO_1. The river boundary in the model is the location of Datong hydrographic station, which is 630 km upstream from the river mouth. Over the period 1950–2010, water levels are driven by daily river discharge on meteorological time scales and by mean monthly values on climatological time scales.

Wind fields, used to calculate sea surface momentum, were obtained from either the National Center for Environmental Prediction (NCEP) reanalysis dataset with a spatial resolution of 0.5° × 0.5° and a temporal resolution of 6 h, or from hydrological stations in the Changjiang Estuary. Figure 2.5 shows the semimonthly mean wind vectors in the Changjiang Estuary and illustrates that northerly wind prevails from September to March; southerly wind prevails from April to August; and wind speed is much stronger in winter than in summer.

The initial salinity and temperature are from data collected inside the river mouth and from the Ocean Atlas (Editorial Board for marine Atlas 1992). The velocity and elevation were initially set to zero. A detailed description of the model configuration and validation can be found in Wu et al. (2010, 2011) and Li et al. (2012).

2.2.2.2 Residual Water Transport in Winter

The NUWF for spring and neap tides was calculated using the numerical model for a winter case study. In this case, the mean monthly river discharge was 11 × 10^3 m^3/s, the mean northerly wind is 5 m/s, and a baroclinic force is induced by the salinity gradient in January.

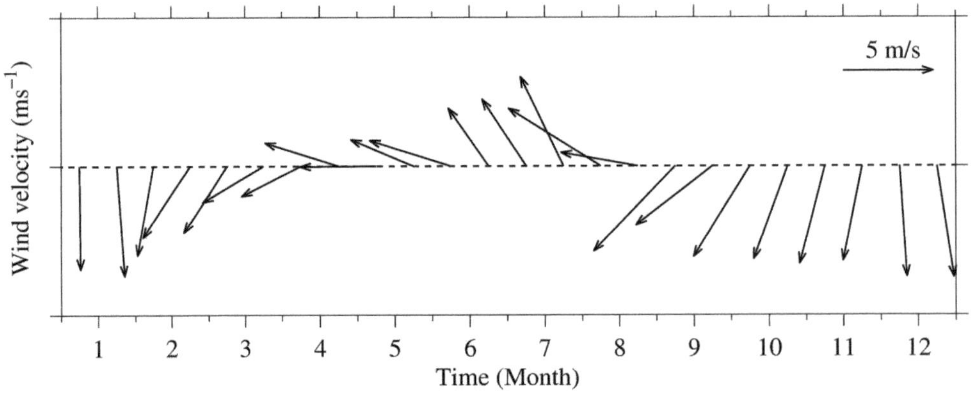

Fig. 2.5 Mean semimonthly wind vectors in the Changjiang Estuary

Figure 2.6a shows the winter NUWF during spring tides. In the SB, SC, NC, and NP, river discharge causes the NUWF to move seaward, and it has a higher magnitude in these channels than in the tidal flats (Fig. 2.6a). In the SP, the NUWF is landward because of the funnel shape produced by the south dyke and the strong salinity front. In sand bar areas, water is transported north across the tidal flats, which is consistent with observational results (Wu et al. 2010). The NUWF in the tidal flat east of Chongming Island is northward due to the tide-induced Stokes drift. In the NB, the NUWF is weaker and landward, and there is net water transport toward the SB.

Figure 2.6b shows the winter NUWF during neap tides. This has similar characteristics to spring tides (Fig. 2.6a) in the SB, SC, NC, and NP, but the magnitude is smaller. Because the tide is weaker, there is net seaward water transport in the southern tidal flats of the SP, and there is no northward water transport across the tidal flats at the river mouth. The effect of the northerly wind stress becomes obvious during neap tides, as the NUWF adjacent to the river mouth flows south. In the NB, the NUWF is even weaker, and there is no net water transport into the SB from the NB.

2.2.2.3 Residual Water Transport in Summer

The NUWF for spring and neap tides was calculated using the numerical model for a summer case study. In this case, the mean monthly river discharge was 49×10^3 m^3/s, the mean southerly wind was 4 m/s, and a baroclinic force was induced by the salinity gradient in July.

Figure 2.7a shows the summer NUWF during spring tides. In the SB, SC, NC, and NP, the NUWF is seaward and has a greater magnitude than for neap tides due to the high river discharge. In the SP, the NUWF is landward and has a lower magnitude than in other channels. The NUWF in the tidal flat east of Chongming Island is northward due to tidal pumping and southerly winds. In the NB, the NUWF is weaker and almost seaward, and the net water transport is into the NB from the SB.

Figure 2.7b shows the summer NUWF during neap tides. In the SB, SC, NC, SP, and NP, the pattern and magnitude of the NUWF are similar to spring tides (Fig. 2.7a), and most of the river water flows into the East China Sea and Yellow Sea through the SB, SC, SP, and NP.

2.3 Saltwater Intrusion

Saltwater intrusion into the Changjiang Estuary is controlled mainly by river discharge, which has a distinct seasonal variability, and by the tidal range (Shen et al. 2003; Wu et al. 2006; Li et al. 2010; Zhu et al. 2010), but is also influenced by wind stress and shelf circulation (Xiang et al. 2009; Wu et al. 2010; Li et al. 2012). Yang et al. (see Chap. 4) report that increased water consumption and dam construction have reduced the annual water discharge by 10 % in the last 150 years. This decrease is primarily attributable to the sharp reduction between August and November when the reservoirs primarily store water. In contrast, the water discharge in the dry season (January and February) has been increasing significantly due to water release from reservoirs (Qiu and Zhu 2013), and the decrease in river discharge in these months intensifies saltwater intrusion into the estuary.

The tide also controls saltwater intrusion into the estuary. On intertidal timescales, the semidiurnal tide drives saltwater into the estuary during flood tides and out of the estuary during ebb tides. The fortnightly spring tide generates greater saltwater intrusion than during the neap tide. Saltwater intrusion is also enhanced by the seasonal variability of tides. The tidal range reaches its maximum during March, and northerly winds produce landward Ekman water transport, which intensifies saltwater intrusion into the NB and the NC (Wu et al. 2010; Li et al. 2012). The Taiwan Warm Current (hereafter referred to as the TWC) and the Chinese Coastal Current (hereafter referred to as the CCC) (Fig. 2.2) also carry high-salinity water into the estuary (Xiang et al. 2009).

Because the Changjiang Estuary is bifurcated, the pattern of saltwater intrusion is different in the NB and SB. In the dry season, when river discharge is low, the NB almost always contains high-salinity water when tides are strong (Shen et al. 2003; Wu et al. 2006). In the lower reaches of the SB, saltwater mainly intrudes landward in a wedge-like manner, as is seen in other partially mixed estuaries (Shen et al. 2003).

The most remarkable feature of saltwater intrusion in the Changjiang Estuary is the SSO from the NB into the SB in the dry season (Box. 2.1). During the spring tide, the water level rises considerably in the upper reaches of the NB due to its funnel shape, leading to a massive amount of saltwater-spilling-over the shoals into the SB. Only a small amount of the saltwater returns to the NB because the shoals are exposed to the air during the ebb tide. The saltwater that is spilled into the SB is transported downstream along with the river discharge and arrives in the middle reaches of the SB during the subsequent neap tide. This effect impacts the Chenhang and Qingcaosha reservoirs and threatens Shanghai's water supply (Box 2.1).

Fig. 2.6 Distribution of the NUWF during winter for: **a** spring tides; and **b** neap tides

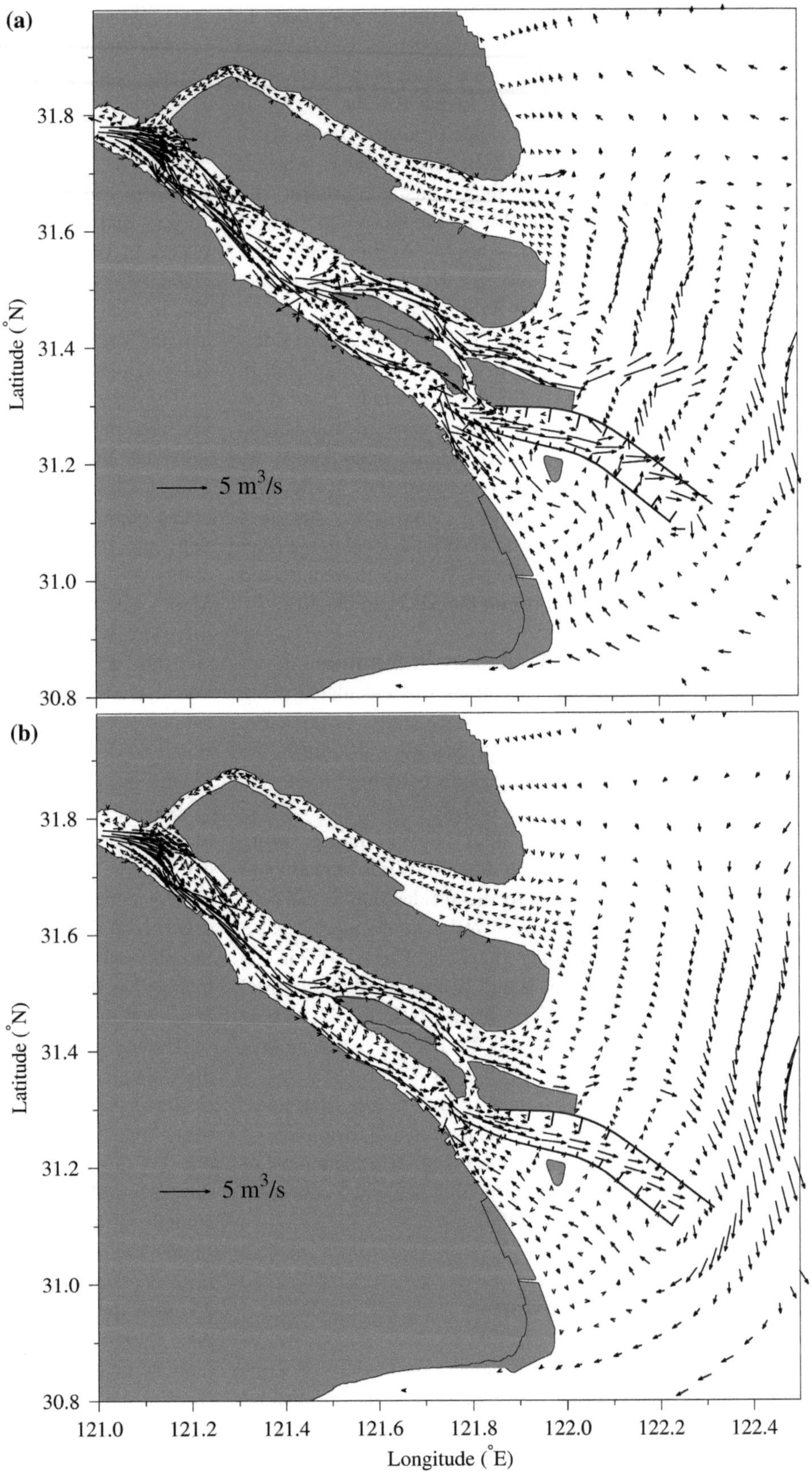

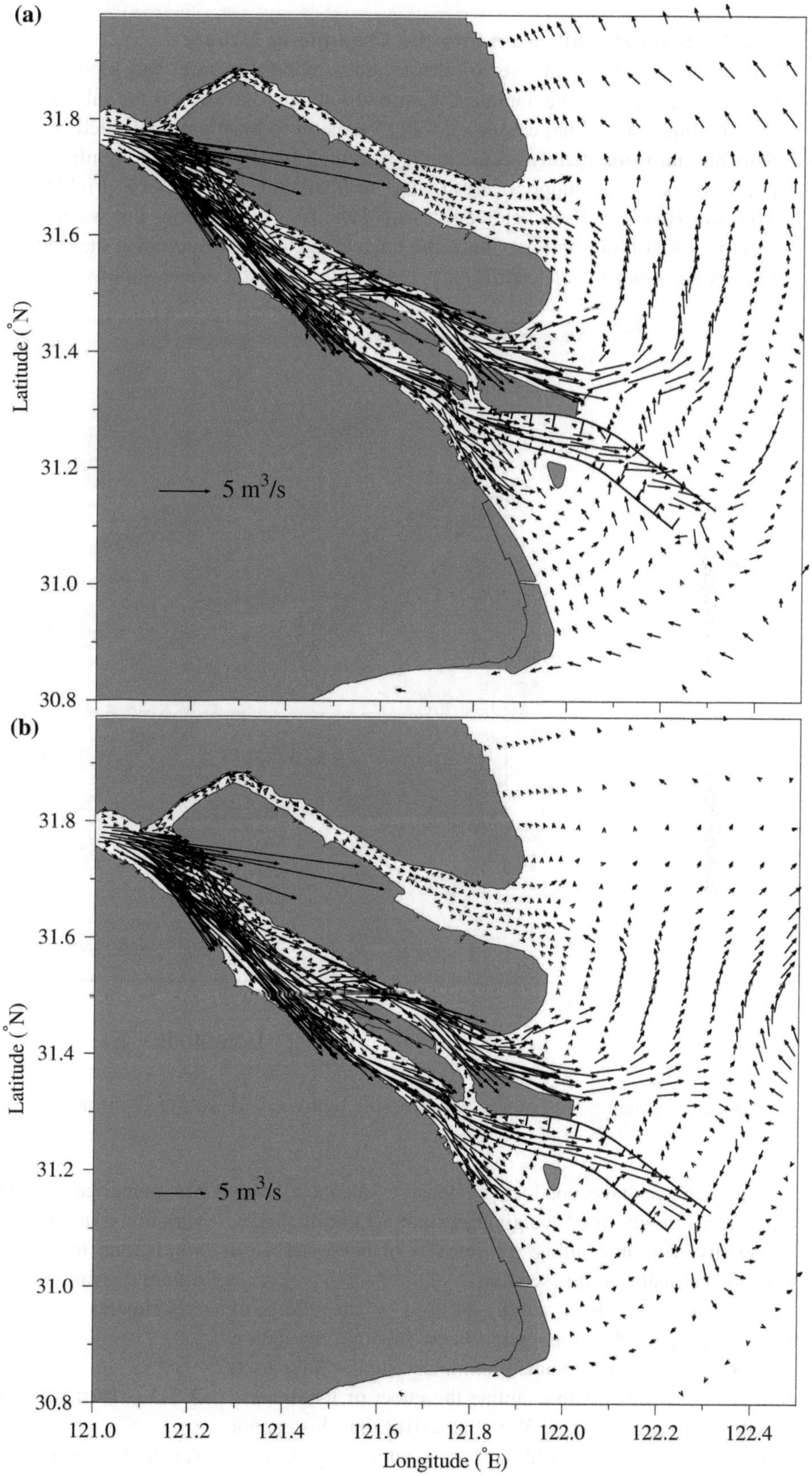

Fig. 2.7 Distribution of the NUWF during summer for: **a** spring tides; and **b** neap tides

Box 2.1 Saltwater Intrusion into the Changjiang Estuary

The combined effects of river discharge, tides, winds, mixing, and topography control saltwater intrusion into the Changjiang Estuary. The dominant feature of saltwater intrusion is the saltwater-spilling-over (SSO) from the NB into the SB (Fig. 2.1), which causes saltwater intrusion to have a complicated spatial and temporal variability in the SB. Saltwater intrusion mainly occurs in dry season. This is not only of significant scientific interest, but also has serious implications for the supply of fresh water to Shanghai. Before 2010, drinking water for Shanghai was mainly taken from the Huangpu River. However, with increasing population, the water quantity became insufficient and water quality was also not optimal. Now, the huge Qingcaosha Reservoir in the estuary supplies about 80 % of Shanghai's fresh water, but saltwater intrusion is a direct threat to the water supply in the dry season.

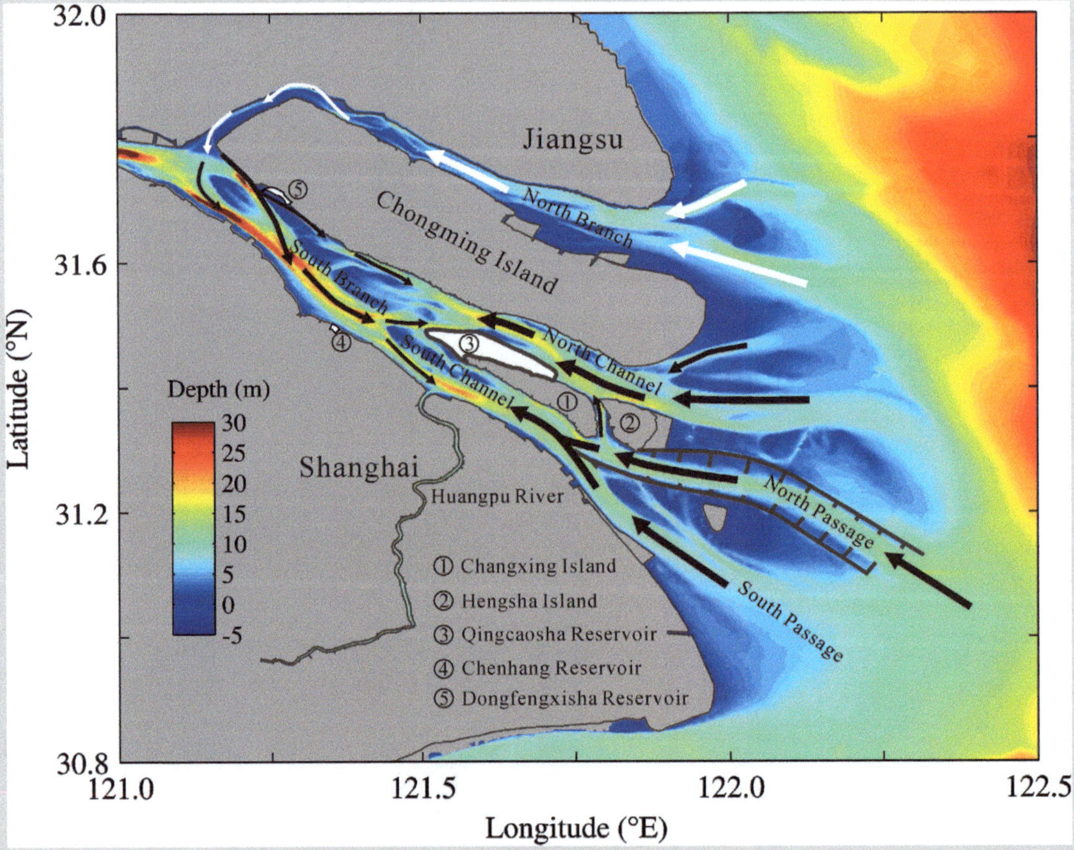

Map of the Changjiang Estuary, showing the pathways for saltwater intrusion (arrows) and the locations of reservoirs.

Although we have conducted intensive studies of saltwater intrusion into the Changjiang Estuary (Li et al. 2012), for brevity, we only outline the impacts of wind stress on saltwater intrusion in this section.

Previous studies have mainly focused on the effects of river discharge, tides, topography, and subtidal circulation (e.g., Shen et al. 2003). Some preliminary studies have used numerical experiments to examine the effect of wind stress on saltwater intrusion (e.g., Wu et al. 2010), but this cannot be thoroughly examined without long-term salinity data.

This subsection describes a detailed analysis of long-term salinity data from Chongxi station. A well-validated 3D numerical model was used to reproduce the salinity variability at Chongxi station during strong northerly winds, and to study the response of saltwater intrusion to different wind stresses based on a series of numerical experiments.

2.3.1 Impacts of Wind Stress on the SSO

A salinity gauging station was established at Chongxi (S2 in Fig. 2.8) in February 2009 to monitor the SSO, and an OBS-3A conductivity sensor was deployed for salinity

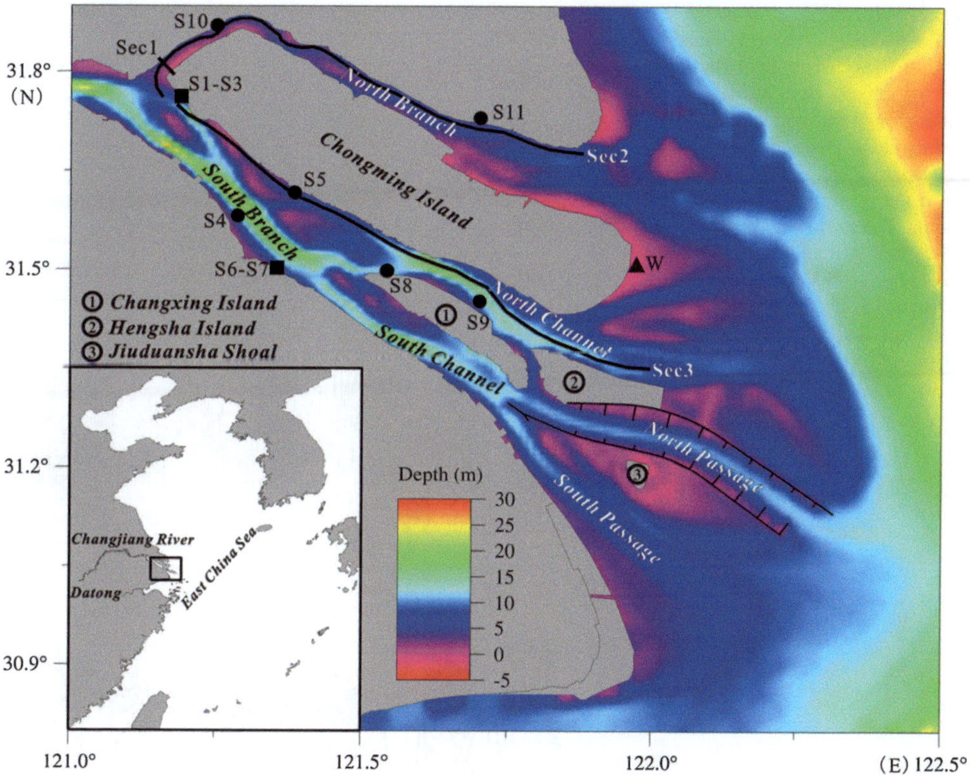

Fig. 2.8 Locations of observational stations and profiles. *S1* is Chongtou; *S2* is Chongxi; *S3* is Xinjian; *S4* is Taicang; *S5* is Nanmen; *S6* is Baogang; *S7* is Chenhang; *S8* is Qingcaosha A; *S9* is Qingcaosha B; *S10* is Qinglonggang; and *S11* is Qidong. The weather station in the eastern Chongming tidal flat is marked by a *black triangle*, and the three profiles (*Sec1*, *Sec2*, and *Sec3*) are indicated by *black lines*

measurements. The instrument was fixed at a depth of about 1 m below the water surface, and the sampling rate was 2 min.

The surface salinity was recorded between November 2009 and February 2010. During this period, the discharge of the Changjiang (Fig. 2.9a) was recorded at Datong station (Fig. 2.8); the water level (Fig. 2.9b) was recorded at Qinglonggang station (S10 in Fig. 2.8); and the wind speed and direction (Fig. 2.9c) were measured at the weather station in the eastern tidal flat of Chongming Island ("W" in Fig. 2.8). The surface salinity is shown in Fig. 2.9d.

Chongxi station is situated at the tip of the bifurcation of the SB and NB, where the salinity exhibits a semimonthly variation. In general, the salinity was higher during spring tides with a maximum value near 4 and was lower during neap tides with a minimum value below 0.5.

The results indicate that the tide plays an important role in determining the degree of saltwater intrusion at Chongxi station. Figure 2.9d shows that the SSO in December 2009 was weaker than in January 2010 during a spring tide, because the river discharge was higher in December 2009 than in January 2010. From November to early December in 2009, the river discharge was more than 13 × 10^3 m^3/s, which was greater than for the period from late December in 2009 to early January in 2010, when it was 12 × 10^3 m^3/s. This indicates that river discharge is another key control on the SSO, which is consistent with previous observations and studies (Shen et al. 2003; Wu et al. 2006).

It is also interesting that the salinity at Chongxi station increased to 2 during a neap tide on November 11–12, 2009. The river discharge of 14 × 10^3 m^3/s was constant over this period, but strong northerly winds occurred with a maximum wind speed of 12 m/s. A similar situation was repeated during February 11–12, 2010. At this time, the tide was moderate, but the salinity at Chongxi station increased anomalously to 3.7. This salinity value was higher than those measured either before or after spring tides. The river discharge was 14.5 × 10^3 m^3/s, but again, strong northerly winds prevailed with a maximum wind speed of 10 m/s. These results lead us to speculate that strong northerly winds tend to intensify the SSO.

2.3.2 Numerical Simulations

The 3D numerical model described in Sect. 2.2.2.1 was used to simulate the abnormal variations of salinity at Chongxi station during the neap tide on November 10–12, 2009, and during the moderate tide on February 11–12, 2010. The model was run from October 1, 2009, and January 1, 2010, respectively, with river discharge data from Datong station and wind speed from the weather station in the eastern tidal flats of Chongming Island.

The model successfully simulated the salinity variation during the neap tide at Chongxi station during November

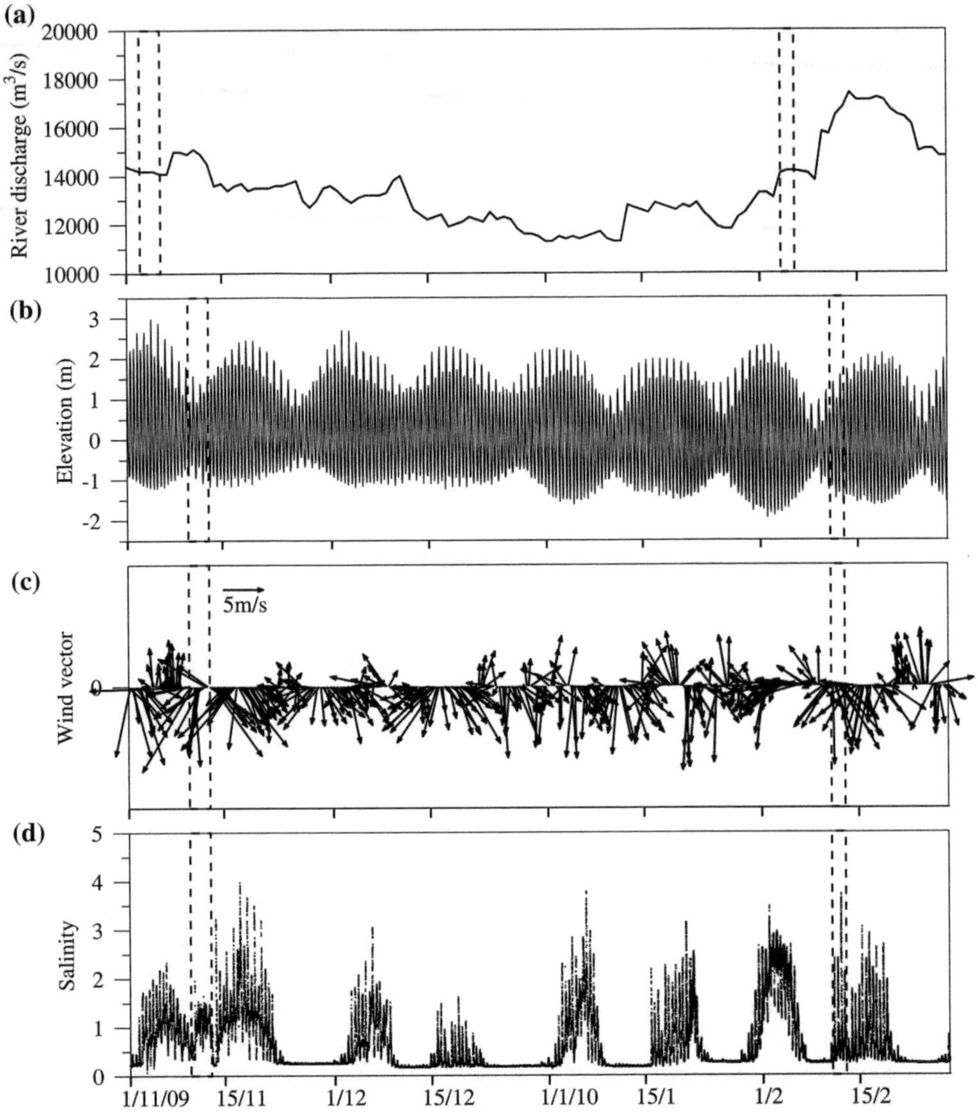

Fig. 2.9 In situ data measured from November 2009 to February 2010: **a** river discharge at Datong hydrographic station; **b** tidal level at Qinglonggang station; **c** wind vectors at the weather station in the eastern Chongming tidal flat; and **d** surface salinity at Chongxi station. The *dashed boxes* in panels **b–d** denote the periods of November 10–12, 2009, and February 11–12, 2010. *Dashed boxes* in panel **a** indicate the period seven days before that shown in panels **b–d**

10–12, 2009, as well as salinity during the spring tide of November 16–20, 2009 (Fig. 2.10d1). To examine the influences of wind stress on the SSO, the model was run again with half of the measured wind speed. The resulting salinity was correspondingly lower than for the measured data during the spring and neap tides (Fig. 2.10d2). For example, on November 10 and November 17, 2009, the simulated daily maximum salinity decreased from 2 and 3.5 to 1.5 and 3, respectively.

The model also reproduced the salinity variation at Chongxi station on February 11–12, 2010, during a moderate tide, and February 15–18, 2010, during a spring tide (Fig. 2.11d1). Simulated salinity that was generated by half the measured wind speed was lower than for in situ station data (Fig. 2.11d2). For example, on February 11–12, 2010, the simulated daily maximum salinity decreased from 3 and 3.5 to 2 and 2.5, respectively. These results demonstrate that strong northerly winds have a pronounced effect on enhancing the SSO.

2.3.3 Mechanisms of Wind Impact on Saltwater Intrusion

The results presented in the previous section show that an abnormal increase in salinity at Chongxi station during neap and moderate tides can be attributed to strong northerly winds. The mechanisms for the impact of wind speed/direction on saltwater intrusion as well as the SSO were investigated with a series of numerical experiments. The results

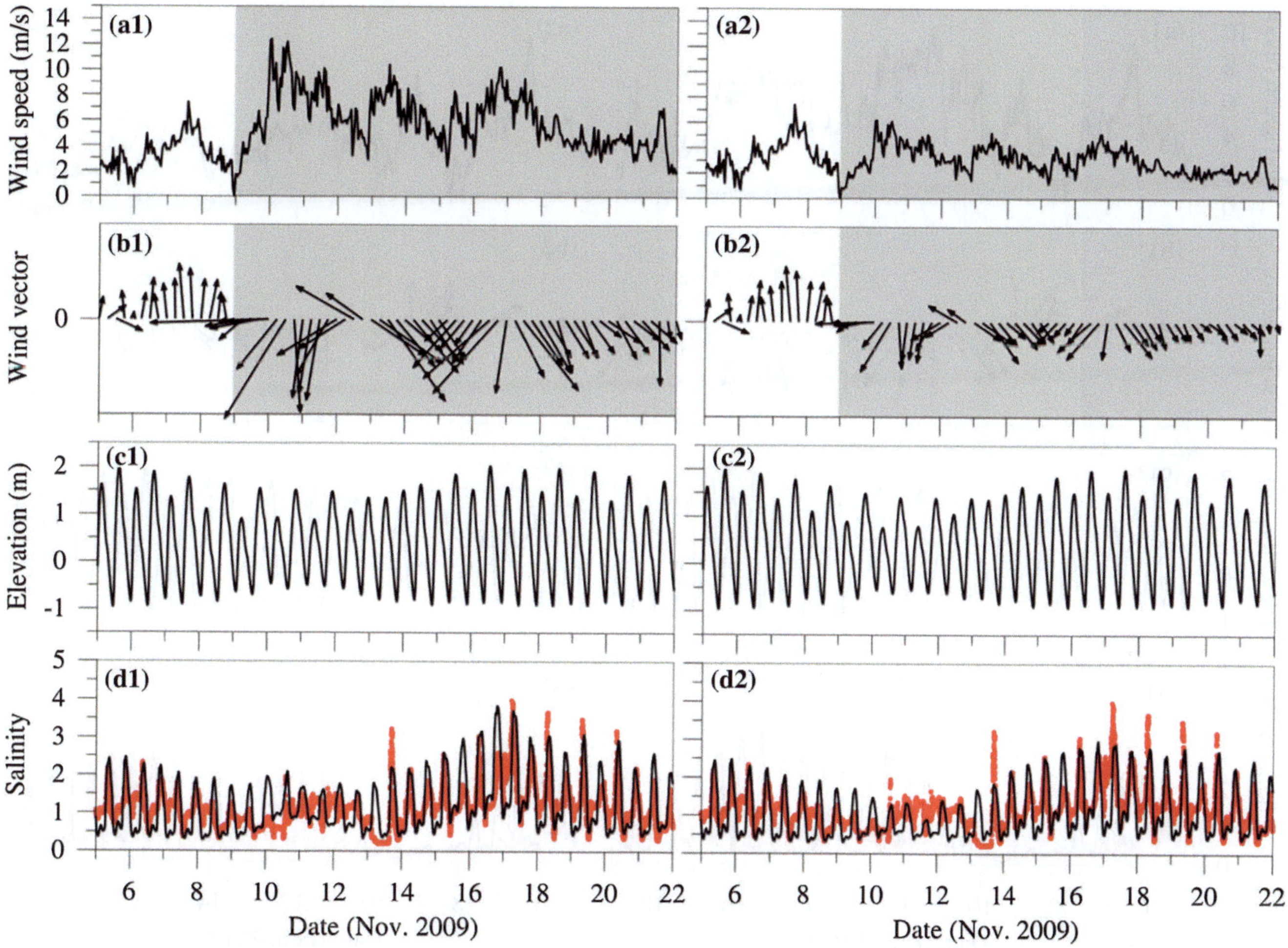

Fig. 2.10 Wind data, and simulated and measured surface salinity for November 5–22, 2009: **a1** measured wind speeds; **b1** wind vectors from the weather station in the eastern Chongming tidal flat; **c1** simulated tidal elevations; **d1** comparison of simulated surface salinity (*black solid line*) and measured data (*red dots*) at Chongxi station; **a2** half of the measured wind speeds; **b2** half of the measured wind vectors; **c2** simulated tidal elevations using half of the measured wind speeds and vectors; and **d2** comparison of simulated surface salinity (*black solid line*) and measured data (*red dots*) at Chongxi station using half the measured wind speeds and vectors

show that northerly and northwesterly winds have significant impacts on the enhancement of wind-driven circulation, causing saltwater intrusion into the NB and NC flows out of the SC, SP, and NP. For further details, see Li et al. (2012).

2.4 Dispersal of Riverine Plumes in the East China Sea

Dispersal of the plume off the Changjiang in summer is one of the most important physical processes in the East China Sea and Yellow Sea (e.g., Mao et al. 1963; Zhu and Shen 1997; Moon et al. 2010). Previous observations and studies have indicated the CDW riverine plume extends to the northeast in summer and flows south in a narrow band confined to the coast in winter (e.g., Mao et al. 1963; Beardsley et al. 1985; Zhu and Shen 1997).

The dispersal of the riverine plumes is strongly influenced by river discharge, the East Asia monsoon, tidal mixing, coastal currents, and continental-shelf circulation. The seasonal cycle of freshwater discharge from the Changjiang dominates the distribution of riverine plumes, especially in summer when peak river output occurs. Strong surface cooling and winds disrupt the vertical salinity stratification during winter, and it is re-stratified by strong heating during summer (Beardsley et al. 1985). Strong semidiurnal tidal currents over the bottom in the shallower regions can cause both strong tidal mixing and generate subtidal residual flow, which contributes to the plume distribution (Wu et al. 2011).

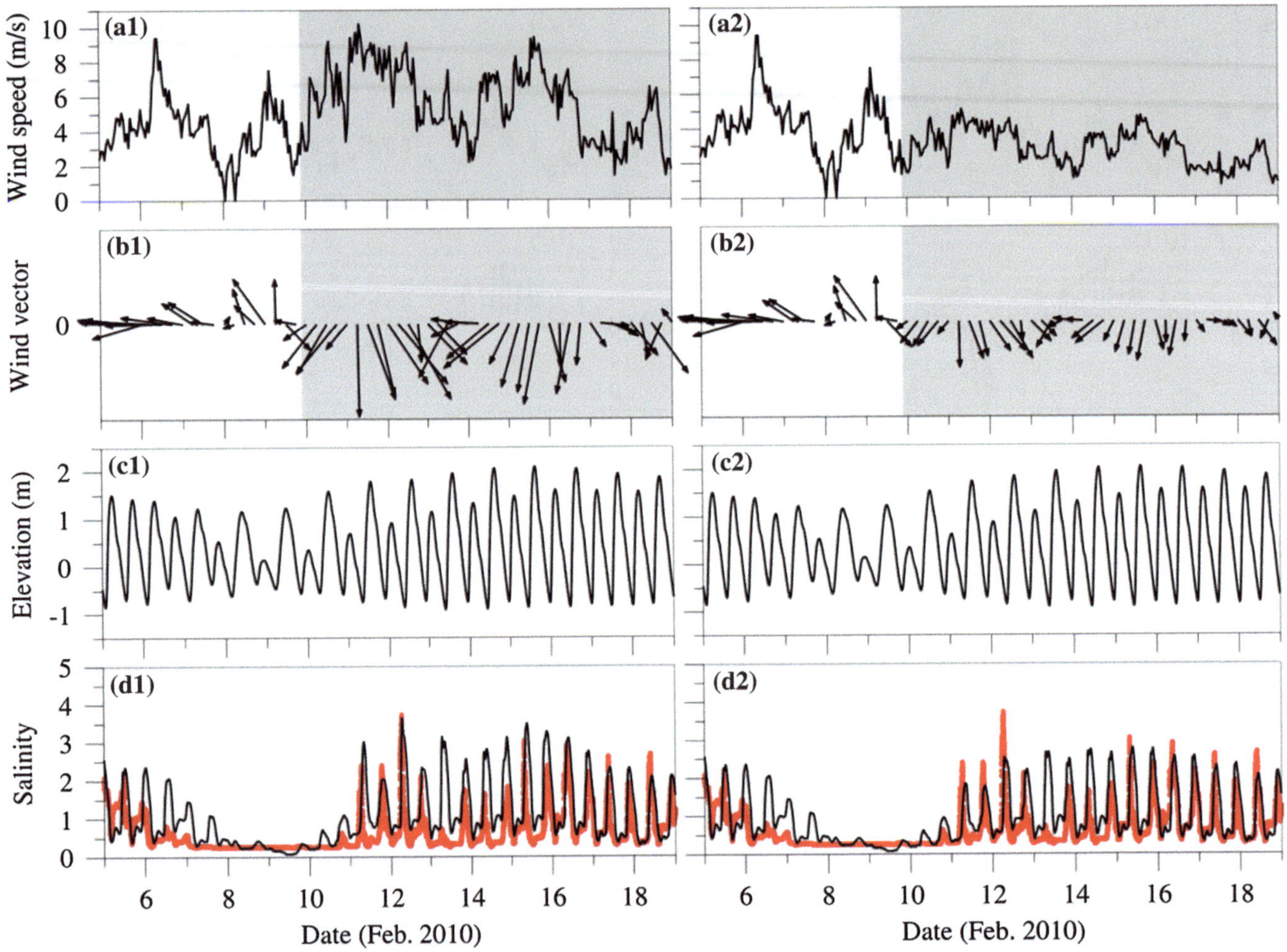

Fig. 2.11 Wind data, and simulated and measured surface salinity for February 5–19, 2010: **a1** measured wind speeds; **b1** wind vectors from the weather station in the eastern Chongming tidal flat; **c1** simulated tidal elevations; **d1** comparison of simulated surface salinity (*black solid line*) and measured data (*red dots*) at Chongxi station; **a2** half of the measured wind speeds; **b2** half of the measured wind vectors; **c2** simulated tidal elevations using half the measured wind speeds and vectors; and **d2** a comparison of simulated surface salinity (*black solid line*) and measured data (*red dots*) at Chongxi station using half the measured wind speeds and vectors

Mao et al. (1963) used salinity data to show, for the first time, that the CDW turns to the northeast in summer, but not to the south as would be expected from the Coriolis force. The southerly summer monsoon is the dominant factor in the northeastward extension of the plume (Zhu and Shen 1997; Chang and Isobe 2003). When the southerly wind speed exceeds 3 m/s, the extension of the plume is skewed to the northeast due to Ekman transport. The far-field diluted water joins the northeastward shelf-current system and flows out of the East China Sea and Yellow Sea region, through the Tsushima–Korea Strait where the year-round freshwater outflow is estimated to be at least 70 % of the total Changjiang river discharge (Isobe et al. 2002; Chang and Isobe 2003).

However, the plume pattern is highly variable in a large region near the Changjiang mouth, illustrating the very dynamic nature of CDW (Box 2.2). Although the plume eventually spreads to the northeast under the influence of the southerly monsoon in summer, the underlying mechanisms controlling its initial development and the variation near the river mouth are still poorly understood. Factors that are thought to be important are southerly winds (Zhu and Shen 1997); the high Changjiang runoff in summer (Le 1984); vortex stretching that results from increased depths offshore

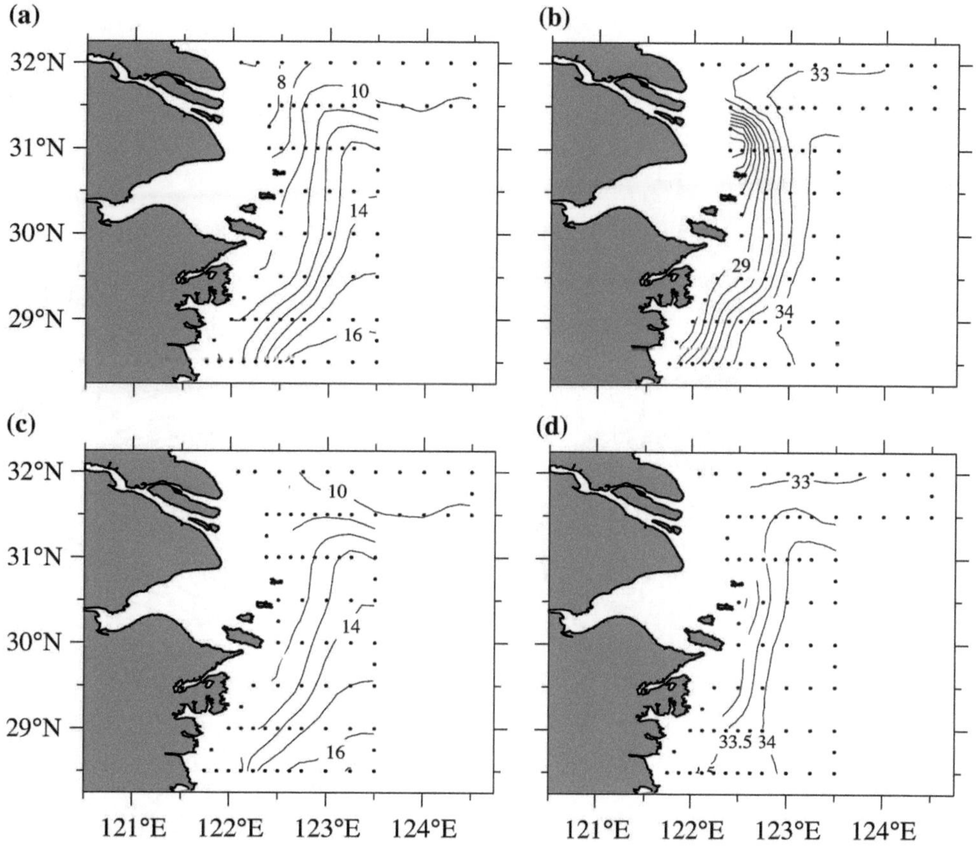

Fig. 2.12 Distribution of temperature and salinity from February 24 to March 10, 2001, for **a** temperature at 3 m depth; **b** salinity at 3 m depth; **c** temperature at 30 m depth; and **d** salinity at 30 m depth. The *dots* mark the observation sites

(Beardsley et al. 1985); basin-scale secondary cyclonic flow produced by the cold water dome in the Yellow Sea (Zhu et al. 1998); and the elevated sea surface that results from the TWC (Zhao 1991). Based on satellite images of surface suspended-sediment concentrations, Pu et al. (2002) reported that the distance and direction of the plume can change completely over a period ranging from less than one week to ten days.

Box 2.2 Extension of the Changjiang Riverine Plume

The Changjiang riverine plume extends to the northeast in summer and south along the coast in a narrow band in winter. The seasonal cycle of freshwater discharge from the Changjiang dominates the distribution of the riverine plume, especially in summer when peak river output occurs. The vertical salinity stratification is mixed in winter by strong surface cooling and wind-mixing and is re-stratified by strong heating during summer. The extension of the riverine plume is mainly controlled by river discharge, winds, and continental-shelf circulation. However, strong semidiurnal tidal currents in shallow regions can cause strong tidal mixing and generate subtidal residual flow, which also influences the plume distribution.

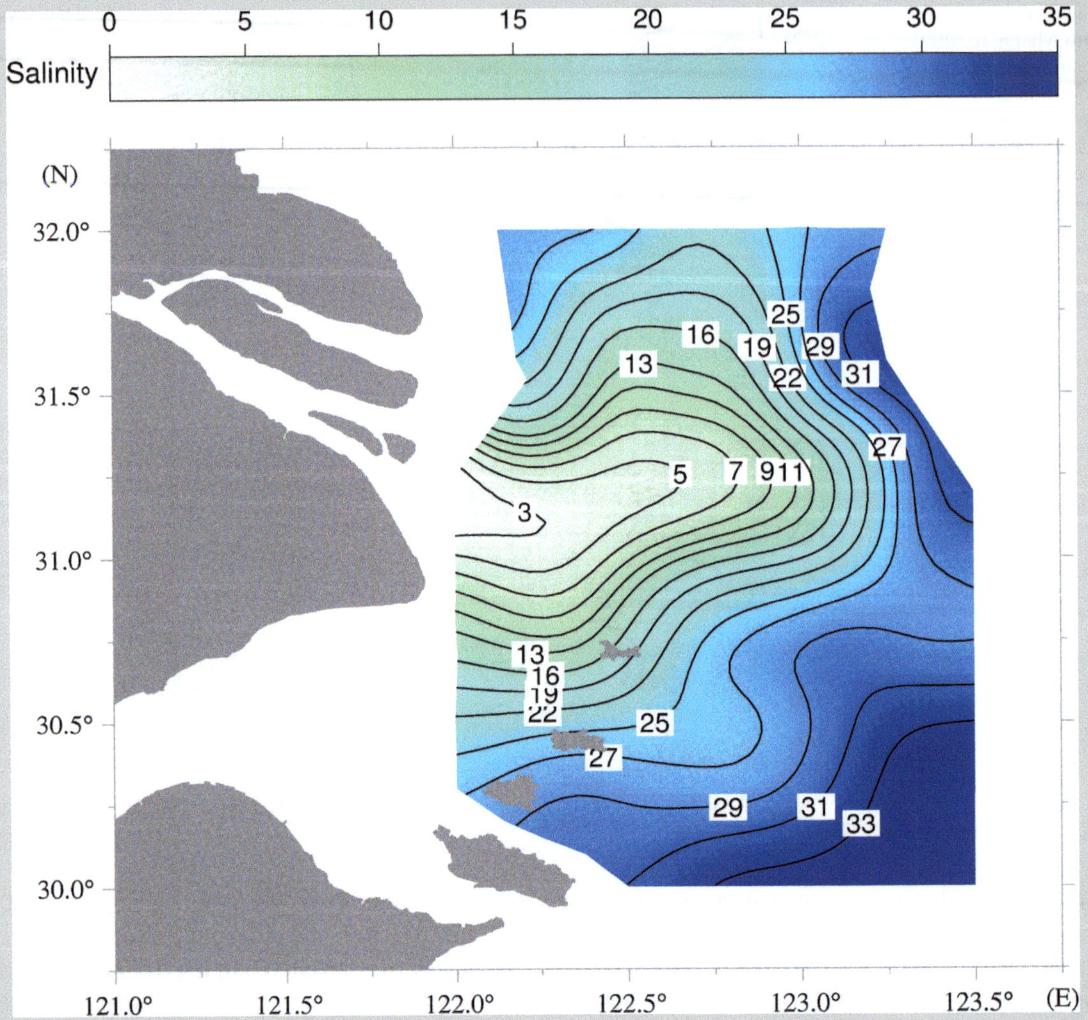

Extension of the Changjiang riverine plume in August 2008.

The dispersal of the plume is closely related to the circulation off the Changjiang Estuary, which includes the wind-driven current, Changjiang river discharge, the TWC, the East China Sea Coastal Current (ECSCC), and the Chinese Coastal Current (CCC) (Fig. 2.2). Due to the East Asia monsoon, which is southerly and stronger in winter, and northerly and weaker in summer, the southward Jiangsu Coast Current is stronger in winter and weaker in summer. Furthermore, under the influence of the monsoon, the Zhejiang Coast Current flows south in winter and north in summer (Beardsley et al. 1985).

A northwestward inflow of relatively warm and saline TWC water is found in the submerged valley located just offshore from the Changjiang mouth. This inflow is the source of the near-bottom saline water found in the lower part of the Changjiang Estuary. This near-bottom saline water is a permanent circulation feature in both winter and summer and has an important influence on dynamic processes in the estuary (Xiang et al. 2009).

For this study, ocean cruises were conducted off the Changjiang mouth almost every year since 2000. The Seabird CTD (Conductivity–Temperature–Depth) was used to measure salinity, temperature, and water depth. In this subsection, we use water temperature and salinity data to analyze the mechanisms that control the extension of the CDW. For brevity, we present winter observations from: (i) February 24 to March 10, 2001, in Sect. 2.4.1, and (ii) summer observations for June 16–30, 2003, in Sect. 2.4.2. Section 2.4.3 describes numerical simulations of the tidal modulation on the extension of the CDW.

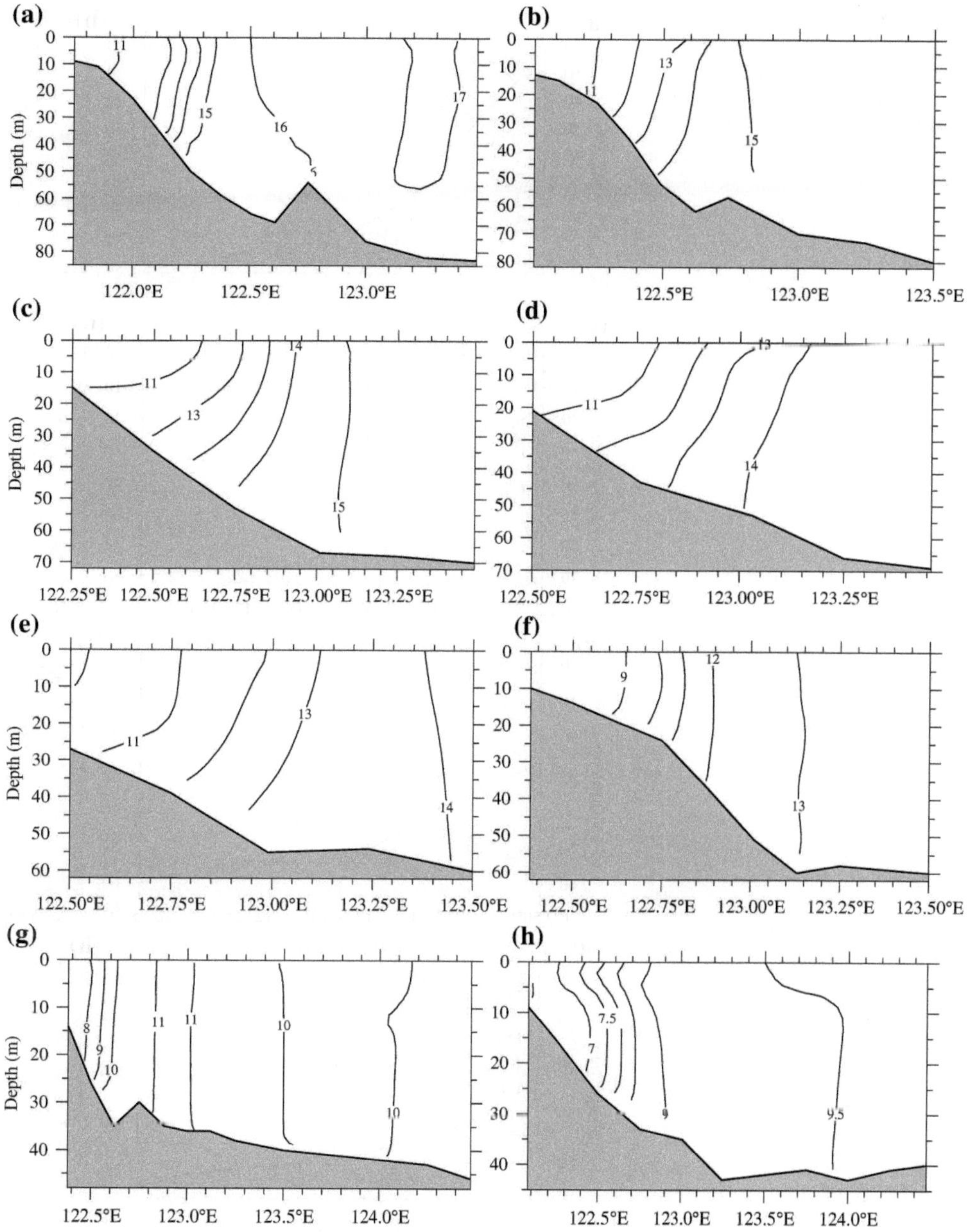

Fig. 2.13 Distribution of temperature along East–West transects for latitudes: **a** 28.5 °N; **b** 29.0 °N; **c** 29.5 °N; **d** 30.0 °N; **e** 30.5 °N; **f** 31.0 °N; **g** 31.5 °N; and **h** 32.0 °N

2.4.1 Extension of the CDW in Winter

The survey was conducted from February 24 to March 10, 2001, in the area bounded by 28.5–32.0 °N and 124.5 °E (Fig. 2.12). There were 79 grid stations and 8 East–West transects, with a high station resolution in the salinity front area just off the river mouth.

Figure 2.12 shows the horizontal distributions of temperature and salinity in winter at depths of 3 and 30 m. The CDW extended south from the river mouth in a narrow band along the coast. As the Changjiang water is fresher and colder than seawater in winter, the water farther from the coast was more saline with higher temperatures. A tongue of this water, brought by the northward-flowing TWC, extended to the northeast and entered the submerged valley off the Changjiang. Mixing between the CDW and the TWC water formed a narrow and strong temperature and salinity front along the coast.

The vertical profiles of temperature and salinity (Figs. 2.13 and 2.14) along the East–West transects show that they were well mixed and lower near the coast. Along the transect at 28.5 °N, the temperature and salinity ranged from 11 to 17 °C and from 28 to 34, respectively. The vertical temperature profile was relatively uniform, but the salinity profile had more vertical structure. Along the transect at 29.0 °N, temperature and salinity were also well mixed and ranged from 11 to 15 and from 27 to 34, respectively.

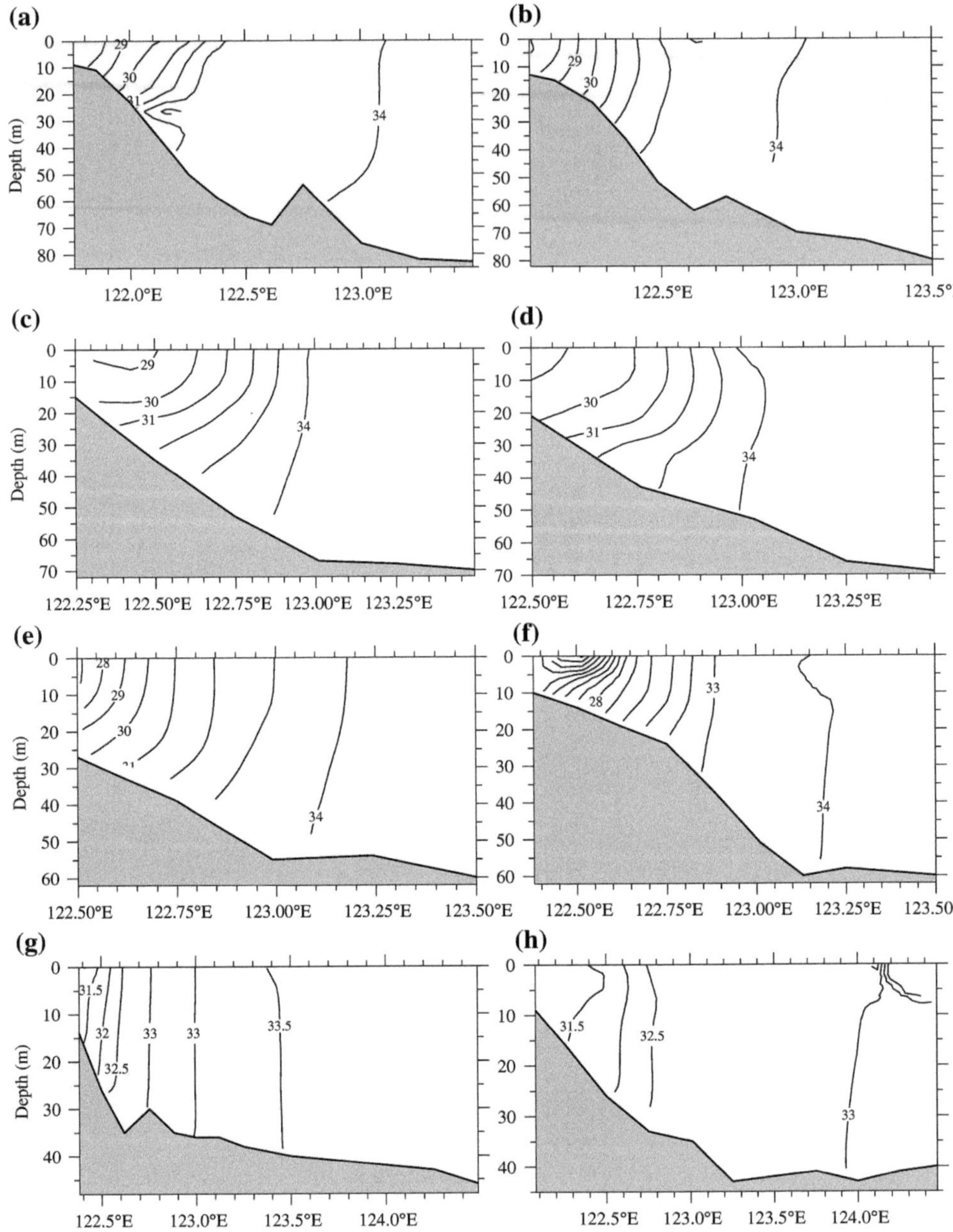

Fig. 2.14 Distribution of salinity along east–west transects for latitudes: **a** 28.5 °N; **b** 29.0 °N; **c** 29.5 °N; **d** 30.0 °N; **e** 30.5 °N; **f** 31.0 °N; **g** 31.5 °N; and **h** 32.0 °N

Along the transects at 29.5, 30.0, and 30.5 °N, the temperature and salinity contours show a seaward bend along the slope bottom, indicating downwelling along the coast. The dynamic mechanism for the downwelling was that strong northerly winds drove the surface water onshore by Ekman transport and, to conserve mass, produced an offshore current along the slope bottom.

Along the transect at 31.0 °N, there was a small domain plume near the river mouth where the Changjiang water empties into the sea. Along the transect at 31.5 °N, there was a strong temperature and salinity front along the coast, which was vertically uniform. Along the northern transect at 32.0 °N, the temperature was lower than along the southern transects.

2.4.2 Extension of the CDW in Summer

A survey off the Changjiang mouth was conducted during June 16–30, 2003, in the region bounded by 29.5–32.5 °N and 124.5 °E. The cruise completed a total of 6 East–West transects and collected data at 80 different locations (Fig. 2.15). Higher resolution observations were taken in the thermal front and in areas with frequent high red tides or algae blooms.

The salinity at a depth of 1.5 m below the surface (Fig. 2.16) shows that the CDW formed a bimodal shape as it extended away from the river mouth, with one limb moving to the southeast, and another larger limb moving to the northeast.

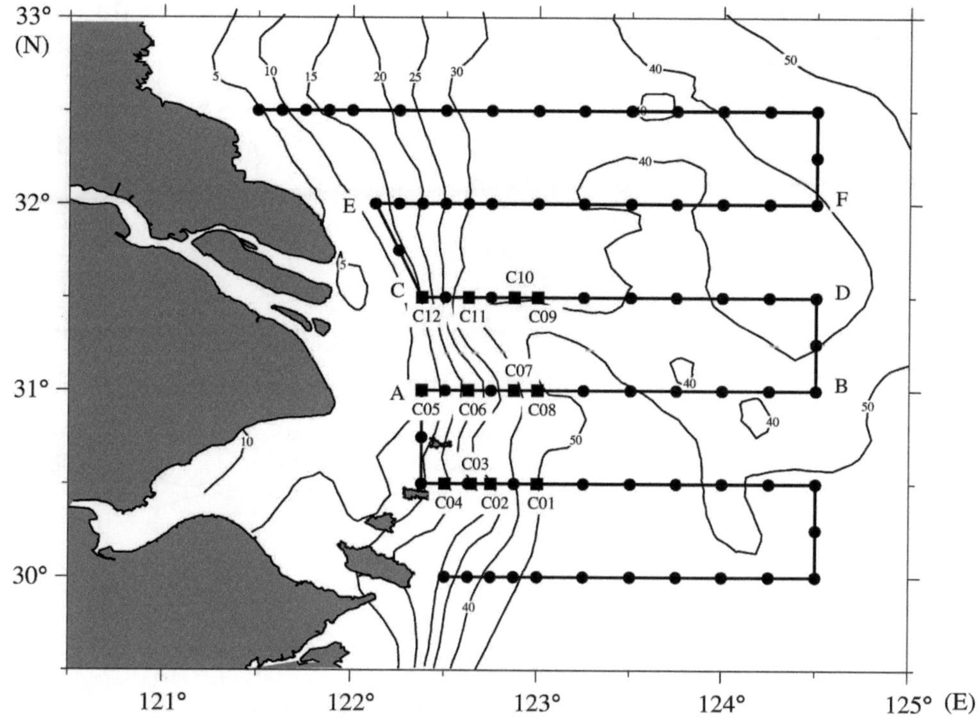

Fig. 2.15 Cruise observations off the Changjiang mouth during June 16–30, 2003. The *thick solid line* is the cruise route, and *thin lines* are the isobaths (in meters). The *blank circles* show the positions of the measurement stations, and *black squares* are the water-sample stations. The labels "AB," "CD," and "EF" mark transects located at latitudes 31, 31.5, and 32 °N, respectively

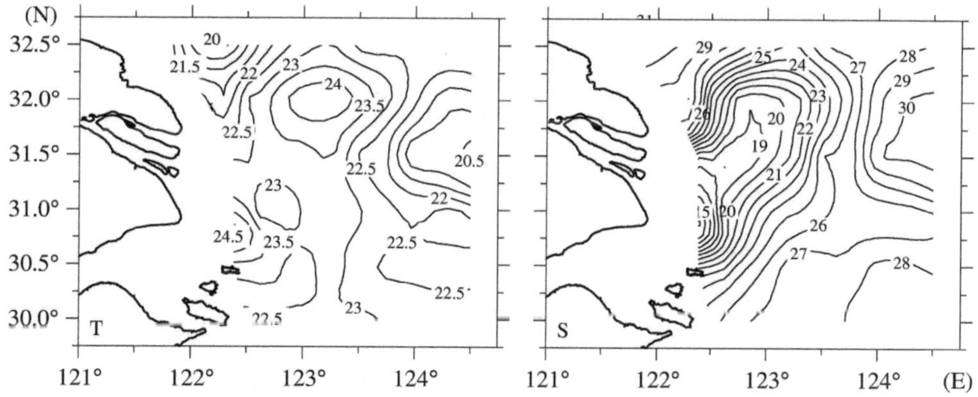

Fig. 2.16 Horizontal distribution of temperature (*left*) and salinity (*right*) at a depth of 1.5 m below the surface. The unit for temperature (*T*) is °C, and the unit for salinity (*S*) is psu

The Changjiang river discharge at Datong hydrographic station during the observational period was between 32×10^3 and 40×10^3 m^3/s. This was lower than the mean value of 43×10^3 m^3/s in mid- and late June 2003. This suggests that lower river discharges promote the northeast extension of the CDW.

The temperature of the Changjiang estuary water was higher than the seawater in summer. After entering the sea, this plume of high-temperature water had the same bimodal shape, but an area of colder water with temperatures of between 22.5 and 23 °C appeared on the west side of the submerged valley. Previous observations and analysis indicated that this was produced by the bottom Ekman-layer effect on the north-flowing TWC, and a baroclinic effect induced by freshwater–saltwater mixing (Zhu 2003).

In view of the distribution of temperature and salinity along transects AB, CD, and EF in Figs. 2.15, 2.16, and 2.17, the plume produced by the CDW was very significant and located above a depth of approximately 10 m. The CDW gradually extended eastward from the southern transact, AB, to the northern transect, EF. The high-temperature water in the plume indicates that the Changjiang Estuary waters flowing into the East China Sea had relatively high temperatures and low salinities, with an obvious thermocline and halocline.

The temperature and salinity contours bend downward at 123.5 °E along the section AB, implying downwelling,

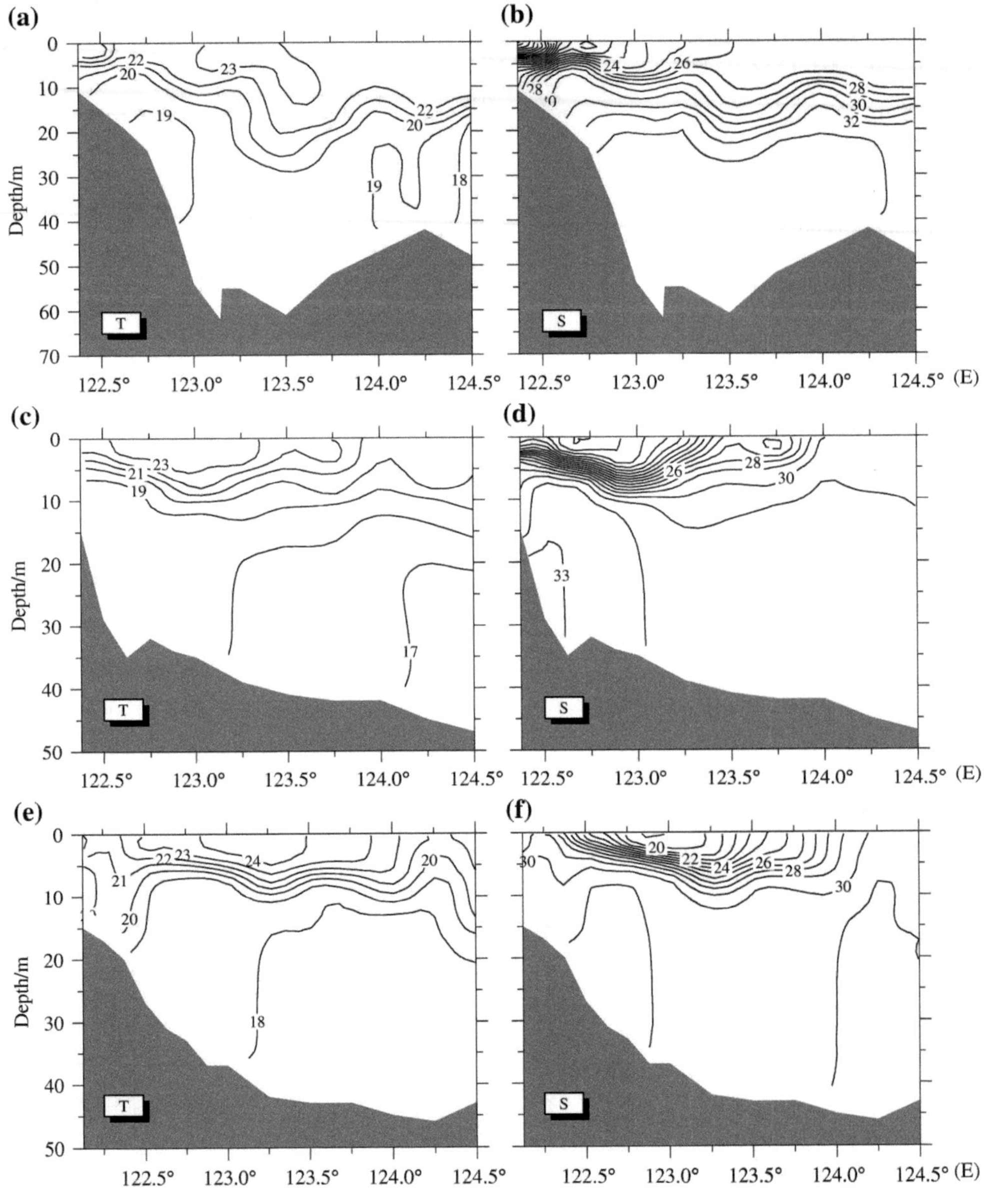

Fig. 2.17 Vertical distribution of temperature (*left panel*) and salinity (*right panel*) along transects AB, CD, and EF (shown in Fig. 2.15)

while they bend upward on the west side, indicating upwelling with a clockwise circulation. The bottom of the thermocline was at approximately 15 m depth, with the exception of the downwelling region where it deepened to 30 m. The seawater in the submerged valley had low temperatures and high salinities, and the upward-bending contours in the west of the valley indicated that upwelling also occurred in this area (Zhu 2003).

2.4.3 Numerical Simulation of Tidal Modulation on the Extension of the CDW

The tidal modulation of the CDW is introduced briefly in this subsection, and further details can be found in Wu et al. (2011). Simulations of the tidal modulation of the CDW were made using a numerical model that encompasses the entire East China Sea and Yellow Sea, with higher resolution grids around the river mouth. Skill assessments, based on the model simulations forced by real observations, indicated that the model can reproduce the tides, shelf currents, and temporal–spatial variations of the CDW. Comparisons of the model-validation simulation, both with and without the tidal forcing, further confirmed the critical role played by tides in modulating the CDW (Wu et al. 2011).

Two experiments were initially conducted in which the wind forcing and the shelf currents were excluded. Without the tidal forcing, we found that the CDW extends strongly to the north along the Jiangsu Coast, a phenomenon that has rarely been observed, but frequently occurs in model studies that do not include tidal forcing. When the tide is superimposed, however, the northward extension of the CDW along the coast disappears. The strong tide along the Jiangsu Coast is sufficient to mix the water column, resulting in the formation of a strong along-coast salinity gradient to the north of the Changjiang mouth, which obstructs the northward extension of the plume (Wu et al. 2011).

The model results also showed that spring–neap tidal variations result in a distinguishable spatial pattern. During the spring tide, there is a near-field plume located inside the 30-m isobath. At its frontal edge, the subtidal momentum is balanced by momentum advection and the Coriolis force via a mechanism of tidal rectification, which drives a residual current mainly to the southeast. Around the head of the submarine valley, an anticyclonic rotating plume-bulge occurs, as a result of the local high-subtidal sea level that can result from tidal asymmetry. The bulge rotates the CDW to the northeast at about 122.5 °E (Wu et al. 2011).

Considering that the wind and the shelf currents were not included in this experiment, this indicates that the north-eastward extension is an intrinsic property of the CDW. In addition, the near-field plume and the bulge illustrate a bi-directional plume structure separated from the head of the submarine valley. During the neap tide, because of the reduced tidal energy, the effects of tidal rectification and tidal asymmetry become small, and the near-field plume therefore extends farther east and the bulge is less evident.

Further experiments, conducted by adding 4 m/s southerly winds and shelf currents to the model, indicated that although the wind forcing ultimately drives the CDW to the northeast in the far field, the dynamics of the tidal modulation are robust near the river mouth. The near-field plume is almost identical to the plume formed without wind forcing. Although the bulge merges into the far-field plume due to the intensified mixing, the CDW rotates to the northeast in almost the same location as the plume formed without wind forcing. Furthermore, the bi-directional plume structure is maintained and is even more pronounced. If the tidal forcing is excluded, however, the southeast branch of the CDW disappears.

It was also found that, during the spring tide, the bottom saline water can penetrate the river plume by convective transport along the slope, thereby cutting off the CDW and forming a detached zone of freshwater. In summary, these results show that the tide has an important influence on the extension of the CDW in the near field, while the wind has a dominant effect in the far field (Wu et al. 2011).

2.5 Summary

The temporal–spatial variations of tides, as well as the tide's harmonic constituents, form, and distortion in the Changjiang Estuary were analyzed based on in situ hourly data from tidal gauging stations. Significant semidiurnal, fortnightly, and seasonal tidal variations were found. Mean monthly water levels varied seasonally in concert with river discharge, which reached its peak during the flood season, and dropped to its lowest values during the dry season. The maximum monthly tidal range exhibited two peaks, in March and September, and fell to its lowest values in June and December each year.

As the tidal wave propagates up-estuary, the tidal range in the SB becomes smaller than that in the lower reaches due to friction and reduced river discharge. In the NB, the tidal forcing is amplified due to the funnel-shaped topography and the weaker influence of river discharge.

Harmonic analysis showed that tides around the Changjiang Estuary are dominated by four semidiurnal constituents (M_2, S_2, N_2, and K_2); four diurnal constituents (K_1, O_1, P_1, and Q_1); and three shallow-water constituents (M_4, MS_4, and M_6). The tidal form, distortion, and asymmetry numbers were calculated using harmonic constants, so that the key tidal features could be characterized.

Numerical simulations of the circulation (i.e., residual current) in the Changjiang Estuary show that flow in the SB has a speed of approximately 0.1 m/s and 0.5 m/s when the river discharge is 11×10^3 and 50×10^3 m^3/s, respectively. Only a small amount of river water flows into the NB due to its shallow topography.

The salinity measured at Chongxi gauging station during a neap tide and moderate tide rose anomalously when northerly winds were strong. This demonstrated that wind stress enhances the SSO. To further investigate this phenomenon, a 3D numerical model was used to reproduce the anomalous salinity at Chongxi station. The results showed that when the wind speed was reduced by half, the salinity was significantly reduced. Additional numerical experiments, using different wind speeds and directions, also indicated that the wind has an important effect on saltwater intrusion.

Data collected off the Changjiang mouth in winter show that the CDW extended south in a narrow band along the coast. A tongue of relatively high-temperature and high-salinity water, brought by the northward-flowing TWC, flows into the submerged valley and can reach the area just off the Changjiang mouth. Mixing between the relatively low-temperate and low-salinity CDW water and the TWC water formed a strong temperature and salinity front along the coast. A small domain plume front was also observed near the river mouth, and downwelling occurred along the coast due to strong onshore winds, which produced Ekman transport in the surface waters and an offshore current along the middle and bottom layers.

A survey off the Changjiang mouth showed that the CDW extended offshore with a significant thermocline and halocline. It had a bimodal structure in summer, with one main

Table 2.5 Mean monthly river discharge (RD, m^3/s) and water volume (WV, 10^8 m^3) of the Changjiang from 1950 to 2010 at Datong hydrographic station

Month	1	2	3	4	5	6	7	8	9	10	11	12
RD	11184	12058	16300	24061	33488	40454	49848	44149	39909	32332	22552	14063
WV	299.6	291.7	436.6	623.7	896.9	1048.6	1335.1	1182.5	1034.4	866.0	584.5	376.7

branch extending to the northeast and another more minor branch extending to the southeast. There was upwelling along the west slope of the submerged valley due to the baroclinic effect and the bottom Ekman effect of the northward-flowing TWC. Numerical simulations indicated that the tide has an important effect on the dispersal of the riverine plume through tidal mixing.

The extension of the CDW is controlled mainly by the river discharge, winds, and tides; however, as observations of the CDW generally take one survey vessel between five to eight days, the wind and tide change during the observation period. For this reason, more than one vessel is needed to synchronize the data. For numerical simulations of the dispersal of the riverine plume, the combined effects of river discharge, wind, tide, and continental-shelf current are important, and each dynamic factor should be correctly parameterized. The numerical simulation showed that the tide has an important influence on the extension of the CDW in the near field, whereas the wind has a distinct effect in the far field (Table 2.5).

Acknowledgments This work was based on research started in the early 1990s, with financial support from the Natural Science Foundation of China (Grant numbers: 40376027, 40776012, 40976056, 40721004, 41176071, and 41021064); the Major Program of Shanghai Science and Technology Committee (Grant numbers: 05XD14006 and 08231200102); and the Ministry of Science and Technology of China (Grant numbers: G1999043803, 2010CB951201, and 2011CB409801). We thank colleagues and students from the State Key Laboratory of Estuarine and Coastal Research (ECNU) for support in collecting field observations, numerical simulations, and laboratory analyses.

References

Beardsley R, Limburner R, Yu H, Cannon GA (1985) Discharge of the Changjiang (Yangtze River) into the East China Sea. Cont Shelf Res 4:57–76

Blumberg AF (1994) A primer for ECOM-si. Technical report of HydroQual, Mahwah

Blumberg AF, Mellor GL (1987) A description of a three-dimensional coastal ocean circulation model. In: Heaps NS (ed) Three-dimensional coastal ocean models. American Geophysical Union, Washington D.C, pp 1–16

Chang PH, Isobe A (2003) A numerical study on the Changjiang diluted water in the yellow and East China Seas. J Geophys Res 108 (C9):3299. doi:10.1029/2002JC001749

Chen ZY (1980) Tidology. Science Press, Beijing, p 301

Chen C, Zhu JR, Ralph E, Green SA, Budd JW, Zhang FY (2001) Prognostic modeling studies of the Keweenaw current in lake superior, part I: formation and evolution. J Phys Oceanogr 31:379–395

Choi BH (1980) A tidal model of the Yellow Sea and Eastern China Sea. Korea Ocean Research and Development Institute (KORDI) report, 72 pp

Editorial Board For Marine Atlas (1992) Ocean atlas in Huanghai Sea and East China Sea (hydrology). China Ocean Press, Beijing

Galperin B, Kantha LH, Hassid S, Rosati A (1988) A quasi-equilibrium turbulent energy model for geophysical flows. J Atmos Sci 45:55–62

Geyer WR (1993) The importance of suppression of turbulence by stratification on the estuarine turbidity maximum. Estuaries 16:113–125

Ianniello JP (1977) Tidally induced residual currents in estuaries of constant breadth and depth. J Mar Res 4:754–786

Isobe A, Ando M, Watanabe T, Senjyu T, Sugihara S, Manda A (2002) Freshwater and temperature transport through the Tsushima-Korea Straits. J Geophys Res 107(C7):3065. doi:10.1029/2000JC00702

Large WS, Pond S (1981) Open ocean momentum flux measurements in moderate to strong winds. J Phys Oceanogr 11:324–406

Le K (1984) A preliminary study of the path of the Changjiang diluted water. Oceanologia Et Limnologia Sinca 15(2):157–167 (in Chinese)

Li L, Zhu JR, Wu H, Wang B (2010) A numerical study on the water diversion ratio of the Changjiang Estuary in the dry season. Chin J Oceanol Limnol 28(3):700–712

Li L, Zhu JR, Wu H (2012) Impacts of wind stress on saltwater intrusion in the Yangtze Estuary. Sci China Earth Sci 55(7):1178–1192. doi:10.1007/s11430-011-4311-1

Lü X, Qiao F, Xia C, Zhu J, Yuan Y (2006) Upwelling off the Yangtze river estuary in summer. J Geophys Res 111:C11S08. doi:10.1029/2005JC003250

Mao H, Gan Z, Lan S (1963) Preliminary study on the Changjiang diluted water and its mixing natures. Oceanologia Et Limnologia Sinca 5(3):183–206 (in Chinese)

Mellor GL, Yamada T (1974) A hierarchy of turbulence closure models for planetary boundary layers. J Atmos Sci 33:1791–1896

Mellor GL, Yamada T (1982) Development of a turbulence closure model for geophysical fluid problem. Rev Geophys 20:851–875

Moon JH, Hirose N, Yoon JH, Pang IC (2010) Offshore detachment process of the low-salinity water around Changjiang bank in the East Chinese Sea. J Phys Oceanogr 40:1035–1053

Pritchard DW (1956) The dynamic structure of a coastal plain estuary. J Mar Res 15:33–42

Pu Y, Huang W, Xu J (2002) The spreading direction change of the Changjiang River diluted water in 7–10 days. J East China Sea 20:1–5 (in Chinese)

Qiu C, Zhu JR (2013) Influence of seasonal runoff regulation by the three gorges reservoir on saltwater intrusion in the Changjiang River Estuary. Cont Shelf Res 71:16–26

Shen HT, Mao ZC, Zhu JR (2003) Saltwater intrusion in the Changjiang Estuary. China Ocean Press, Beijing (in Chinese)

Simpson JH, Brown J, Matthews J, Allen G (1990) Tidal straining, density current, and stirring in the control of estuarine stratification. Estuaries 13:125–132

Smagorinsky J (1963) General circulation experiments with the primitive equations., I. The basic experiments. Mon Weather Rev 91:99–164

Su J, Yuan Y (2005) Oceanography of China Seas. China Ocean Press, Beijing, p 367

Wu H, Zhu JR (2010) Advection scheme with 3rd high-order spatial interpolation at the middle temporal level and its application to saltwater intrusion in the Changjiang Estuary. Ocean Model 33:33–514

Wu H, Zhu JR, Chen BR, Chen YZ (2006) Quantitative relationship of runoff and tide to saltwater spilling over from the North Branch in the Changjiang Estuary: A numerical study. Est Coast Shelf Sci 69:25–132

Wu H, Zhu JR, Choi BH (2010) Links between saltwater intrusion and subtidal circulation in the Changjiang Estuary: A model-guided study. Cont Shelf Res 30:1891–1905

Wu H, Zhu JR, Shen J, Wang H (2011) Tidal modulation on the Changjiang River plume in summer. J Geophys Res 116:C08017. doi:10.1029/2011JC007209

Xiang YY, Zhu JR, Wu H (2009) The impact of the shelf circulations on the saltwater intrusion in the Changjiang Estuary in winter. Prog Nat Sci 19(2):192–202 (in Chinese)

Zhao B (1991) On the extension of Changjiang diluted water. Acta Oceanol Sin 13(5):600–610 (in Chinese)

Zhu JR, Shen HT (1997) Extension mechanisms of Changjiang diluted water. East China Normal University Press, Shanghai (in Chinese)

Zhu JR, Xiao CY, Shen HT (1998) Numerical model simulation of expansion of Changjiang diluted water in summer. Acta Oceanol Sin 20:13–22 (in Chinese)

Zhu JR (2003) Dynamic mechanism of the upwelling on the west side of the submerged river valley off the Changjiang mouth in summertime. Chin Sci Bull 48(24):2754–2758

Zhu JR, Wu H, Li L, Wang B (2010) Saltwater intrusion in the Changjiang Estuary in the extremely drought hydrological year 2006. J East China Normal Univ (Nat Sci) 4:1–6, 25 (in Chinese)

Changjiang Estuary Sediment Transport Dynamics

3

Qing He, Leicheng Guo, Hong Liu, and Ya Wang

Abstract

This chapter deals with river flow and sediment transports in the Changjiang River continuum. We first briefly introduce riverine flow and sediment discharges acting on the estuary and their variations in the past half century. There is an ongoing decrease in sediment discharge since the 1980s, which has accelerated since 2003 when the Three Gorges Dam (TGD) commenced operation. In situ flocculation of suspended sediment is detected in the freshwater river and the blackish estuary, affecting sediment transport in the system. In the estuary, changes in sediment concentrations and grain sizes are observed in response to reduced river-borne sediment load. Analysis of sediment grain sizes, suspended sediment concentration (SSC), and the clay–silt–sand content of the sediments in the estuary reveals the presence of a nearshore depocenter featured by high sediment exchange ratio off the South Passage. Numerical experiments of estuarine circulation and water age suggest that large-scale engineering works have had marked impact on the regional hydrodynamics. Overall, the river, the estuary, and the marginal sea are inherently connected through the water and sediment linkage; therefore, systematical insights on their hydro-, morpho-, and eco-dynamics are required.

Keywords

River flow • Sediment discharge • Sediment transport dynamics • Flocculation • Water age

3.1 Introduction

The Changjiang (Yangtze River) is one of the world's longest rivers, which delivers huge amount of water and sediment to the marginal East China Sea (ECS), thus providing a significant contribution to the global terrestrial flux (Syvitski et al. 2005). The Changjiang basin is home to 36.9 % of the China's population, or 0.48 billion people in 2010 (Zhao et al. 2013). Industrial activities in the Changjiang basin are also economically important, contributing 41.19 % of the national GDP in 2010 (Zhao et al. 2013).

In this chapter, we review and analyze long-term hydrological data from government bulletins (CWRC 2005, 2010, 2011) and scientific literatures (see Chap. 4) to interpret the temporal and spatial changes of river flow and sediment discharges. Mountainous regions in the upper reaches of the river are excluded from the analysis due to data scarcity. As sediment load variability is discussed in detail elsewhere in this volume (see Chap. 4), we only briefly describe sediment load changes.

The Changjiang Estuary was formed from persistent accumulation of riverine sediment. Its morphology is mainly governed by river discharge and tides, while waves and winds are of secondary importance (Chen et al. 1985; Yun

Q. He (✉) · L. Guo · H. Liu · Y. Wang
State Key Laboratory of Estuarine and Coastal Research, East China Normal University, 3663 Zhongshan Road North, Shanghai 200062, China
e-mail: qinghe@sklec.ecnu.edu.cn

L. Guo
e-mail: candleguolc@gmail.com

H. Liu
e-mail: Liuhong007@yahoo.com.cn

Y. Wang
e-mail: ricowangya@gmail.com

J. Zhang (ed.), *Ecological Continuum from the Changjiang (Yangtze River) Watersheds to the East China Sea Continental Margin*, Estuaries of the World,
DOI 10.1007/978-3-319-16339-0_3,

2004). The estuarine morphology, and associated hydrodynamics and sediment transport processes are highly complex. Here, we describe the spatial pattern of suspended sediment concentration (SSC, which is equivalent to the TSM) and the grain sizes of both suspended and bed sediments. Sediment transport and deposition are described based on a large body of researches that have considered the exchange between suspended and bed sediments using grain size analysis.

To assess the impacts of large-scale navigation channels in the North Passage (NP) of the estuary, we use a numerical model to trace the water age. This enables us to examine estuarine circulation and transport timescales, define the water age distribution across the estuary under different river and tidal forcing scenarios, and consider the sensitivity of these processes to engineering works. Overall, it is concluded that the Changjiang system, including the fluvial, estuarine, and adjacent coastal areas, are linked by water motion and sediment transport processes.

3.2 Riverine Sediment Transport

3.2.1 Water and Sediment Discharge

Water and sediment discharge data have been recorded at Datong station, about 650 km upstream from the river mouth (Fig. 3.1), since 1950. The mean annual river discharge is 28.9×10^3 m^3/s or 903.4 km^3/year (1950–2005) and exhibits no significant trends over the past 60 years (Fig. 3.2). This indicates that the hydroclimatology of the basin has not changed over this time period. Extremely low-flow years were 1978 and 2006, and high-flow years were 1954 and 1998. It is noted that seasonally low water conditions occurred more frequently in the past decade in the Changjiang basin, and further investigation is required to understand this variability (Chen and He 2009; Xu and Ma 2009).

The mean annual sediment load exhibited low variability from 1950 to 1985 (with an average value of 470 million tons) and then decreased between 1986 and 2000 (Fig. 3.2). An abrupt sediment load reduction is observed around 2000–2003, after which the mean sediment discharge decreased to 152 million t/year (CWRC 2010). The gradual decrease in sediment discharge from the mid-1980s onward is attributed to the combined effects of land-use changes, anti-deforestation measures, and dam construction, such as the Danjiangkou Dam in 1973 and the Gezhou Dam in 1981 (Yang et al. 2006; Wang et al. 2007). The dramatic decrease in sediment discharge since 2003 is mainly attributed to the construction of the Three Gorges Dam (TGD) (see Chap. 4).

Due to the tropical monsoon climate, there is strong seasonal variation in both river and sediment discharge from the Changjiang. The wet season, from May to October each year, contributes 71 and 87 % of the annual river discharge and sediment load, respectively (Chen et al. 2007). Maximum discharge occurs in July when 14 and 21 % of the water and sediment are discharged, respectively, and minimum discharge occurs in January when only 3.2 and 0.7 % of the water and sediment are discharged. The most recent dataset, collected from 2006 to 2009, shows that the seasonal distributions of water and sediment are changing compared with the distributions before 2005. The water and sediment

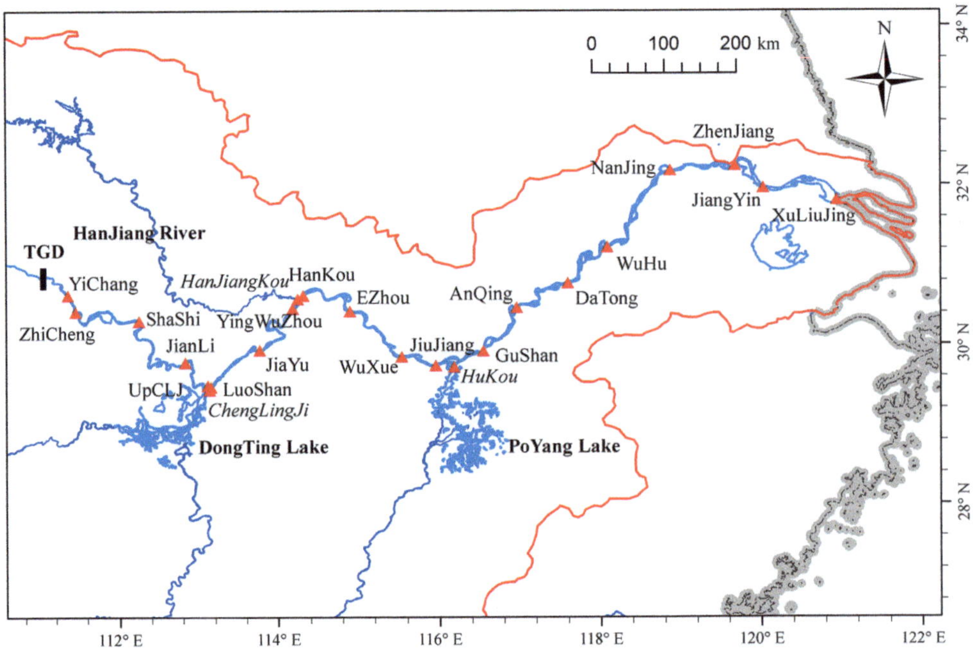

Fig. 3.1 The middle and lower Changjiang basin and the location of the major gauging stations

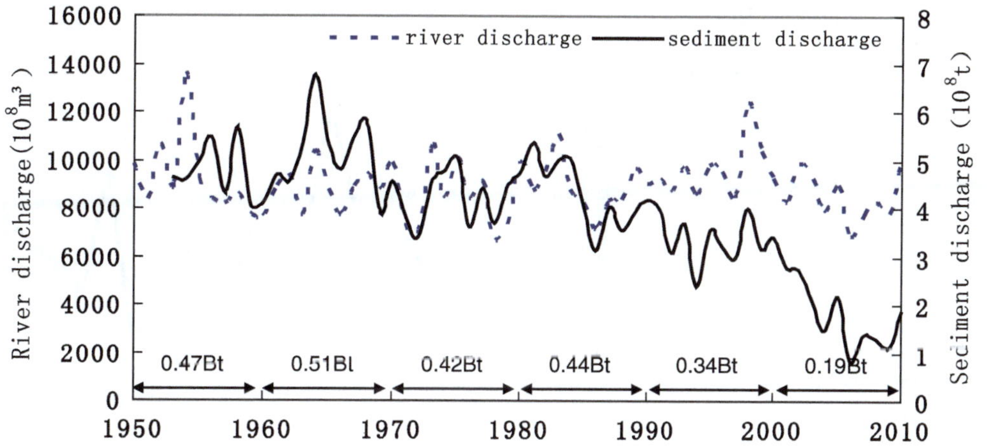

Fig. 3.2 Mean annual river and sediment discharge at Datong station from 1950 to 2010 (sediment discharge data are missing in 1950 and 1952) (data from Yun 2010)

discharge in the wet season (May–October) decreased by 68 and 81 %, respectively, while the discharge in the dry season (November–April) increased by 32 and 19 %, respectively (Fig. 3.3). This change is also attributed to the operation of the Three Gorges Reservoir (TGR), which reduced the peak flow in the wet seasons and increased the discharge in the dry seasons (Chen et al. 2010).

Sediment transported along the Changjiang can be carried as either bed load or suspended load. However, more than 90 % of the sediment transport at Datong station is carried as suspended load, while the bed load is relatively minor (Chen et al. 2007). Due to the relatively high suspended sediment load, the Changjiang is very turbid. The multi-year (1950–2002) average total solid matter (TSM) in the suspended sediment load at Yichang and Datong stations is 1.11 and 0.48 kg/m^3, respectively (CWRC 2005).

The time series of TSM at Datong station shows that it has decreased since the 1980s. It was 0.486 kg/m^3 between 1950 and 2000 and declined by around half to 0.241 kg/m^3 between 2001 and 2007 (Guo and He 2011). Between 2003 and 2011, the average TSM was 0.12 and 0.17 kg/m^3 at Yichang and Datong stations, respectively (CWRC 2011). Field measurements in January 2008 also showed that the TSM was low in the reaches just downstream the TGD (~0.0002 kg/m^3) and progressively increased up to about 0.19 kg/m^3 downriver (Guo and He 2011). The TGR started to store water in 2003, trapping much of the upstream sediment, and discharging clean water downstream. This is responsible for the remarkable decrease in sediment concentration downstream the dam, while upstream concentrations remained stable (CWRC 2010). Downstream of the TGD, along-channel degradation and the delivery of sediment from tributaries acted to offset sediment reductions due to the TGD, causing the downstream increase in TSM.

3.2.2 Sediment Grain Sizes

Suspended sediments in the Changjiang are relatively fine, ~10 μm, without significant spatial variations (CWRC

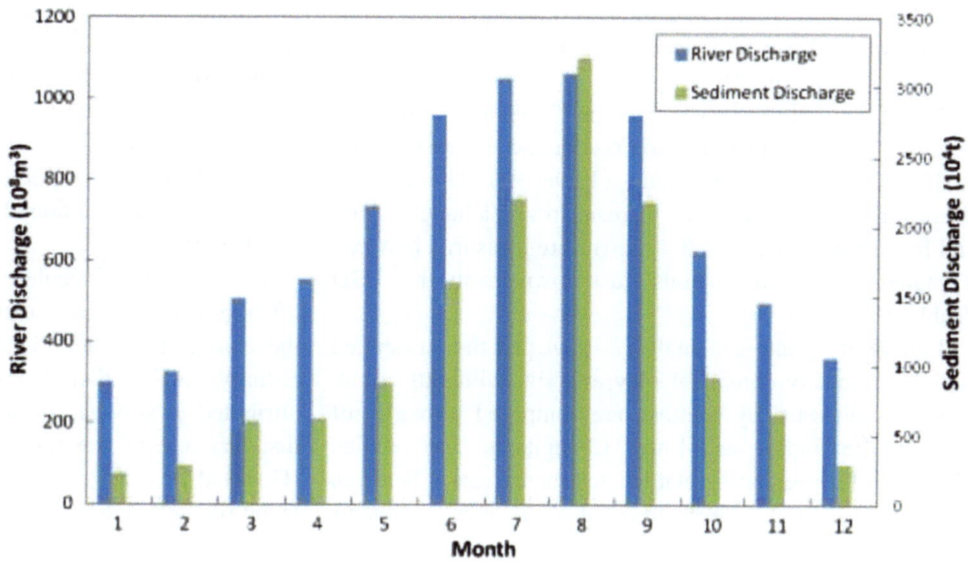

Fig. 3.3 Mean monthly river and sediment discharge at Datong station from 2006 to 2009 (data from CWRC 2010)

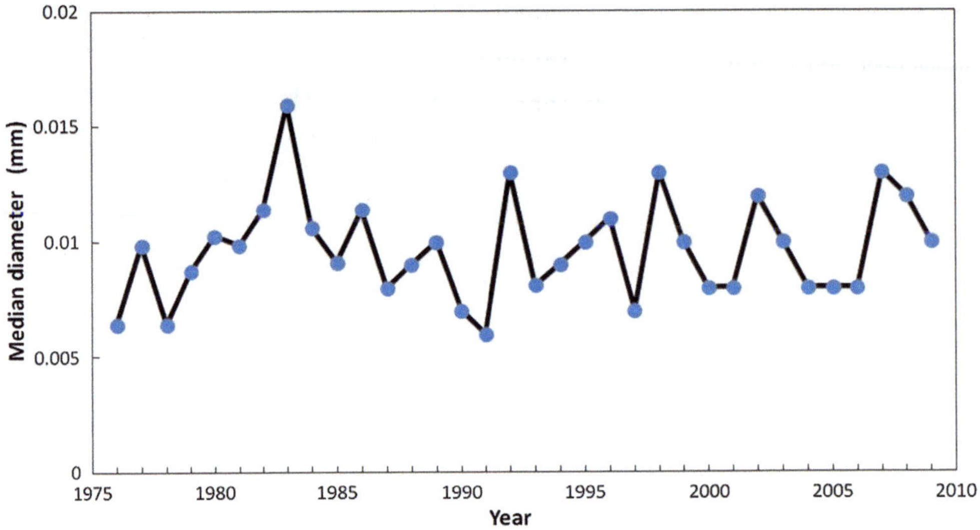

Fig. 3.4 The median diameters of suspended sediments from 1976 to 2009 at Datong station

2010), which contributes to the high turbidity, and exhibit little spatial variability. In January 2008 (i.e., the dry season), the median diameter of suspended sediments varied between 4 and 10 μm in the middle and lower Changjiang (Guo and He 2011). Field data from September to October 2009 also show that the suspended sediments are 4–7 μm in median diameter in the upper Changjiang (Chen et al. 2011). Temporally, the multi-year (1950–2000) average median diameter of suspended sediment was 17 μm, compared with approximately 10 μm between 2000 and 2010 at Datong station (CWRC 2010). However, no significant long-term changes have occurred in the past three decades (Fig. 3.4).

The bottom sediments in the Changjiang are relatively coarse in the middle and lower reaches, with a median diameter range of 50–300 μm. The diameter has a downward decreasing trend due to sorting and mixing processes in the river channel (Luo et al. 2012). In the past decade, the bottom sediments have been coarsening downstream of Yichang station. This is attributed to the operation of the TGR, which reduces downstream sediment, resulting in bed scour and associated bottom sediment coarsening in the downriver direction (Luo et al. 2012).

Based on field data collected in 2008 and 2009, the vertically averaged median grain size diameter varies from 4 to 11 μm along the Changjiang (Table 3.1). The section-averaged TSM increases downstream from Yichang station, and low TSM values (1.9–3.4 mg/l) were measured between Yichang and Shashi stations, just downstream the TGD (Table 3.1).

Laboratory grain size analyses show that the suspended sediments mostly consist of clay and silt, with little sand. Bottom sediments, by contrast, are composed primarily of sand in the middle and lower Changjiang. The median diameter of dispersed suspended sediments varies between 4.4 and 9.4 μm, with a trend toward finer sediments downstream (Table 3.1). Suspended sediments become coarser in the reaches between Nanjing and Xuliujing stations (Fig. 3.5), which implies that coarser sediments are brought into suspension downstream Nanjing station due to the influence of incoming tides. Suspended sediments are a little coarser in the tributaries, especially the Hanjiang River (CWRC 2010; Guo and He 2011).

3.2.3 Flocculation of Suspended Sediment

Flocculation of fine suspended sediment is widely reported in the Changjiang Estuary (He et al. 2004; Tang 2007), but is less studied along the Changjiang. With the help of a LISST-100X (Laser In Situ Scattering and Transmissometry, Sequoia Scientific Inc., Bellevue, WA, USA), significant in situ flocculation of suspended sediment was observed along the Changjiang (Chen et al. 2011; Guo and He 2011). In situ floc sizes vary in mean diameter from 22 to 182 μm (Fig. 3.5), which is a comparable range to floc sizes in the estuary (i.e., 50–120 μm). This indicates that freshwater flocs are not necessarily smaller than those found in saline environments (Guo and He 2011). There is evidence that riverine flocs are transported into the estuary, thus acting as 'parent flocs' for continued flocculation in the estuary (Guo and He 2011).

By examining flocculation in the river, estuary, and in the offshore marine environment, Guo and He (2011) found that the characteristic floc sizes are smaller in the Changjiang Estuary than in either the river or the offshore area. This was attributed to stronger shear stress in the estuary that to the disadvantage of the formation of macro-flocs (>128 μm) (Guo and He 2011). Biochemical processes are also thought to be partially responsible for floc formation in the riverine

Table 3.1 Measured river discharge, TSM, median diameter of the dispersed particles, and mean diameter of the flocs in the upper Changjiang in September 2009 (data from Chen et al. 2011) and in the middle and lower Changjiang in January 2008 (data from Guo and He 2011)

	Stations	Discharge (m^3/s)	SSC (mg/l)	D_{50} (μm)	Mean floc diameter (μm)
Yangtze River in January 2008	YiChang	4510	1.9	8.80	171.1
	ZhiCheng	–	2.2	8.30	181.9
	ShaShi	5450	3.4	9.40	152.4
	JianLi	–	18.8	6.40	130.9
	UpCLJ	–	32.5	8.20	129.1
	ChengLingJi	2190	69.2	6.70	95.0
	LuoShan	7430	45.5	7.70	141.4
	JiaYu	–	44.3	8.90	93.7
	YingWuZhou	–	58.3	5.90	87.0
	HanJiangKou	731	76.1	11.40	73.7
	HanKou	8980	41.6	7.30	98.3
	Ezhou	–	30.6	5.10	96.7
	WuXue	–	33.1	5.10	94.3
	JiuJiang	8730	26.4	5.50	99.4
	HuKou	1340	190.9	7.00	22.5
	GuShan	10,870	126.8	4.70	43.6
	AnQing	–	87.7	5.00	58.0
	DaTong	11,400	78.5	4.90	55.6
	WuHu	–	174.3	4.40	54.2
	NanJing	–	34.2	4.40	120.4
	ZhenJiang	–	44.0	4.60	89.4
	JiangYin	–	41.1	5.50	77.4
	XuLiuJing	–	50.4	8.20	86.9
Yangtze River in September 2009	WanZhou	–	29.8	4.16	35.8
	YiChang	8560	16.9	1.87	7.1
	HongHu	11,500	56.9	3.72	37.4
	HuangShi	13,700	57.2	5.30	67.9
	AnQing	15,600	78.9	3.68	50.0
	WuHu	–	62.2	7.51	39.7
	YangZhou	–	58.5	7.84	71.5
	XuLiuJing	–	32.1	4.85	57.8

and offshore waters (Droppo et al. 1997; Liu et al. 2007). Another field survey in the Changjiang showed that flocs in the wet season are smaller than in the dry season (Chen et al. 2011). This may be attributed to higher river discharge, stronger river currents, and thus higher shear stress during the wet season (Chen et al. 2011; Guo and He 2011).

3.2.4 Discussion

The exceptionally high river runoff and sediment discharge of the Changjiang is one of the major forces responsible for building up the Changjiang Delta, shaping the estuary, and driving nearshore processes along the marginal ECS (Chen et al. 1985; Yun 2004; Liu et al. 2006). In that sense, an understanding of water source, sediments, and nutrients in the river basin are of high importance. In the past two decades, human activities such as land-use changes and dam construction has become more intensive in the Changjiang basin.

The significant freshwater flocculation of suspended sediment in the Changjiang is of fundamental significance to understanding of riverine sediment transport dynamics. Aggregated sediments settle faster than dispersed particles, and floc settling velocities are estimated to be 0.07–1.59 mm/s, which is an order of magnitude larger than

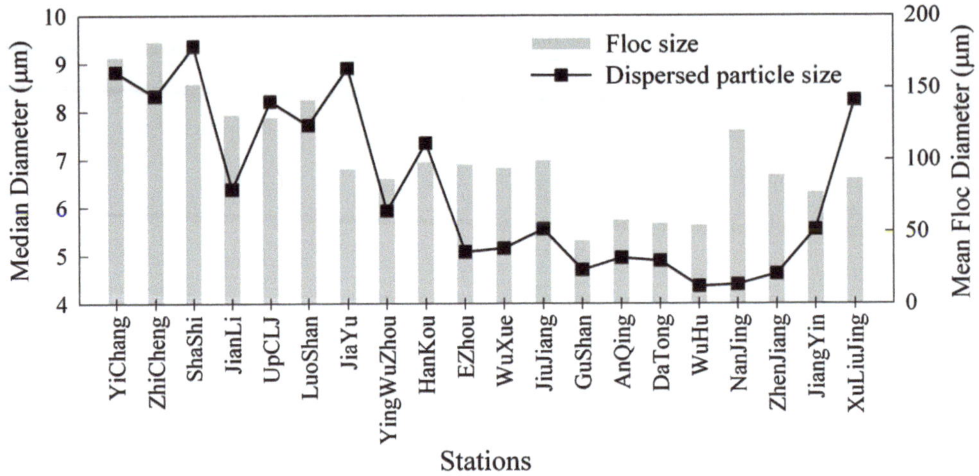

Fig. 3.5 Comparison of the median dispersed particle sizes and mean floc diameters in January 2008 along the middle and lower Changjiang (from Guo and He 2011). Station locations are shown in Fig. 3.1

primary particles (Guo and He 2011). Failure to consider flocculation processes may, therefore, lead to underestimation of sediment deposition in the river, and also in lakes and reservoirs.

Moreover, flocs are composed of both inorganic particles and organic matter (Winterwerp and van Kesteren 2004). The presence of organic matter, whether in a dissolved or particulate state, stimulates the flocculation processes due to the adhesion (Droppo et al. 1997), and causes flocs to settle faster in the water column. Thus, the formation of flocs and their transport processes may play a significant role in the transport of nutrients from the river to the sea (Bainbridge et al. 2012). However, these physical–chemical–biological processes are still poorly understood, and further research is required.

3.3 Estuarine Sediment Transport

3.3.1 Suspended Sediment

The Changjiang Estuary covers a large area stretching from Datong to the nearshore zone (50 m isobath), with a core zone between Jiangyin and the 10 m isobaths (Fig. 3.6). Within this large area, river and tidal influences lead to strong spatial variations in sediment transport dynamics. For instance, the grain size of suspended sediment exhibits a fine–coarse–fine trend from the upper estuary (between Jiangyin and Xuliujing), to the Turbidity Maximum (TM) zone (mouth bar area), and to the nearshore shelf zone (seaward of the 10-m isobath) (Fig. 3.7). These zones have median grain sizes (D_{50}) of 8.9,

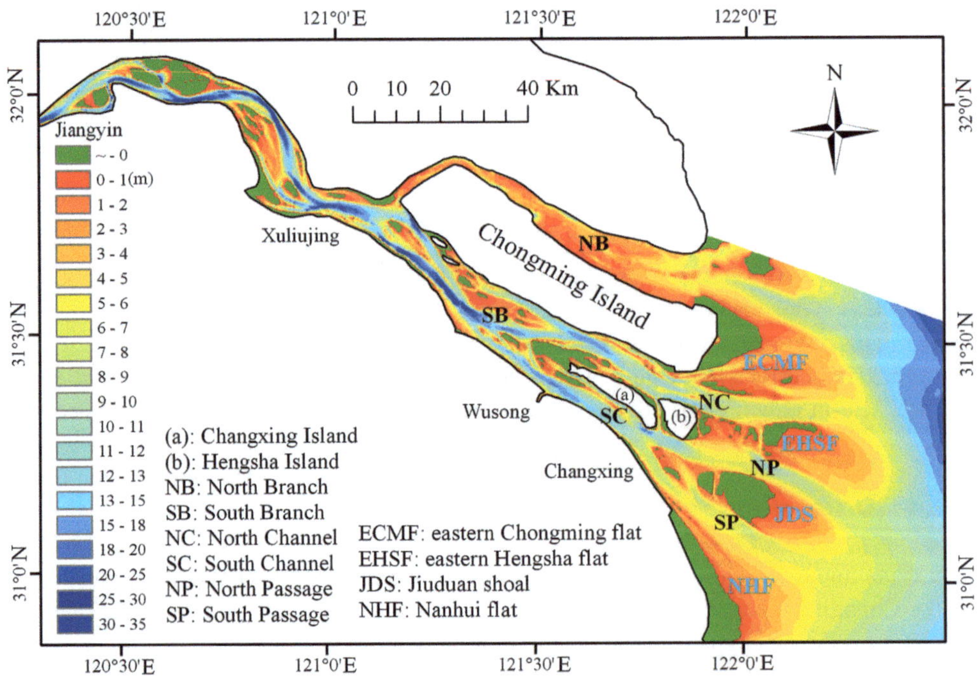

Fig. 3.6 Geometry and bathymetry of the Changjiang Estuary in 1997. Jiangyin, Xuliujing, Qiyakou, Wusong; upper estuary, TM, outer mouth bar, inner shelf zone

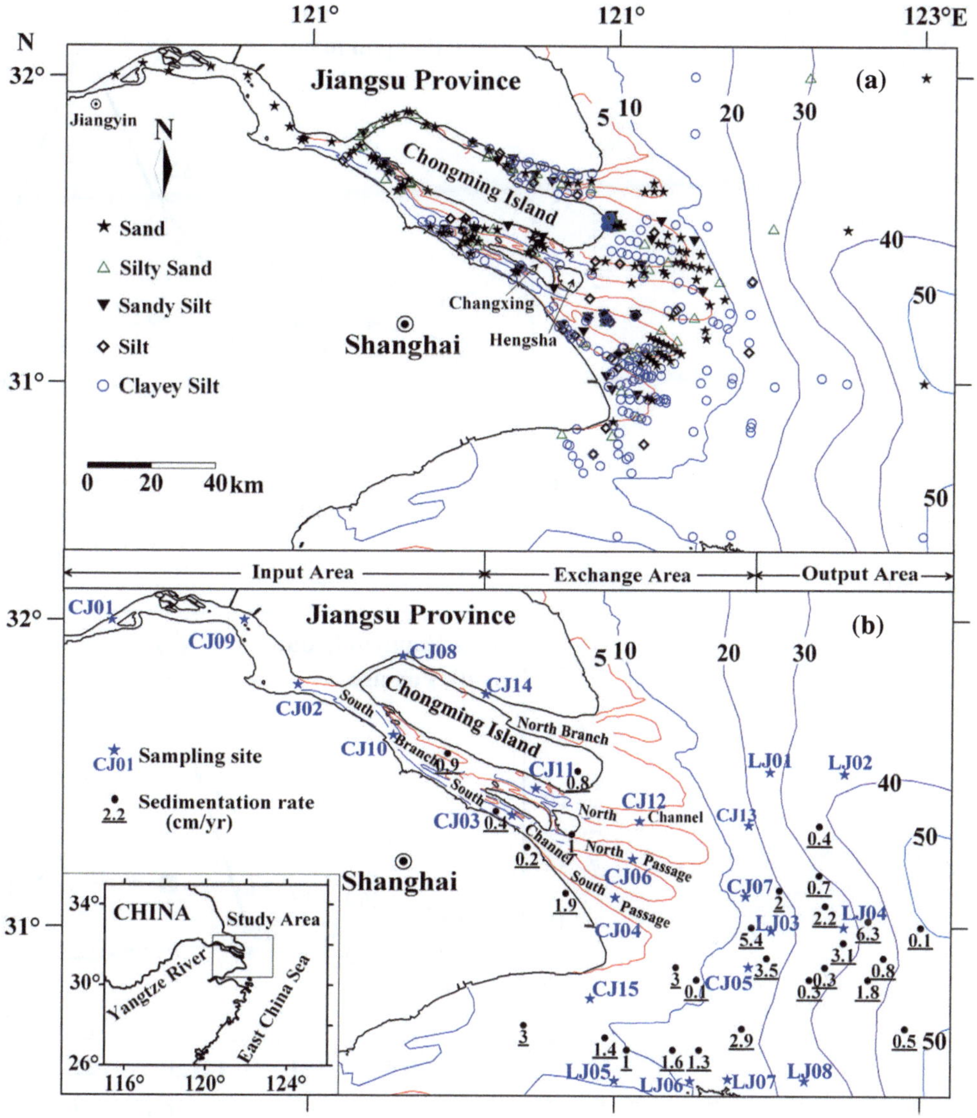

Fig. 3.7 Study area and sampling sites for surface and suspended sediments. **a** isobath and sampling sites for surface sediments (subdivided into five types); **b** sampling sites for suspended sediments and sedimentation rate (cm/year) based on ^{210}Pb dating of sediment cores. Samples of sites CJ01–CJ15 were collected in February 2003, while samples from sites LJ01–LJ08 were collected in July 2006. Sedimentation rate data are from Duan et al. (2005) and Liu et al. (2006)

10.5, and 4.5 μm, respectively. Sedimentary processes in the North Branch (NB) of the estuary and Hangzhou Bay (HZB) to the south of the estuary are dominated by tides (with less fluvial influence), with grain size D_{50} values of 9.9 and 5.6 μm, respectively (Fig. 3.7).

In the upper estuary, sediments mainly originate from riverine sources and, as shown by the vertical distribution of D_{50}, the grain sizes are relatively uniform. In contrast, the depth-averaged D_{50} of suspended sediment in the TM zone is 10.5 μm with obvious vertical differences (Liu et al. 2012) (Fig. 3.8).

The river mouth behaves as a 'filter' for terrestrial sediment input. It retains part of the coarse sediment by strong mixing and sorting processes (Shen 2001), but fine particles are transported long distances into the offshore zone and over the continental shelf (Shen 2001; Yun 2004). As a result, suspended sediment around the 20-m isobath is much finer (~4.5 μm) than in the inner estuary (~10 μm) (Liu 2009).

The major component of the suspended sedimentary material inside the estuary is silt (4–63 μm) (Table 3.2), while silt and clay are the main components over the inner continental shelf. The Changjiang Estuary and adjacent ECS shelf can be divided into five zones: the upper estuary, the TM, the inner continental shelf, HZB, and the NB, on the basis of their hydrodynamics and suspended sediment properties (Liu 2009). Mixing and sorting processes in the estuary result in substantial changes to the characteristics of suspended sediments, such as seaward fining, an increasing clay component, and a decreasing silt component. The suspended sediment over the inner continental shelf consists of silt and clay and is much finer than in HZB where it is close in size to estuarine sediments (Table 3.2). This indicates that sediment is transported from the Changjiang Estuary to the HZB.

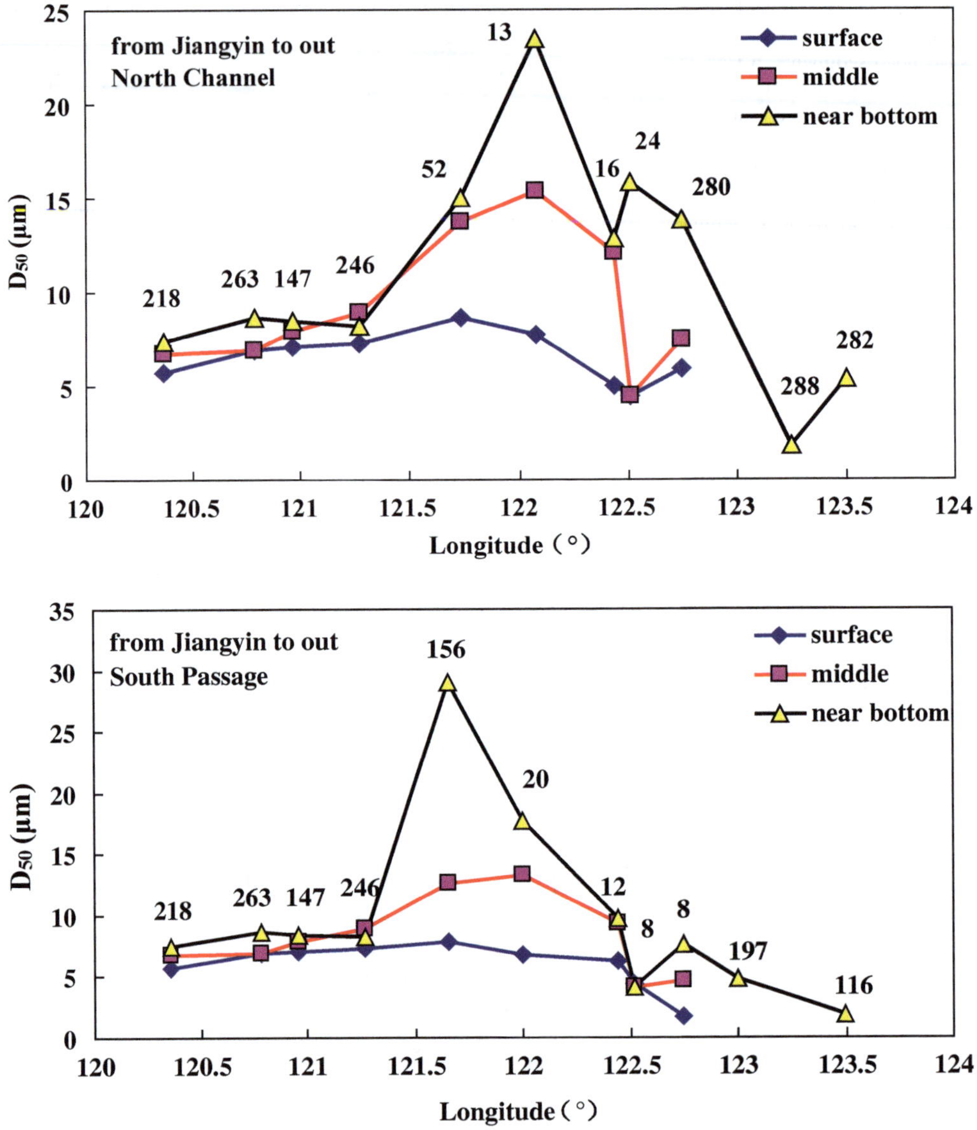

Fig. 3.8 Distribution of suspended sediment D_{50} along the channels

Table 3.2 Sand, silt, and clay content of suspended sediment in different regions of the Changjiang Estuary

	D_{50} (μm)	Content of clay (%)	Content of silt (%)	Content of sand (%)	Sediment
Upper estuary	8.9	30.2	65.9	3.9	Clayey silt
TM	10.5	27.7	67.2	5.1	Clayey silt
NB	9.9	28.4	66.7	4.9	Clayey silt
HZB	5.6	39.4	60.1	0.5	Clayey silt
Inner continental shelf	4.5	48.5	51.4	0.1	Clayey silt, silt

To explore the variation of suspended sediment grain size in the Changjiang Estuary before and after the TGD, suspended sediment grain size diameter (SSD) and the SSC were compared using data from 2003 to 2007 (Liu 2009). This analysis shows that the D_{50} of SSD and SSC were 11 and 21.9 % higher, respectively, in 2003 (before the TGD) compared with 2007 (after the TGD) (Table 3.3).

The different sedimentary characteristics of the wet and dry seasons were also considered both before and after the TGD commenced operation. About 81 % of suspended sediment is transported into the ECS in the wet season at Datong, which represents the upstream boundary for changes to the estuary. SSD in the dry season in Changjiang Estuary was a little higher in 2007 compared with 2003, but SSD in the wet season was between 28 and 39 % lower in

Table 3.3 Comparison of D_{50} and SSC during the wet and dry seasons in the Changjiang Estuary. Positive ratios mean increasing and negative means decreasing

Regions	D_{50}/μm						SSC/kg · m^{-3}					
	2003 dry	2007 dry	Ratio	2003 wet	2007 wet	Ratio	2003 dry	2007 dry	Ratio	2003 wet	2007 wet	Ratio
Upper estuary	9.1	11.2	0.23	8.5	5.7	−0.33	0.17	0.20	0.15	0.53	0.19	−0.65
TM	11.2	11.3	0.01	9.9	6.0	−0.39	0.79	0.64	−0.19	0.63	0.62	−0.02
Inner continental shelf	8.2	9.9	0.22	6.0	4.3	−0.28	0.38	0.36	−0.05	0.24	0.12	−0.50
North Branch	9.5	7.8	−0.18	10.3	7.7	−0.25	1.47	1.04	−0.29	0.73	0.67	−0.09

2007 compared with 2003 (aside from in the NB). SSD decreased much more between 2003 and 2007 in the TM than in other regions (Table 3.3). Overall, SSC decreased more in the wet season (32–65 %) than in the dry season (5–19 %), and SSC decreased more between 2003 and 2007 in the upper estuary and inner continental shelf than in other regions.

In summary, the SSC was more sensitive to the reduction of sediment loads from the river basin in the upper estuary and TM area compared with other regions of the Changjiang Estuary. The variations of SSD and SSC are attributed to the decline in riverine sediment transport and the dynamic adjustment of sediment in the estuary.

In the dry season, the sediment in the TM mainly originates from resuspension on the tidal flat and from inner continental shelf by wave dynamics. These sediment sources compensated for the decrease in riverine sediment and buffered the effects of sediment load reduction (Chen et al. 1988a, b), but also led to grain size changes.

3.3.2 Bottom Sediment

Bottom sediments are classified into clay (0.5–4.0 μm), silt (4.0–62.5 μm), and sand (62.5–2000 μm) according to the method proposed by Shepard (1954). Five primary sediment types occur in the Changjiang Estuary and the adjacent coastal region: clayey silt, sand, silty sand, sandy silt, and silt (Fig. 3.9). Clayey silt is the dominant sediment type (~46 % of all 570 sediment samples over the Changjiang Estuary and its adjacent area in Fig. 3.7) and is primarily deposited in the muddy lower reaches of the Changjiang Estuary (the TM zone), and in the outer estuary (Fig. 3.10a, b). Sand (24 % of all the samples) is present in the upper reaches of the estuary, on the shoals, and in the primary channels of the TM (Fig. 3.10c). The relict deposits over the Holocene subaqueous delta are also composed of sand. Silty sand (15 % of all the samples), sandy silt (11 %), and silt (4 %) are present in the estuarine mouth bar area and in the upper NB (Fig. 3.10b, c).

The spatial distribution of the mean diameter of surface sediment (Fig. 3.10d) shows a clear coarse–fine–coarse trend from the upper estuary to the outer estuary. On average, the diameters of surface sediment in the South Branch (SB) vary from 150 to 220 μm and display a general seaward-fining trend. The mean grain size in the NB is finer than that of the SB, with values in the range of 50–120 μm in the upper reaches, and 15–60 μm in the lower reaches.

The spatial pattern of bottom sediments in the mouth bar area is more complex. Coarser sediments are found in the North Channel (NC), and the finer sediments are present in the South Passage (SP) (Fig. 3.10d).

In the southeastern portion of the estuary, surface sediments are also fine, with a mean diameter of less than 20 μm. The mean diameter of the outer continental shelf deposits ranges from 150 to 300 μm (Fig. 3.10d).

Fine particles (grain size <32.5 μm) are transported further seaward than coarser particles and are deposited in the outer estuary (Fig. 3.10a). The silt component is transported eastward and southward through the estuary and deposited in the SP and further southward along the coast (Fig. 3.10b). The sand component is primarily deposited in the upper reaches of the estuary (Fig. 3.10c).

3.3.3 Exchange Between Suspended and Bottom Sediments

3.3.3.1 Grain Size Distributions

To examine the sediment exchange between suspended and bottom sediments, we used grain size distribution curves from the following locations: four stations located in the upper estuary (between Jiangyin and Wusong), the TM (between Wusong and the nearshore 10-m isobaths), the outer mouth bar (seaward of the 10-m isobaths), and the inner continental shelf (shoreward of the 50-m isobaths)

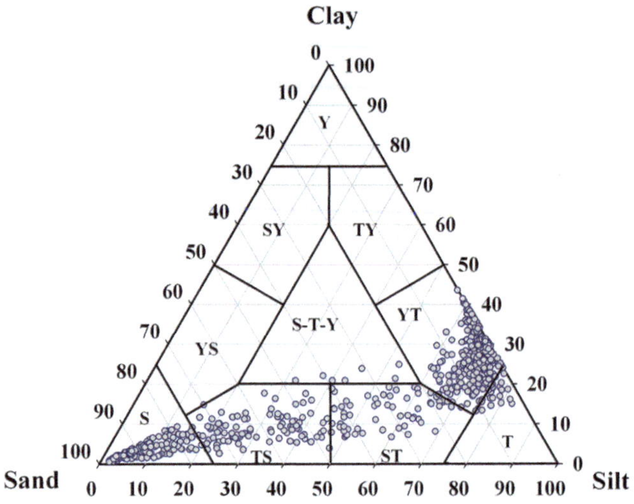

Fig. 3.9 Ternary diagram of sand/silt/clay proportions in the Changjiang Estuary and adjacent coastal zones (*Y* clay; *T* silt; *S* sand; *TY* silty clay; *SY* sandy clay; *YT* clayey silt; *ST* sandy silt; *YS* clayey sand; *TS* silty sand; *S-T-Y* sand–silt–clay)

(Fig. 3.7). We separated the cohesive and non-cohesive sediment using a threshold of 32.5 μm according to the methods of Zhang (1996) and Tang (2007).

In the upper estuary, we found a peak suspended sediment grain size of around 8 μm (Fig. 3.11a), indicating that fine particles dominate in this region. The fine material (grain size < 32.5 μm) accounted for 90 % of the total sediment volume, and the bottom sediment was much coarser, accounting for only 3 % of the sediment volume. This implies that there is little chance of sediment exchange between suspended grains and bottom material (Fig. 3.11a; Table 3.4).

In the TM, the surface suspended sediment grain size is similar to that in the upper estuary, while the near-bed suspended sediment is coarser with a peak at 40 μm (Fig. 3.11b). This indicates that resuspended coarse material from bottom sediments is contributed during transport processes. Also, two peaks are observed in the grain size distribution, which implies that the finer peak is related to riverine sediment, while the coarse peak is caused by resuspended bed sediment. The components of the near-bed suspended sediment are similar to those of the bottom sediment, suggesting both the suspension of bottom sediment and the settling of suspended sediment. In summary, the possibility for sediment exchange between the suspended load and bed load is much higher in the TM than that in the upper estuary.

In the nearshore zone, there are two peaks in the suspended sediment grain size distribution at 8 and 20 μm (Fig. 3.11c) and the vertical differences are smaller than those in the TM. In contrast to the upper estuary and the TM, the cohesive content increases, while the coarse components decrease. The

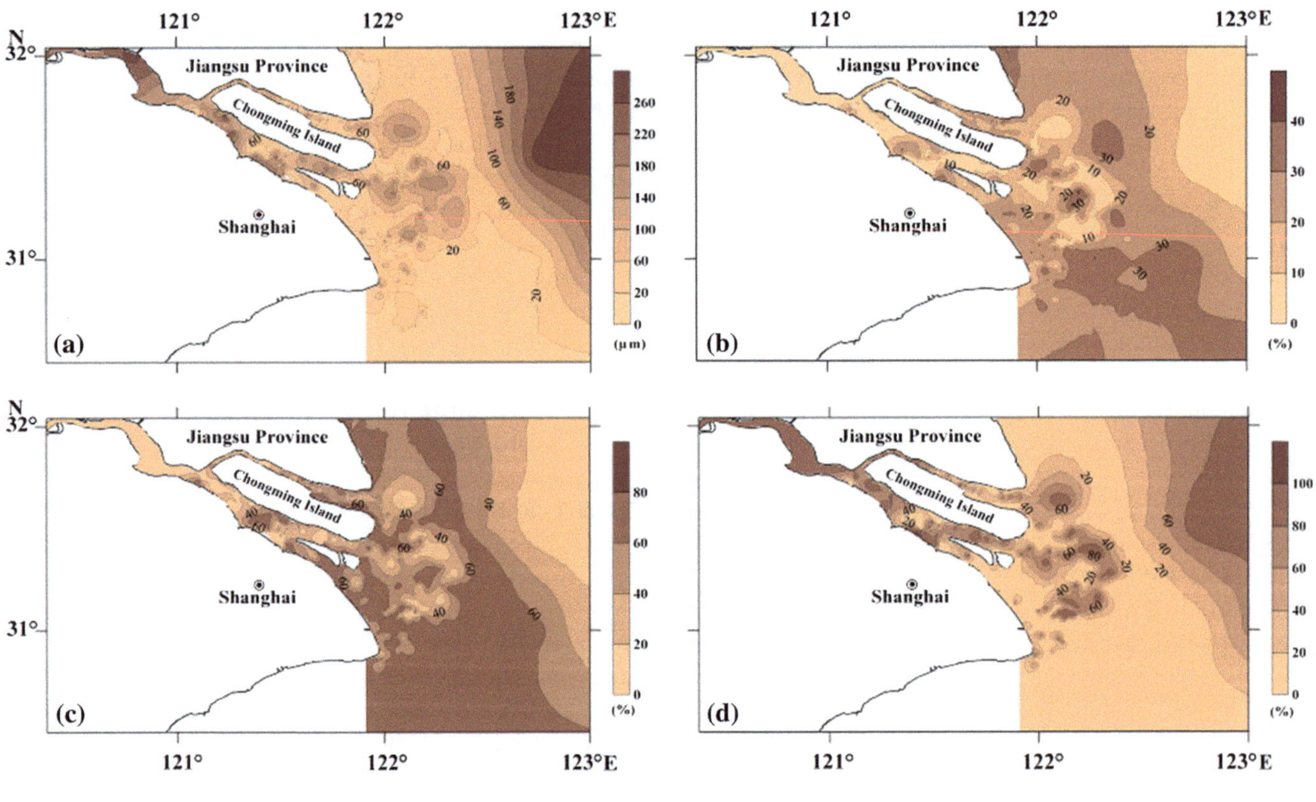

Fig. 3.10 Spatial distribution of the content (%) of **a** clay, **b** silt, **c** sand, and **d** mean diameter (μm) of surface sediment

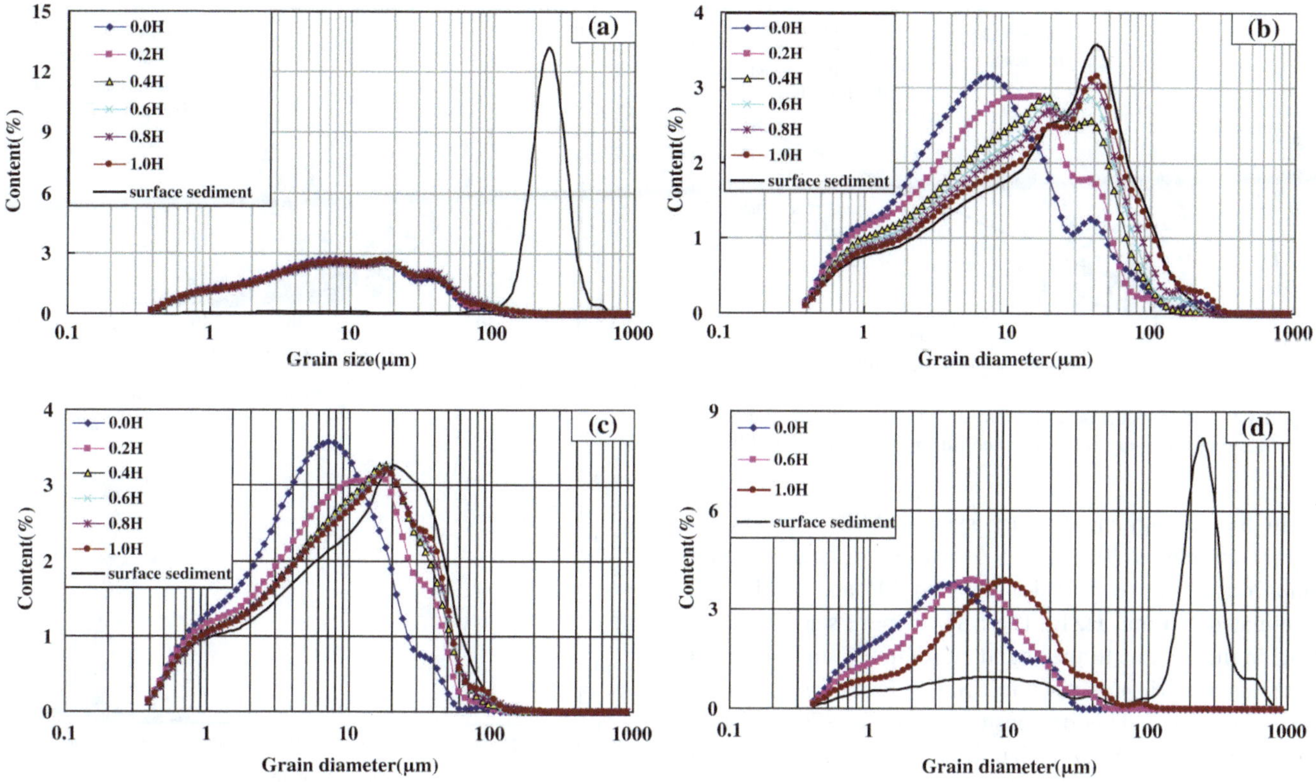

Fig. 3.11 Grain size distribution of suspended and surface sediments in the estuary and continental shelf system. **a** CJ10, in the upper reaches of estuary; **b** CJ04, in the turbulence maximum; **c** CJ05, in the outer mouth bar; and **d** LJ10, in the inner continental shelf. *H* represents water depth, and locations are shown in Fig. 3.7

Table 3.4 Cohesive sediment content (<32.5 μm) in the estuary and nearshore coastal areas (inner shelf of the ECS)

Regions	Percentage of cohesive sediment content (%)						Surface sediment
	0.0H	0.2H	0.4H	0.6H	0.8H	1.0H	
Upper estuary	89.0	87.0	86.3	83.4	84.8	86.3	3.0
TM	89.4	88.5	81.2	77.0	73.4	68.1	65.3
Outer mouth bar	97.0	92.0	89.0	88.1	87.2	86.7	82.6
Inner continental shelf	100.0			98.5		94.8	31.1

grain size distributions of suspended sediments and bottom sediments are similar and have equal silt and clay contents, which suggest that the exchange between suspended sediments and the bottom sediments is high in this region.

The cohesive content of suspended sediments increases to 95 % over the ECS shelf. The peak sizes are 4, 5 and 10 μm in the surface, middle, and bottom layers, respectively (Fig. 3.11d). The grain size distributions of the bottom sediments are characterized by two peaks, with 66 % of the particles being coarser than 100 μm. These coarser particles are thought to be relict sediments left over from lower Late Holocene sea levels in the nearshore shelf zone (Chen et al. 1985). Finer grain sizes, from 4 to 100 μm, make up 34 % of shelf sediments and originate from the settling of the suspended sediments that have escaped estuarine trapping. These results demonstrate that the exchange between suspended sediment and bottom sediment becomes lower over the nearshore shelf region.

The above description indicates that the surface suspended sediment carries a signature from river-borne sediments, which is imparted during transport processes from the Changjiang Estuary to the ECS. Exchanges between the near-bed suspended load and bottom sediments in the TM zone alter the grain size distribution of the suspended sediments. This is further illustrated by the increasingly coarse grain size of the near-bed suspended sediments. However, coarse sediment mainly settles inside the estuary and is rarely transported to offshore areas. As a result, exchange processes become less important in a seaward direction.

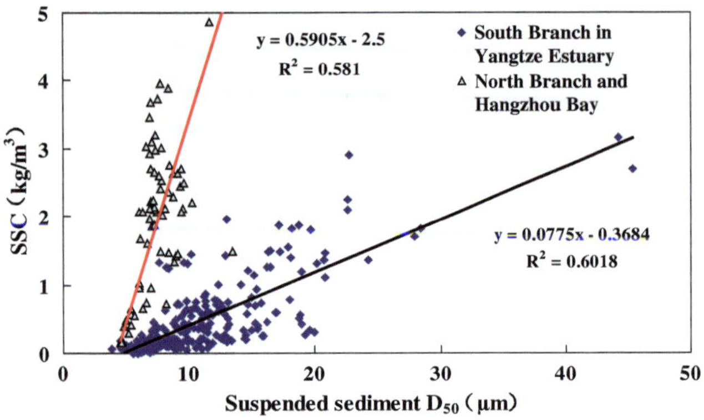

Fig. 3.12 Relationship between suspended sediment D_{50} and SSC in the Changjiang Estuary (data sampled during spring tide in 2007)

In the Changjiang Estuary system, SSD and SSC are highly correlated (Fig. 3.12). However, different relationships occur in the fluvial-dominated SB and in the tide-dominated NB and HZB. The SSD is more sensitive to the SSC in the SB due to the abundant sediment supply. In tide-dominated environments, the SSD is less sensitive to the SSC (Fig. 3.12). This demonstrates that the SSC variability is related to the SSD variability in different ways under different hydrodynamic conditions. Field data show that the clay content decreases exponentially with increasing SSC, and to the contrary, the silt content increases exponentially with increasing SSC (Fig. 3.13).

Hydrodynamics are the primary factor that trap sediment inside the estuary, but sediment mixing and exchange processes between suspended and bottom sediments are also important. The turbulence intensity is highly related to the velocity gradient, which is given by $\sigma = f(u^m \frac{du}{dz})$, where u is current velocity (m/s), z is water depth (m), m is the order of the power, and f means an implicit function. Additionally, the shear stress is calculated by $\tau = \mu \frac{du}{dz}$, where μ is the dynamic viscosity of water (N · s/m^2), and τ is the shear stress (N/m^2). Thus, the vertical velocity gradient $\frac{du}{dz}$ is one of the key parameters representing the turbulence strength and shear stress (Zhang 1998).

The vertical velocity gradient is relatively small in the reaches between Jiangyin and Qiyakou, with values between 0.03 and 0.05 s^{-1}. It begins to increase from the upper reaches of the South and North channels and reaches a maximum in the TM zone, before decreasing again further offshore. The presence of shear stress is the main factor inducing water and sediment mixing, resulting in high sediment concentrations in the TM zone. The settling velocity of cohesive sediment is enhanced by flocculation (Tang 2007; Guo and He 2011), which in turn intensifies the stratification effect. At the same time, vertical exchange processes between the surface and bottom sediments are also enhanced. Our results show that the vertical gradients of grain sizes and sediment concentration

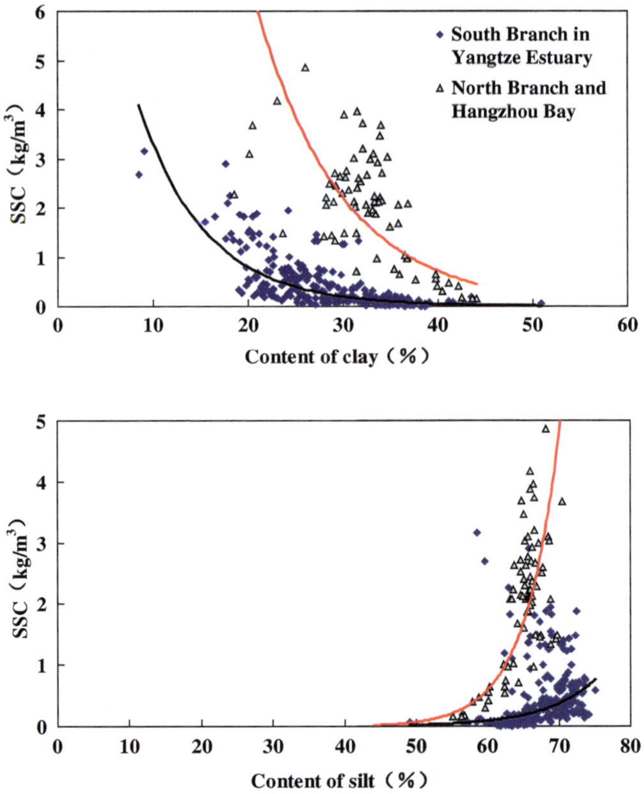

Fig. 3.13 Relationship between suspended sediment compositions and SSC in the Changjiang Estuary during a spring tide

increase with increasing velocity gradients with a strong linear relationship (Fig. 3.14).

High sediment concentrations can change the flow structure, such as damping turbulence, and thus influences the vertical velocity and sediment concentration profiles (Zhang 1998; Toorman et al. 2002). Given that the grain sizes of the suspended sediment vary little longitudinally, the increasing grain sizes in the bottom suspended sediment are mainly induced by high shear stress and resuspension. This suggests that resuspension induced by high shear stress is the main reason for the increasing grain size and concentration of coarser particles in the bottom suspended sediments.

3.3.3.2 Quantification Methods of the Sediment Exchange

A quantitative analysis of the vertical sediment exchange process was conducted by evaluating the constraints imposed by the sediment budget, according to which the sum of the masses of the exchange and output sediments should equal the mass of the input sediments (Liu et al. 2010):

$$G_I = pG_E + (1 - p)G_O \tag{3.1}$$

where G_I and G_O are the grain size distributions (GSDs) of the input and output sediments, and G_E represents a mixture

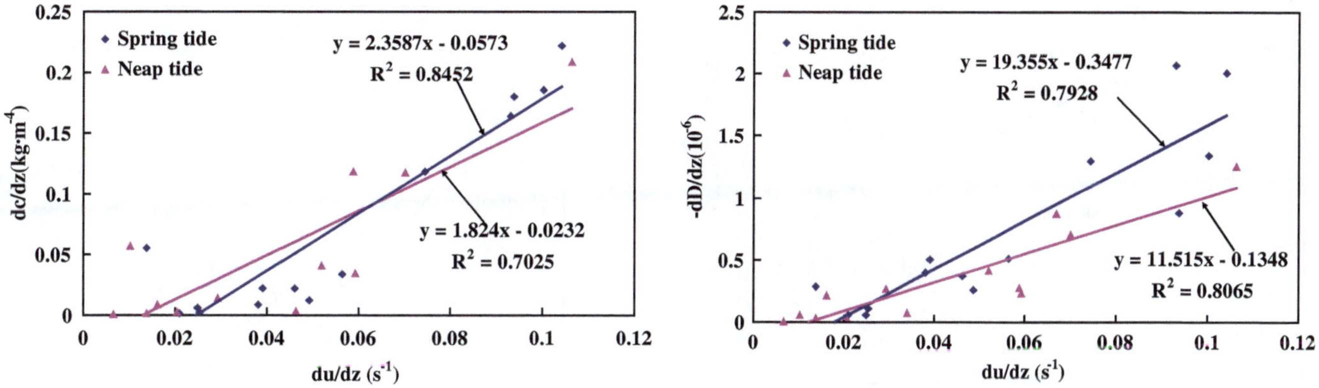

Fig. 3.14 Relationship between the current velocity gradient, the SSC gradient, and the surface sediment diameter

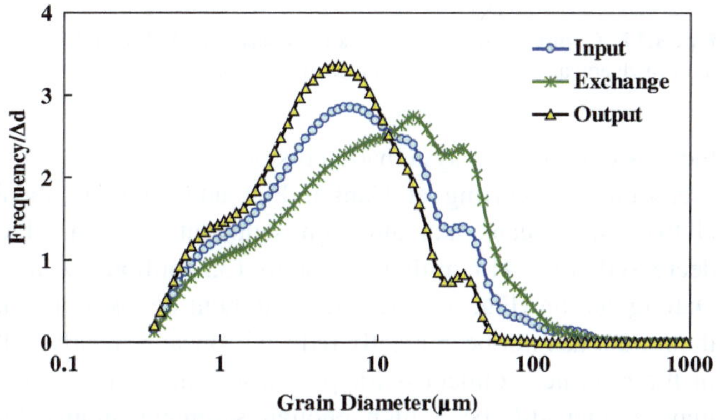

Fig. 3.15 Average grain size distributions of near-bottom suspended sediments in the Changjiang Estuary

of G_I and G_O by a portion p and $1 - p$, respectively. Each GSD is a vector of k elements (mass fractions of grain size classes), which sum to unity (or 100 %). The unknown, p, is a proportion ($0 \leq p \leq 1$), termed the exchange ratio. The least squares solution to this constrained linear mixing problem is given by

$$p = \frac{\sum_{j=1}^{k} (G_{Ej} - G_{Oj})(G_{Ij} - G_{Oj})}{\sum_{j=1}^{k} (G_{Ej} - G_{Oj})^2} \quad (3.2)$$

To convert these proportions to rates of sediment exchange between the water column and the bed (V_E), independent information regarding the bulk sediment accumulation rates (V_B) is required:

$$V_E = pV_B \quad (3.3)$$

where V_E is the rate of sediment exchange, p is the exchange ratio, and V_B is the sediment accumulation rate (see Fig. 3.7).

The GSDs of suspended and surface sediments illustrate the vertical sediment exchange processes in different sections of the Changjiang Estuary (Fig. 3.15). The suspended sediment content, with a range of 18.0–110 μm in the exchange region, increases by 16.6 % compared with the input area due to the exchange between the suspended and bottom sediments (Fig. 3.15). However, the suspended sediment content, with a range of 0.4–13.6 μm in the output region, increases by 15.5 % compared with the input area. This increase is due to the selective deposition of coarse components, so that fine fractions are relatively higher.

According to above analysis, the suspended sediment grain size distributions change remarkably due to sediment exchange and selective deposition during transport to the sea. The bulk suspended sediment exchange ratio in the Changjiang Estuary can be calculated if the average GSD of suspended sediments in the exchange area (Fig. 3.9) is substituted into Eq. 3.1 as G_E. The results of this calculation give an exchange ratio of p = 49.4 %. This means that 49.4 % of suspended sediments supplied from the river basin deposited into the TM zone and contribute to the morphological evolution of the estuary. These results are consistent with studies in Changjiang Estuary conducted by Milliman et al. (1985), Shen (2001), and Liu et al. (2006), which provided estimates of 40, 42, and 47 %, respectively. Hence, ca. 50 % of suspended sediments escape from the estuary and are transported south where they are deposited on the outer estuary.

More detailed information regarding sediment exchange in the study area may be obtained by analysis of the spatial variation of sediment exchange ratios. Small amounts of suspended sediments were deposited onto the seabed of the upper estuary (sediment exchange ratio, $p < 0.1$) because the fine-grained suspended sediments in this region were transported to the mouth bar area by the ebb-dominated tidal flow (Fig. 3.16a). The sediment exchange ratios in the outer estuary also have very low values ($p < 0.1$) because of oceanic currents offshore. However, intensive sediment exchange occurred in the inner estuary due to sand–mud mixing, which was controlled by the bidirectional tidal flow (Fig. 3.16a). The most frequent sediment exchange in the muddy area, evidenced by an exchange ratio of about 0.8, was observed at the depocenter

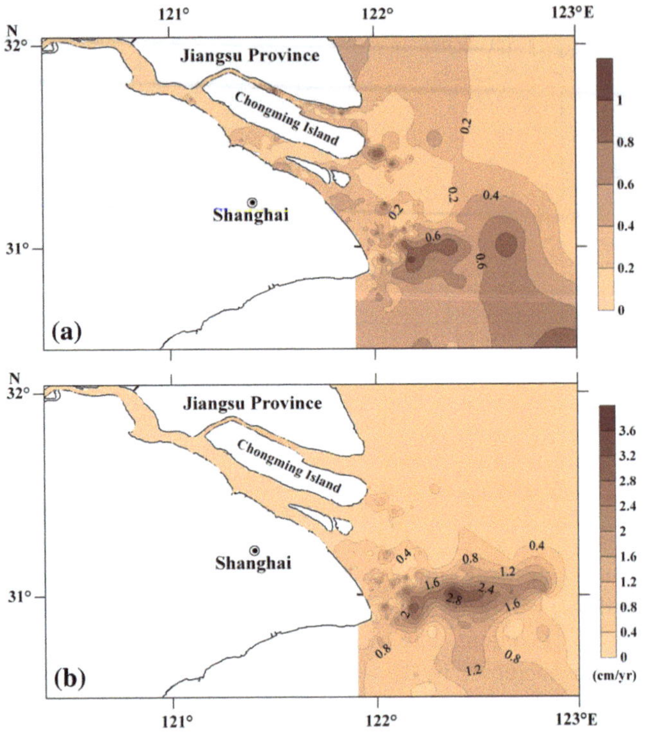

Fig. 3.16 **a** Spatial distribution of the vertical sediment exchange ratio, *p*, and **b** the sediment exchange rate (cm/year)

of the suspended sediment. The results also show that the southward transport pathway of suspended sediments in the Changjiang Estuary is located between 122°E and 123°E longitude along the coast (Fig. 3.16a). This belt of high sediment exchange ratios parallel to the coast is consistent with the pattern of longshore currents.

The sediment exchange ratio, *p*, is a relative measure that describes the local balance between the suspended and surface sediments. The ratio can be transformed to the sediment exchange rate (V_E), using the sedimentation rate, by $V_E = pV_B$, where V_E is the rate of sediment exchange, *p* is the exchange ratio, and VB is the sediment accumulation rate. The low V_E value in the upper part of Changjiang Estuary indicates that a small amount of suspended sediment is deposited in this area (Fig. 3.16b). Conversely, the high sediment exchange rate in the muddy area, with a value greater than 2 cm/year, implies that this area is the depocenter for terrestrial sediment (Liu et al. 2010, 2011). Beyond the depocenter, sediments are transported southward along the coast.

3.3.3.3 Evidence of Sediment Exchange

The subaqueous region of the Changjiang Delta has begun to recede in recent years because of major reductions in river sediment (see Chap. 4). The retreat is particularly evident around the 10-m isobaths near the depocenter. By contrast, the inner estuary and the tidal flat are still advancing, and bathymetrical analysis around the 5-m isobaths shows that

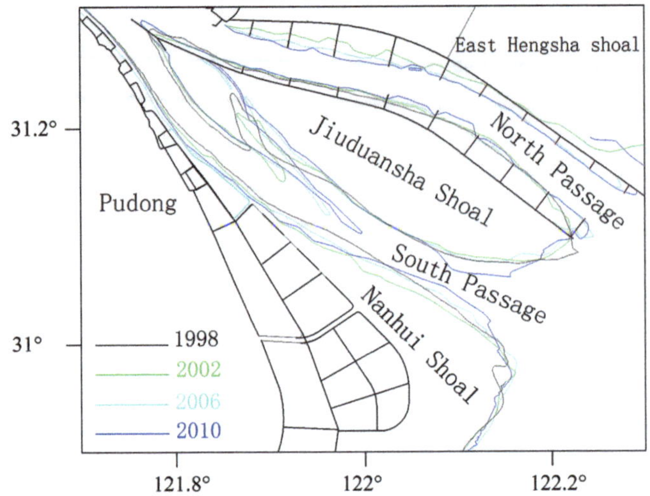

Fig. 3.17 Comparison of the 5 m isobath between 1998 and 2010 and map of deepwater navigation channel in the North Passage

there is currently no systematic erosion (Fig. 3.17) and that accretion is occurring at Nanhui Flat and Jiuduan Shoal (JDS). This means that although sediment discharge has decreased from 414 million t/year to 135 million t/year at Datong station (Fig. 3.2), riverine sediment deposition near the mouth has not been greatly reduced. Because nearly half of the riverine sediment is trapped inside the estuary, there may in fact still be a high enough sediment supply for accretion at the mouth. Moreover, the nearshore depocenter is a source of sediment for the estuary and contributes to building tidal flats, especially during dry season (Liu 2009).

Yun (2004) argued that river-derived sediment is deposited in the nearshore zone seaward of the NC, NP, and SP, where the water depth is between 7 and 25 m. The depocenter for this riverine sediment is located in the nearshore zone seaward of the SP. The seismic profiles also show that there is a 'muddy trough' oriented to the southeast around the depocenter (Chen and Xu 1995, 2000). The trough is located in front of the modern sedimentation center and consists of relict sediment derived from Late Holocene material that has experienced tectonic uplift and partial exposure (Chen et al. 1985; Hori et al. 2001; Liu et al. 2006). The elevated bed over the depocenter demonstrates that it has been a sediment sink over timescales of a decade to a century (Fig. 3.18). In summary, evidence shows that the depocenter is the center of subsidence and also the sink of the river-borne sediments.

There is a dominantly eastward-moving residual current pattern in the Changjiang Estuary (Fig. 3.19). However, in the northern part of the estuary (to the north of eastern Hengsha Flat), the current turns slightly northeast and in the southern estuary it tends to flow southeast. The strength of the residual current is strongest in the NC, where it flows east in the channel and then south along the shallow eastern Hengsha Flat. The residual currents at the mouth of the NP and SP

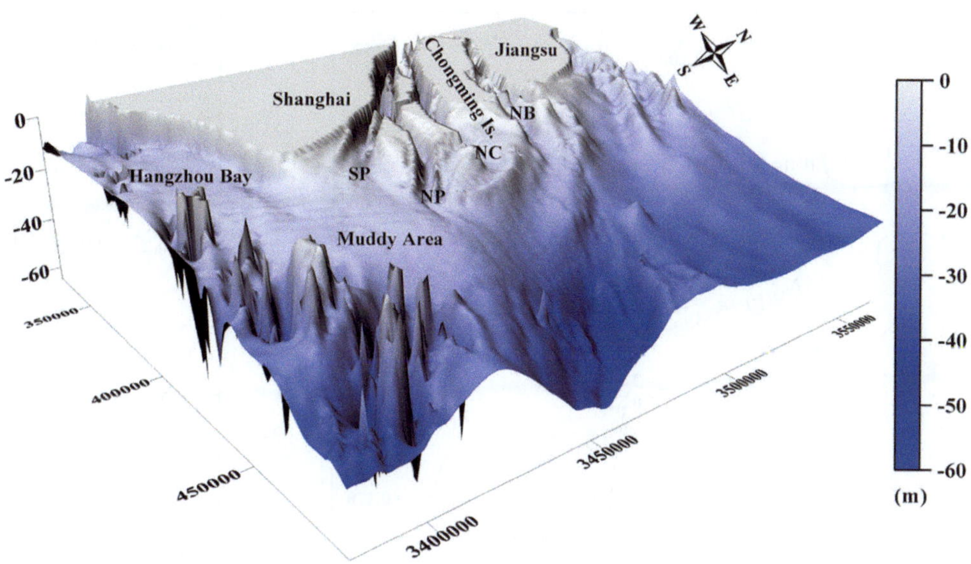

Fig. 3.18 3D topography of the Changjiang Estuary (*SP* is South Passage; *NP* is North Passage; *NC* is North Channel; and *NB* is North Branch)

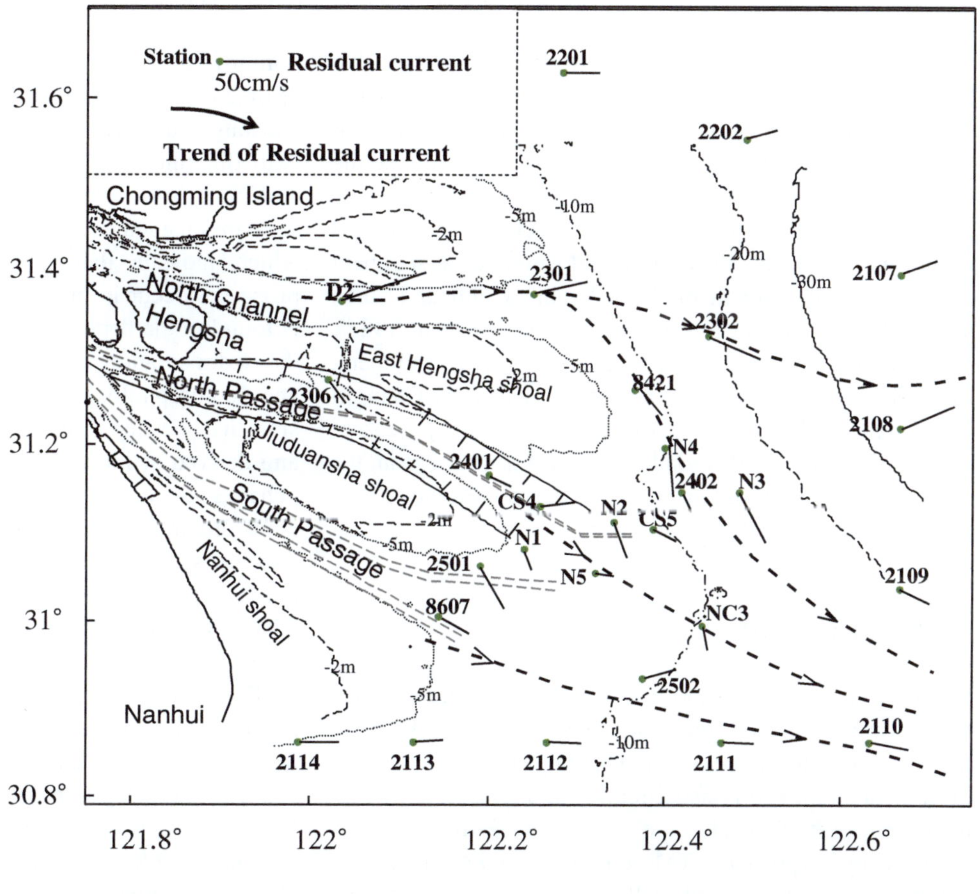

Fig. 3.19 Residual currents in the outer Changjiang Estuary based on measurements in 2004 (Lou 2005)

converge toward the depocenter (Fig. 3.19), and the horizontal residual current structure suggests that the hydrodynamics favor sediment transport toward the depocenter near the SP.

SSC data measured in 1982 shows that the vertically averaged sediment concentration was higher in the TM zone than in the landward and seaward regions (Chen et al. 2006). The SSC is also higher during the dry season than during the wet season, and higher during spring tides than during neap tides (Chen et al. 1988a, b). The area seaward of the SP is also characterized by high sediment concentrations, which can be explained by the seaward-moving Changjiang plume (Fig. 3.19). The spatial distribution of the sediment concentration is in line with the residual current pattern and is also verified by satellite observations (Yun 2004; Lou 2005).

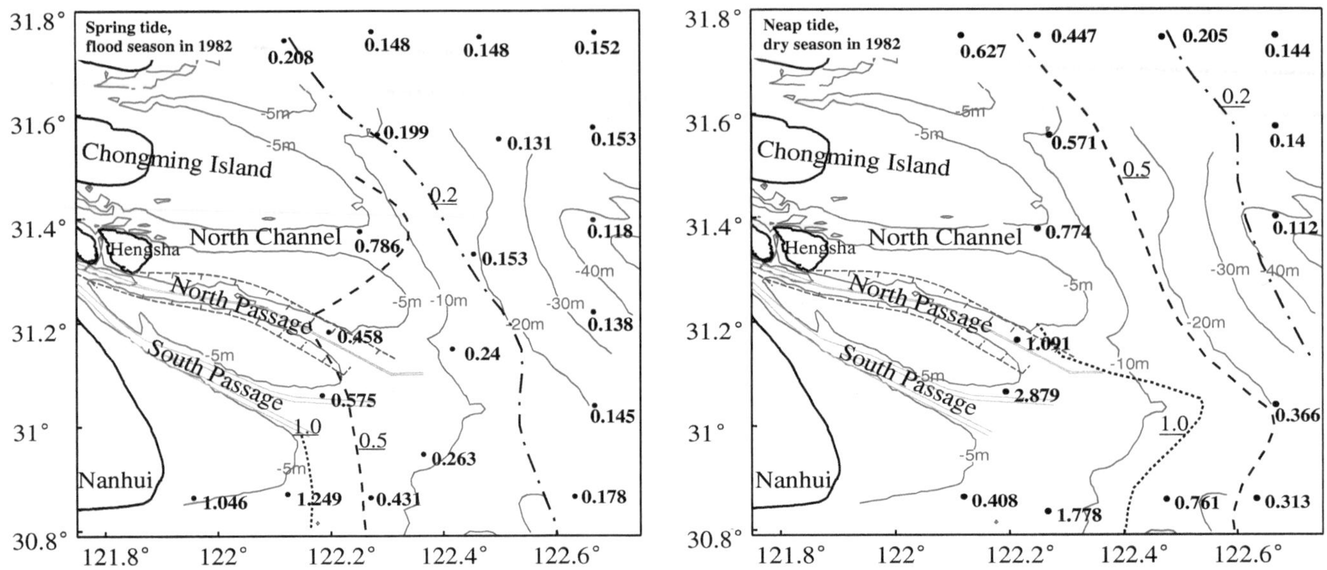

Fig. 3.20 Distribution of average SSC (kg/m^3) in the outer Changjiang Estuary (Lou 2005)

The current field and sediment transport patterns show that both the ebb current and ebb-directed sediment transport move southeast near the mouth of the SP, forming a region of high SSC (Fig. 3.20). The alongshore coastal current and the northward Taiwan Warm Current also play an important role in this sediment transport process by blocking the eastward water and sediment dispersion and promoting sediment deposition at the mouth of the SP. This sediment is then transported south by the alongshore current (Chen et al. 1985).

The depocenter near the SP is a sink of riverine sediment. This sediment can also be transported back to the Changjiang Estuary, HZB, and, particularly in the dry season, southward along China's southern coast. For this reason, the depocenter is also a sediment source that contributes to the estuarine geomorphologic evolution.

3.3.4 Discussion

Estuarine sediment transport processes are far more complex than riverine transport processes. The interaction between the river and tides induces complex hydrodynamics in terms of vertical stratification and circulation, and associated sediment transport. Typically, the Changjiang Estuary is characterized by a very large TM zone in the mouth area, with underwater mouth bars and shallower depths than inland and outer areas. Additionally, both suspended and bottom sediments are much finer over the TM zone than inside the estuary (such as the SB) and the outer estuary (the ECS shelf), which indicates the role of the estuary in trapping and filtering sediment moving from the river to the sea.

Sediment transport is accompanied by along-channel sorting, mixing, and exchange processes between (near-bottom) suspended sediments and bed sediments. We use sediment grain size distributions to quantify the sediment exchange rate. The results show strong spatial variability, with different grain size distributions in the inner estuary, TM zone, and outer estuary. The TM zone and nearshore area are characterized by a higher sediment exchange rates, which indicate that strong deposition and resuspension processes occur in these regions. The highest exchange rate is found in the depocenter zone (southeast of the SP), where the sedimentation rate is also higher. The hydrodynamics and geomorphology of the estuary also support the conclusion that the depocenter is the active zone where Changjiang-derived sediments are deposited. Finally, the TM zone behaves as a 'sediment bank,' providing sediment for landward transport by tidal forcing in times of decreasing riverine sediment input. Based on the sediment exchange rate, it is also estimated that about 47 % of sediment from the Changjiang may deposited in the estuary, whereas the remainder is transported seaward.

There are still many unanswered questions about sediment transport dynamics in the Changjiang Estuary. These questions include the following: How important is fine sediment in transport processes such as fluid mud dynamics? How important are wave dynamics in sediment transport in the mouth area, where broad tidal flats and channels coexist? How are sediment transport processes related to the sedimentary record in different timescales (tidal, seasonal, annual, or decadal)? How much sediment comes from the northern Jiangsu coast (from the abandoned Yellow River delta)? How much Changjiang sediments are transported into HZB and how much is transported southward by the alongshore coastal current? Further systematic studies are needed to address these questions.

3.4 Impacts of Engineering Works

3.4.1 Introduction

The Changjiang is historically termed a 'golden waterway' due to its significance for navigation. However, the mouth bar in the estuary constrains ship movement because of its shallow channel depth (less than 10 m below mean sea level) (Chen et al. 1985). With the aim of achieving a 12.5-m-deep navigation channel, the Deep Waterway Project was initiated in the NP in 1997 (Fig. 3.17). This project involved the construction of two 50-km-long jetties and 19 groins, along with extensive dredging (Fig. 3.17). The construction changed the ebb flow diversion between the SP and the NP, so that less ebb tidal water is carried through the NP compared to the SP (Jiang et al. 2013). Because the engineering work is located in the mouth bar area, where estuarine density circulation is important, it is also anticipated that circulation patterns may also change. High sediment deposition was also observed in recent years over JDS, and strong sediment exchange between the shoal and the navigation channel in the NP is evident (He et al. 2007). Thus, it is of practical interest to know how the navigation channel engineering works affect estuarine dynamics.

Recently, many approaches have been introduced to study transport processes in estuaries, such as an integrated approach, parcel tracking, and dispersion models (Cheng and Casulli 1982; Luff and Pohlmann 1995; Delhez et al. 1999; Deleersnijder et al. 2001; Liu et al. 2004; Shen and Wang 2007; Gong et al. 2008; Wang et al. 2010). To understand the influence of circulation pattern changes on transport processes due to waterway construction in the Changjiang Estuary, we apply a three-dimensional transport model forced by different discharge conditions. A passive tracer model is used to diagnose the variation of transport properties in the Changjiang Estuary, both spatially and temporally. We use a novel method, water age theory, to examine the transport timescales of water over the whole estuary. Modeling scenarios, both with and without engineering works, enable us to assess the impacts of the navigation channel works.

3.4.2 Model Setup

This study uses the Environmental Fluid Dynamics Computer Code (EFDC). The model has a structured orthogonal curvilinear grid in the horizontal and sigma coordinates in the vertical. The modified Mellor and Yamada level 2.5 turbulence closure scheme is implemented in the model (Mellor and Yamada 1982; Galperin et al. 1988). The model resembles the widely used Blumberg–Mellor model in both the physics and the computational scheme utilized (Blumberg and Mellor 1987).

Because of frequent water exchanges between the Changjiang Estuary and HZB, a much larger area is included, which encompasses the Changjiang Estuary, HZB, and their surrounding coastal regions. A curvilinear and orthogonal grid is generated for the Changjiang Estuary and HZB. The seaward boundary is placed offshore at 123°E and prescribed by tidal water levels, and the landward boundary along the Changjiang is 600 km upstream of the estuary to Datong and prescribed by time series river discharges. In the first instance, we use only the hydrodynamic model, while the sediment transport model is ongoing work.

In the numerical model, two transport equations are introduced to calculate the tracer concentration and age concentration. Considering only one tracer discharging into the water body, without additional source and sink terms, the tracer concentration and age concentration are introduced by Delhez et al. (1999) and Deleersnijder et al. (2001) and can be given as:

$$\frac{\partial C(t,\vec{x})}{\partial t} + \nabla(uC(t,\vec{x}) - K\nabla C(t,\vec{x})) = 0 \tag{3.4}$$

and

$$\frac{\partial \alpha(t,\vec{x})}{\partial t} + \nabla(u\alpha(t,\vec{x}) - K\nabla \alpha(t,\vec{x})) = C(t,\vec{x}) \tag{3.5}$$

where C is the tracer concentration, α is the age concentration, u is the velocity, K is the diffusivity tensor, t is time, and $\vec{x}$ is the spatial coordinate. The mean water age, a, can then be calculated as follows:

$$a\left(t,\vec{x}\right) = \frac{\alpha(t,\vec{x})}{C(t,\vec{x})} \tag{3.6}$$

Equations 3.4–3.6 are used to compute the water age using the EFDC model with specified initial and boundary conditions. The tracer with a concentration of 1 (arbitrary unit) is continuously released into the headwater of the estuary. The open boundary condition of the tracer is treated as similar to salinity (conservative), except that the incoming tracer and age concentrations are set to zero. The vertical mean water age is calculated by averaging the water age at each layer. Sensitivity scenarios are run based on high- and low-river discharges, simulating wet and dry seasons in the

Changjiang Estuary, and also based on scenarios with and without engineering works. The bathymetry is fixed for all scenarios using 2003 data. For more detailed information on the model setup, please refer to Wang et al. (2010).

3.4.3 Changes in Estuarine Circulation

The Deep Waterway Project not only altered the gravitational circulation in the mouth bar area by changing the freshwater discharge diversion and associated stratification, but also affected the horizontal water exchange between the SP and NP. Before the dike was built, the water in the NP could flow freely to the SP over the tidal flat during high tide and enter HZB. The dike greatly reduced the horizontal water exchange between the NP and SP. Figure 3.21 shows the horizontal circulation patterns, and Fig. 3.22 shows the surface residual current patterns due to water particle movements, both with and without the dikes. Due to a reduced water flux from the NP downstream of JDS, water transported out of the NP and SP merges downstream of the dike, forming a stronger current that turns right toward HZB. A portion of the water is transported back to the SP along the tidal flat near the southern bank of the SP, resulting in the formation of a strong eddy and high residence time on the tidal flat (Figs. 3.21 and 3.22). The water transported out of the NP turns right and returns through the SP, forming two clockwise eddies. These results show that the dike has a significant effect on the horizontal estuarine circulation.

3.4.4 Changes in Transport Timescales

The model results show changes in both the horizontal and vertical water age distribution (Fig. 3.23). The causes of these changes are largely due to the increase in friction resulting in changes to the barotropic pressure gradient, which leads to the change in flow diversion in the NC and SC, and in the NP and SP. To further diagnose the underlying processes, the water ages downstream of Xuliujing along the NP and SP are plotted in Fig. 3.23, under both high- and low-flow conditions. It can be seen that the water age linearly increases in the first 50 km and there is no difference between the cases with and without the dike. A marked change in water age occurred between 60 and 120 km along the transect after the dike was built (Fig. 3.23). The transport time decreased from peak values in the NP by 16 % and increased in the South Passage by up to 41 %.

The horizontal distribution of vertical mean water age indicates a change in estuary circulation (Fig. 3.24). The water age is almost identical before and after the dike was built due to the distance of the project from the estuary. The difference in water age between the NC and the SC decreased, with water transport becoming faster in the NC and slower in the SC after

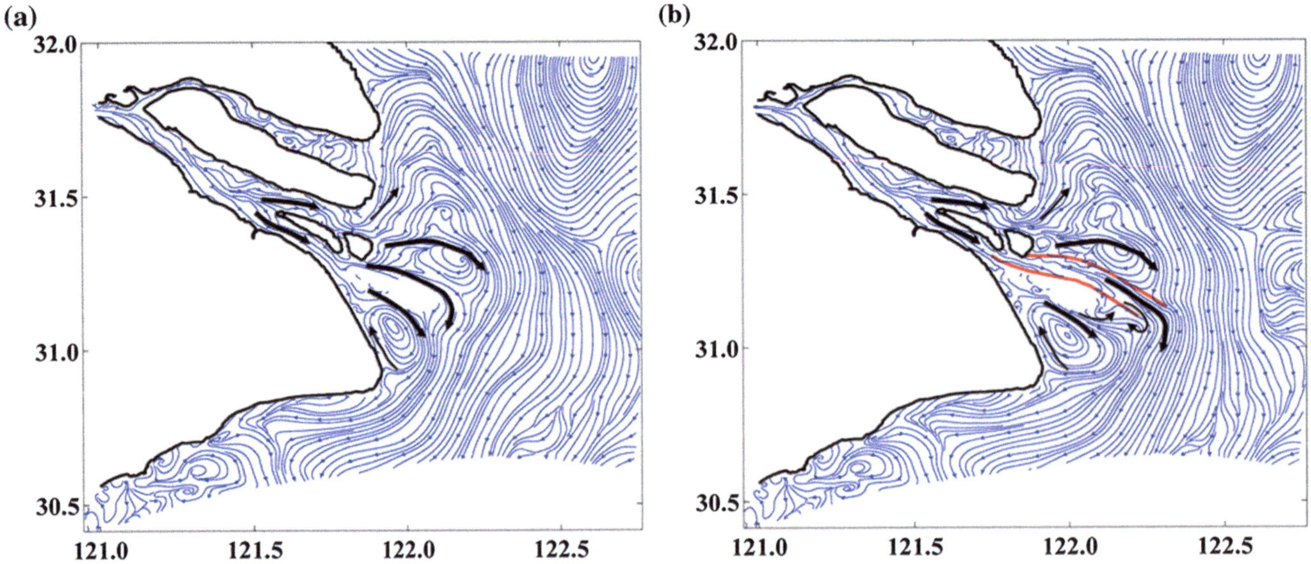

Fig. 3.21 Comparison of the horizontal circulation pattern **a** without the dike and **b** with the dike

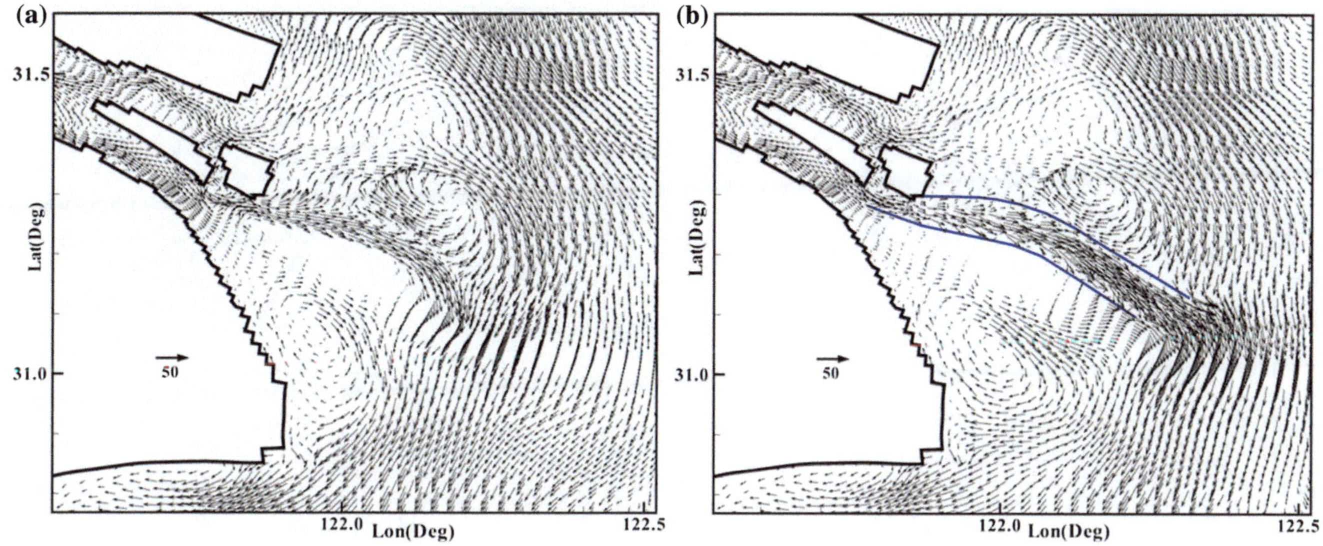

Fig. 3.22 The surface residual current during low discharge: **a** without the dike and **b** with the dike

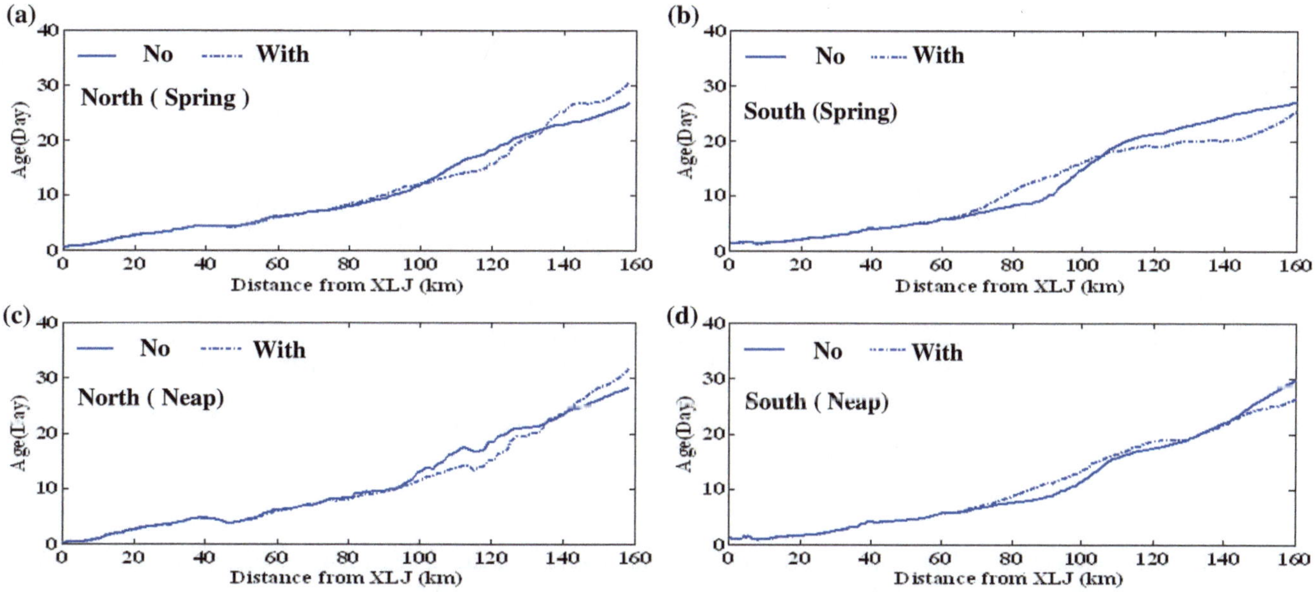

Fig. 3.23 Distribution of depth-averaged water age along the transect under mean flow discharge both with and without waterway project for **a** spring tide in the NP; **b** spring tide in the SP; **c** neap tide in the NP; and **d** neap tide in the SP

the construction of the dike. The lateral difference in water age shows decreasing water age in the SC and increasing water age in the NC. The water exchange between the NP and NC through overwashing of the eastern Hengsha shoal is constrained due to the restrictions of the dike, resulting in a decrease in the water age in the upper portion of the NP. The dike also led to a high water age zone near JDS, with values about 1.3 times larger than in nearby areas.

The engineering structures in the NP also significantly altered the circulation patterns in adjacent areas. After the

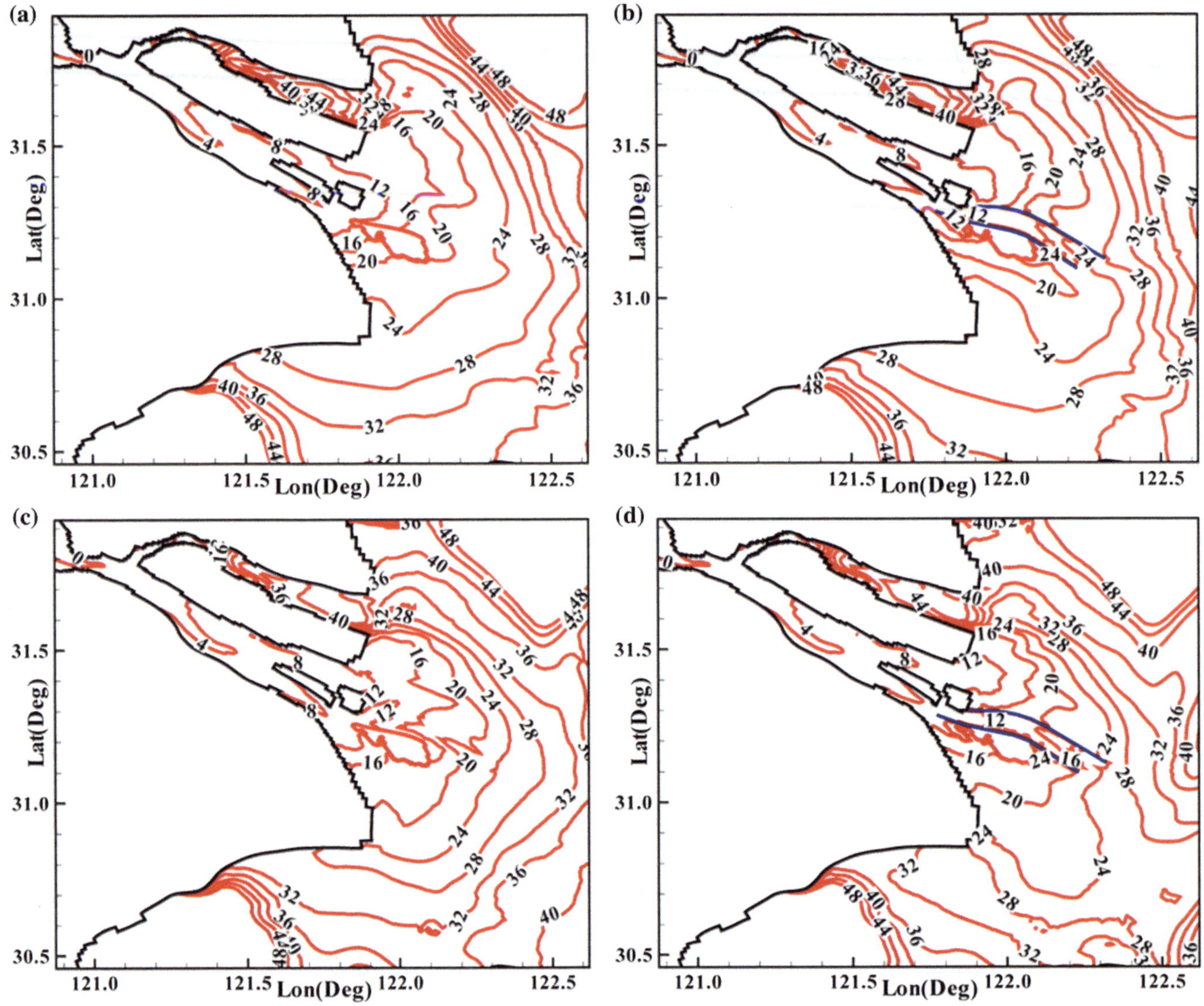

Fig. 3.24 Vertically averaged water age distribution with and without the dike during dry season for **a** spring tide with the dike; **b** spring tide without the dike; **c** neap tide with the dike; and **d** neap tide without the dike

completion of the project, the depth of the channel was altered due to channel adjustment, and the vertical salinity structure changed in the NP. Vertical stratification due to density differences was enhanced after the dike was built in the NP, and salinity intrusion moved 5 km landward during the neap tide in the NP.

The change in the water age can also be used to examine changes in vertical transport. An increase in water age indicates that the speed of water being flushed seaward is decreasing. Overall, the differences between the surface and bottom water increased after the dike was built. Taking the NP as an example, there is less than 2-day difference in water age before the dike project and up to 10-day difference after the dike was built. This result shows that the surface water travels faster, while bottom transport is slower after the dike was constructed (Fig. 3.25).

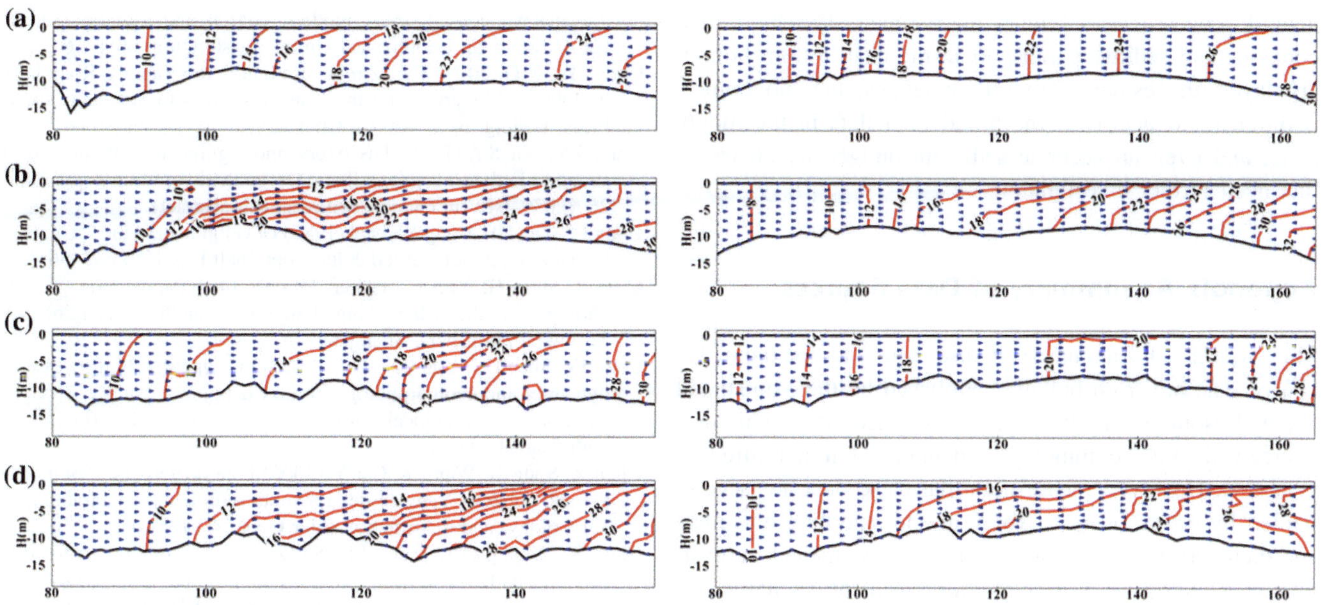

Fig. 3.25 The water age distribution along the transect showing the effect of the dike during dry season mean discharge in the Changjiang Estuary for both the NP (*left panels*) and SP (*right panels*). **a** Spring tide without the dike; **b** neap tide without the dike; **c** spring tide with the dike; and **d** neap tide with the dike

3.4.5 Discussion

The impacts of the navigation channel in the NP on estuarine hydrodynamics were explored using a 3-D numerical model, leading to new insights into transport timescales. Model results show that the transport time decreased from peak values in the NP by 16 %, whereas the transport time increased in the SP by up to 41 % due to the navigation channel. The vertically averaged water age decreased overall, but the transport time increased near the bed and decreased near the surface.

The horizontal water exchange between the NP and SP was significantly reduced due to the engineering works. The water transport out of the SP now has a tendency to flow out along the dike, and some of the water transported out of the NP returns to the SP. This change in horizontal circulation affects the overall circulation pattern in the mouth area and deserves further research attention.

3.5 Summary

The sediment and water transport processes in the Changjiang Estuary have been subject to significant changes in the past two decades as a result of intensified human activity. Firstly, there has been an ongoing decrease in sediment discharge since the 1980s, which has accelerated since 2003 when the TGD commenced operation. At the same time, the sediment properties have adapted, and coarsening riverbed sediments have been observed. Secondly, changes in sediment concentrations and grain sizes are observed in the estuary in response to changing riverine sediment loads. Lastly, the estuarine hydrodynamic and sediment dynamics have changed in response to local human intervention. Research on sediment transport dynamics needs to adapt to these changes and focus on new research questions with updated perspectives.

We observed in situ flocculation of suspended sediment in the Changjiang, which is expected to impact both riverine and estuarine sediment transport dynamics. To better understand the source and depositional history of sediments, we characterized both the suspended and bottom sediment transport processes in the Changjiang Estuary by analyzing the grain size distributions, the SSC, and the clay–silt–sand content of the sediments. One important finding is the unique presence of a nearshore depocenter that may offset riverine sediment reduction. This depocenter can act as a 'sediment bank', supplying sediment for resuspension, shaping the estuary, and building up the delta. By modeling the estuarine circulation and water age, we find that large-scale engineering works have had marked impact on the regional hydrodynamics. The work so far suggests that the river–estuary–sea environments must be studied as an integrated and interconnected system, to better understand the sediment transport dynamics, as well as nutrient and contaminant transport processes.

Sediment transport dynamics are particularly complex in the Changjiang Estuary because of the interactions between fluvial processes (riverine water and sediments) and marine dynamics (tides, waves, and storms); between hydrodynamics, sediment transport, and morphology; and between natural processes and human interference. The fluid mud behavior and sediment transport in the near-bottom boundary layer, the estuarine morphodynamics, the biogeomorphological impacts over the broad tidal flats in the mouth zone, and even an accurate sediment budget are all topics that need further investigation.

Appendix A: Summary of Data Sources

The monthly and annual river discharge and sediment discharge data are from bulletins published by the Ministry of Water Resources of China each year since 2000. This data source is the most commonly used and cited in literature (see Chap. 4). However, please note that these data are accompanied by errors of up to 16 % due to sampling and measurement errors, and standard errors in statistical analysis (Wang et al. 2007). Also, regarding sediment discharge, the bed load is usually excluded from the samples and is therefore less reported and studied. Long-term grain size data are less as available in the Changjiang other than few stations such as Datong.

There are also a number of past field surveys that are described in the literature (e.g., Liu et al. 2008; Guo and He 2011). For instance, the data in Guo and He (2011) were collected during a survey conducted by drifting downstream from Yichang along the Changjiang until Xuliujing. Water samples were collected in six vertical layers for SSC and grain size analysis, and samples of bed load for grain size analysis were taken approximately every 50 km along the channel. The SSC is obtained by filtering, drying, and weighing samples, and grain sizes are measured using Coulter LS-100 instruments. Further details are available in the literature cited above.

There are extensive surveys and sampling of SSC, suspended sediment grain sizes, and bottom sediment in the Changjiang Estuary. Since the hydrodynamics and sediment transport dynamics are characterized by strong variability in seasonal (influence of river discharge and monsoonal winds) and tidal (about 12 h for semi-diurnal tides and 14 days for spring–neap tidal cycles) timescales, the discussion refers to relative timescales, whereas the longer term characteristics are presented by averaging the data over these time periods.

References

Bainbridge ZT, Wolanski E, Alvarez-Romero JG, Lewis SE, Brodie JE (2012) Fine sediment and nutrient dynamics related to particle size and floc formation in a Burdekin River flood plume, Australia. Mar Pollut Bull 65:236–248

Blumberg AF, Mellor GL (1987) A description of a three-dimensional coastal ocean circulation model. In: Heaps NS (ed) Three-dimensional coastal ocean models, coastal and estuarine science vol 4. American Geophysical Union, Washington, pp 1–16

Chen JY, He Q (2009) The extreme drought in the Yangtze River basin in 2006 and its effects on the water resources to Shanghai. Ocean Press, Beijing, pp 38–80 (in Chinese)

Chen ZY, Xu SY (1995) Discovery and significance of mud diapir structure in the Changjiang delta. Oceanologic et Limnologia Sinica 26(6):639–642 (in Chinese with English abstract)

Chen JY, Zhu HF, Dong YF et al (1985) Development of the Changjiang Estuary and its submerged delta. Cont Shelf Res 4(1/2):47–56

Chen J, Shen H, Yun C (1988a) Process of dynamics and geomorphology of the Changjiang Estuary. Shanghai Scientific and Technical Publishers, Shanghai

Chen JY, Shen HT, Yun CX et al (1988b) Introduction. In: Processes of dynamics and geomorphology of the Changjiang Estuary. Shanghai Scientific and Technical Publishers, Shanghai, pp 1–4 (in Chinese with English abstract)

Chen Z, Song B, Wang Z, Cai Y (2000) Late quaternary evolution of the sub-aqueous Yangtze delta, China: sedimentation, stratigraphy, palynology, and deformation. Mar Geol 162:423–441

Chen SL, Zhang GA, Yang, SL, Shi Z (2006) Temporal variation of fine suspended sediment concentration in the Changjiang River estuary and adjacent coastal waters, China. J Hydrol 331:137–145

Chen ZY, Xu KQ, Watanabe M (2007) Dynamic hydrology and geomorphology of the Yangtze river. In: Gupta A (ed) Large rivers: geomorphology and management. Wiley, England, pp 457–469

Chen ZY, Wang ZH, Finlayson B, Chen J, Yin DW (2010) Implications of flow control by the three gorges dam on sediment and channel dynamics of the middle Yangtze (Changjiang) River. China. Geol 38(11):1043–1046

Chen JS, He Q, Guo LC (2011) Flocculation characteristics of suspended particulate matter in Yangtze River. J Sediment Res 10 (5):11–18 (in Chinese with English abstract)

Cheng RT, Casulli V (1982) On lagrangian residual currents with applications in south San Francisco Bay, California. Water Resour Res 18(6):1652–1662

China Water Resources Committee (CWRC) (2005) China river sediment bulletin. Ministry of Water Resources of China (ed) China Water Power Press, Beijing (in Chinese)

China Water Resources Committee (CWRC) (2010) China river sediment bulletin. Ministry of Water Resources of China (ed) China Water Power Press, Beijing (in Chinese)

China Water Resources Committee (CWRC) (2011) China river sediment bulletin. Ministry of Water Resources of China (ed.) China Water Power Press, Beijing (in Chinese)

Deleersnijder E, Campin JM, Delhez EJM (2001). The concept of age in marine modelling: I. Theory and preliminary model results. J Mar Syst 28(3–4):229–267

Delhez EJM, Campin JM, Hirst AC, Deleersnijder E (1999) Toward a general theory of the age in ocean modelling. Ocean Model 1 (1):17–27

Droppo IG, Leppard GG, Flannigan DT, Liss SN (1997) The freshwater floc: a functional relationship of water and organic and inorganic floc constituents affecting suspended sediment properties. Water Air Soil Pollut 99:43–53

Duan LY, Wang ZH, Li MT, Pan JM, Chen ZY (2005) ^{210}Pb distribution of the Changjiang estuarine sediment and the implications to sedimentary environment. Acta Sedimentologica Sinica, 23 (3):514–522

Galperin B, Kantha LH, Rosati A, HASSID. S (1988) A quasi-equilibrium turbulent energy model for geophysical flows. J Atmos Sci 45(1):55–62

Gong W, Shen J, Jia J (2008) The impact of human activities on the flushing properties of a semi-enclosed lagoon: Xiaohai, Hainan, China. Mar Environ Res 65(1):62–76

Guo LC, He Q (2011) Freshwater flocculation of suspended sediments in the Yangtze River, China. Ocean Dyn 61(2–3):371–386

He Q, Cheng J, Wang YY (2004) In situ estimates of floc size and settling velocity in the Changjiang Estuary. In: Proceedings of ninth international symposium on river sedimentation. Yichang, China, pp 416–422

He Q, Wang Y, Yu Z, Jin L (2007) Sediment transport processes in deep water navigation channel, Yangtze estuary, China. In: Smith JM (ed) Proceedings of the international conference on coastal engineering: world scientific, vol 3, pp 3007–3018

Hori K, Saito Y, Zhao Q, Cheng X, Wang P, Sato Y, Li C (2001) Sedimentary facies of the tide-dominated paleo-Changjiang (Yangtze) estuary during the last transgression. Mar Geol 177(3–4):331–351

Jiang CJ, de Swart HE, Li JF, Liu GF (2013) Mechanisms of along-channel sediment transport in the North Passage of the Yangtze Estuary and their response to large-scale interventions. Ocean Dyn 63:283–305

Liu H (2009) Sediment mixing and exchange processes in the Yangtze Estuary. PhD Dissertation in East China Normal University, China pp 82–100

Liu Z, Wei H, Liu G, Zhang J (2004) Simulation of water exchange in Jiaozhou Bay by average residence time approach. Estuar Coast Shelf Sci 61(1):25–35

Liu JP, Li AC, Xu KH, Velozzi DM, Yang ZS, Milliman JD, DeMaster DJ (2006) Sedimentary features of the Yangtze River-derived along-shelf clinoform deposit in the east China Sea. Cont Shelf Res 26:2141–2156

Liu H, He Q, Meng Y, Wang YY, Tang JH (2007) Characteristics of surface sediment distribution and its hydrodynamic responses in Changjiang Estuary. Acta Geographica Sinica 62(1):81–92 (in Chinese with English abstract)

Liu H, He Q, Xu J, Dai Z, Chen J (2008) The effect of the exceptionally low water discharge on sediment characteristics of the middle and lower Yangtze River. Acta Geographica Sinica 63(11):50–64 (in Chinese with English abstract)

Liu H, He Q, Wang ZB, Weltje GJ, Zhang J (2010) Dynamics and spatial variability of near-bottom sediment exchange in the Yangtze Estuary, China. Estuar Coast Shelf Sci 86:322–330

Liu H, He Q, Weltje GJ, Chen JY (2011) Sediment exchange and transport processes in the Yangtze River Estuary: concurrent discuss on the effects of sediment sink in the muddy area. Acta Geographica Sinica 66(3):291–304 (in Chinese with English abstract)

Liu H, He Q, Wang Y, Chen JY (2012) Processes of suspended sediment mixing in the Yangtze River Estuary. Acta Geographica Sinica 67(9):1269–1281 (in Chinese with English abstract)

Lou F (2005) The sedimentary and accretion-erosion environment of the outer navigation channel in Changjiang Estuary. Thesis in East China Normal University, China, pp 8–21

Luff R, Pohlmann T (1995) Calculation of water exchange times in the ICES-boxes with a eulerian dispersion model using a half-life time approach. Deutsche Hydrografische Zeitschrift 47(4):287–299

Luo XX, Yang SL, Zhang J (2012) The impact of the three gorges dam on the downstream distribution and texture of sediments along the middle and lower Yangtze River and its estuary, and subsequent sediment dispersal in the east China Sea. Geomorphology 179:126–140

Mellor GL, Yamada T (1982) Development of a turbulence closure model for geophysical fluid problems. Rev Geophys Space Phys 20 (4):851–875

Milliman JD, Shen HT, Yang ZS, Meade RH (1985) Transport and deposition of river sediment in the Changjiang Estuary and adjacent continental shelf. Cont Shelf Res 4(1/2):37–45

Shen HT (2001) Material flux of the Changjiang Estuary. China Ocean Press, Beijing, p 176 (in Chinese)

Shen J, Wang HV (2007) Determining the age of water and long-term transport timescale of the Chesapeake Bay. Estuar Coast Shelf Sci 74(4):585–598

Shepard FP (1954) Nomenclature based on sand-silt-clay ratios. J Sediment Petrol 24:151–158

Syvitski JPM, Vörösmarty CJ, Kettner AJ, Green P (2005) Impact of humans on the flux of terrestrial sediment to the global coastal ocean. Science 308(5720):376–380

Tang JH (2007) Characteristics of fine cohesive sediment's flocculation in the Changjiang Estuary and its adjacent sea area. Master degree thesis, East China Normal University, Shanghai, May 2007 (in Chinese with English abstract)

Toorman EA, Bruens AW, Kranenburg C, Winterwerp JC (2002) Interaction of suspended cohesive sediment and turbulence. Proc Mar Sci 5:7–23

Wang ZY, Li YT, He PY (2007) Sediment budget of the Yangtze River. Water Resour Res 43:W04401. doi:10.1029/2006WR005012

Wang Y, Shen J, He Q (2010) A numerical model study of the transport timescale and change of estuarine circulation due to waterway constructions in the Changjiang Estuary, China. J Mar Syst 82 (3):154–170

Winterwerp JC, Van Kesteren WGM (2004) Introduction to the physics of cohesive sediments in the marine environment. In: Developments in sedimentology, vol. 56. Elsevier, Amsterdam

Xu M, Ma CD (2009) Yangtze River basin climate change vulnerability and adaptation report. Water Publication Press, Beijing, pp 193–216

Yang ZS, Wang HJ, Saito Y, Milliman JD, Xu KH, Qiao S, Shi G (2006) Dam impacts on the Changjiang (Yangtze) River sediment discharge to the sea: the past 55 years and after the three gorges dam. Water Resour Res 42:W04407. doi:10.1029/2005WR003970

Yun CX (2004) The recent development rules of the Changjiang Estuary. China Ocean Press, Beijing, pp. 234–266 (in Chinese)

Yun CX (2010) A picture guide of the Yangtze River estuary evolution. China Ocean Press, Beijing, pp. 260 (in Chinese)

Zhang ZZ (1996) The characteristics of the cohesive fine sediment in the Yangtze Estuary. J Sediment Res 1:67–73

Zhang RJ (1998) River sediment dynamics, 2nd edn. China Water-Power Press, Beijing, pp 149–169 (in Chinese)

Zhao L, Xu T, Xu C (2013) Spatial and temporal analysis of economic evolution of the Yangtze River basin economic belt. Resour Environ Yangtze Basin 22(7):846–851 (in Chinese with English abstract)

Temporal Variations in Water and Sediment Discharge from the Changjiang (Yangtze River) and Downstream Sedimentary Responses

4

S.L. Yang, and H.F. Yang

Abstract

This chapter focuses on anthropogenic impacts on the water and sediment discharge of the Changjiang over the past 150 years and the sedimentary response of its delta and adjacent waters. Increased water consumption and dam construction have reduced the annual water discharge by ~10 % in the past 150 years. This decrease is mainly attributable to the reduction between August and November each year when the reservoirs store water. In contrast, the water discharge in January and February each year has shown a significant increasing trend, presumably due to water release from reservoirs in the dry season. Human impacts on the riverine sediment load are even more pronounced than impacts on river discharge. The annual sediment load has decreased by more than 70 % since the 1970s and by more than 50 % since the 1990s, due to dam construction and soil conservation in the drainage basin. The decline in sediment discharge has caused a drastic decrease in the progradation of intertidal wetlands on the delta front and has generated erosion on the subaqueous delta. Sediment resuspension from delta erosion partly offsets the decrease in riverine sediment supply, so that the suspended sediment concentration (SSC) in the estuary and adjacent waters only decreased by an average of 26 % over the period 1992–2009. It is predicted that these temporal trends will continue to occur in response to anthropogenic impacts for the remainder of this century.

Keywords

Water discharge • Sediment discharge • Anthropogenic impacts Changjiang (Yangtze River) • Coastal response

4.1 Introduction

In recent decades, many of the world's rivers have shown significant changes in water and sediment discharge in response to anthropogenic impacts and climate change (Walling 2006; Milliman and Farnsworth 2011). For example, in the Nile and Colorado rivers, water and sediment discharge has nearly ceased completely due to dam construction and water diversion (Wiegel 1996; Vörösmarty et al. 2003). The Huanghe (Yellow) River in northern China has experienced a reduced water discharge and sediment load of 85 % (Dai et al. 2009), due to the combined impacts of dam construction, water extraction, soil preservation, and reduced precipitation (Xu 2003; Wang et al. 2007a, b).

With a length of 6.3 × 10^3 km, the Changjiang is the longest river in Asia (Fig. 4.1). It ranks fifth globally in terms of water discharge (~900 km^3/year) and, until recently, ranked fourth in terms of sediment load (~470 Mt/year) (Milliman and Farnsworth 2011). The population in the catchment has increased in conjunction with the world-wide population boom, from 155 million in 1890 to 450 million in 2005 (Fig. 4.2). The increasing population and the

S.L. Yang (✉) · H.F. Yang
State Key Laboratory of Estuarine and Coastal Research, East China Normal University, 3663 Zhongshan Road North, Shanghai 200062, China
e-mail: slyang@sklec.ecnu.edu.cn

H.F. Yang
e-mail: hfyang1991@163.com

J. Zhang (ed.), *Ecological Continuum from the Changjiang (Yangtze River) Watersheds to the East China Sea Continental Margin*, Estuaries of the World,
DOI 10.1007/978-3-319-16339-0_4, © Springer International Publishing Switzerland 2015

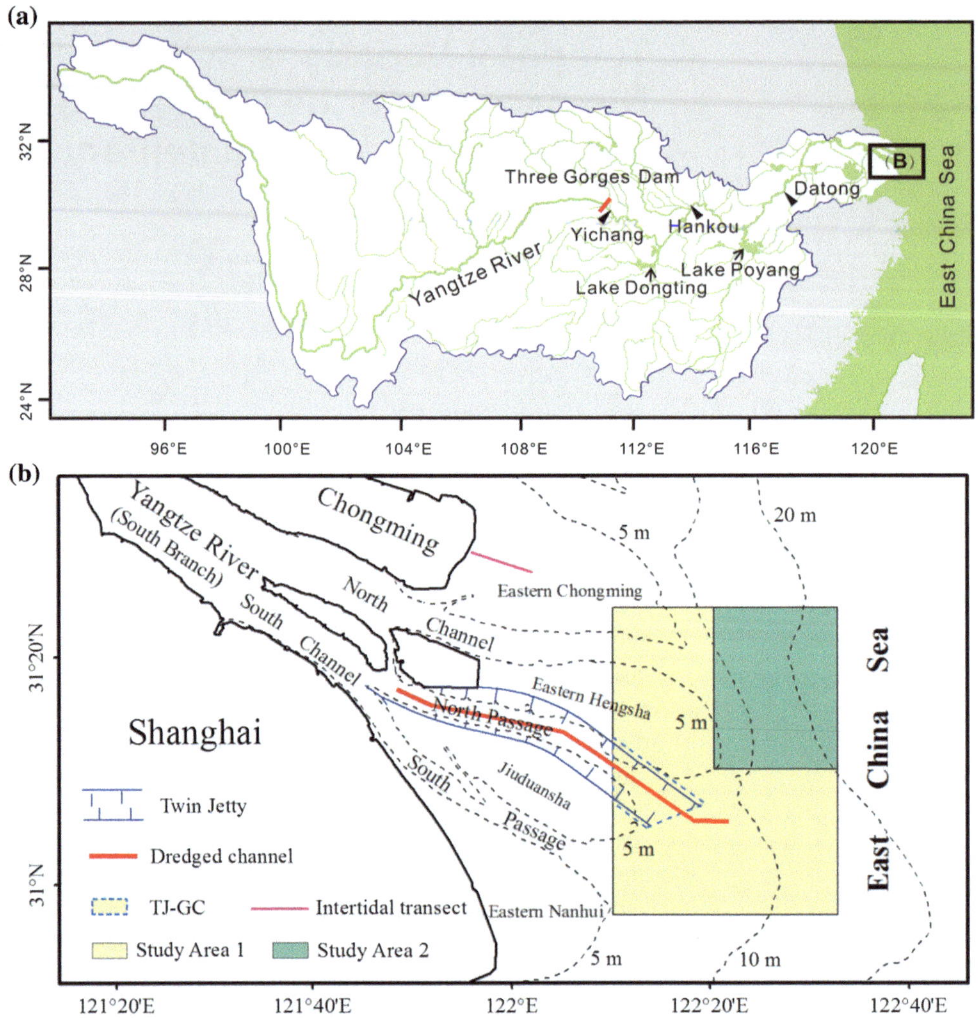

Fig. 4.1 **a** Changjiang watershed, showing the locations of the Three Gorges Dam (TGD) and the Yichang, and Hankou and Datong gauging stations. **b** The subaqueous delta, showing study areas *1* and *2*, the Twin Jetty–Groyne Complex (TJ–GC), and its associated dredged channel (modified from Yang et al. 2011)

rapid development of agriculture and industry in the river basin have greatly increased soil erosion and water consumption. The area of surface erosion in the Changjiang basin has more than doubled in the last century (Yang et al. 2004).

In addition to increased soil erosion and water consumption, nearly 50,000 reservoirs have been constructed in the Changjiang basin since the 1950s. The total storage capacity of reservoirs has increased from <1 km^3 in the early 1950s (Zhu 2000) to more than 200 km^3 in the 2000s (Yang et al. 2005b). Considering that the storage capacity of large-scale reservoirs had reached ~200 km^3 in 2013 (Fig. 4.3) and the storage capacity of large-scale reservoirs contributes two-thirds of the total reservoir storage capacity (Yang et al. 2005b), this suggests a ratio of reservoir storage capacity to annual water discharge of 35 % in the Changjiang, much higher than the global average of 20 % (http://www.seaweb.org/background/book/dams.html).

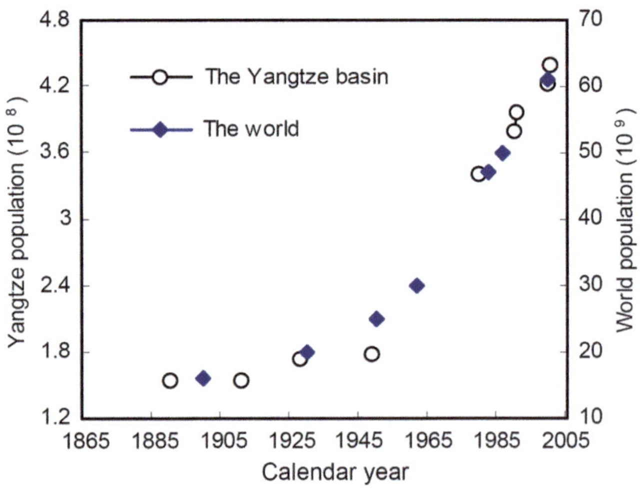

Fig. 4.2 Increasing trends in global population and the population of the Changjiang basin since the late ninetieth century (cf. Yang et al. 2005a)

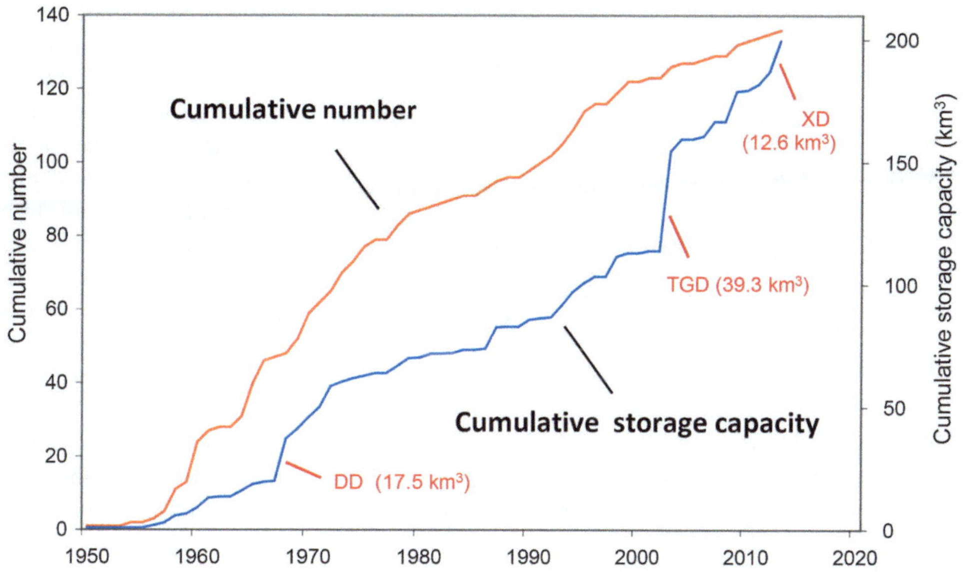

Fig. 4.3 Increase in the number and storage capacity of large-scale reservoirs (>0.1 km^3) within the Changjiang basin since the 1950s, with those larger than 10 km^3 indicated. DD is Danjiangkou Dam; TGD is Three Gorges Dam; and XD is Xiluodu Dam (modified from Yang et al. 2005b)

Changes in the water discharge and sediment load from the Changjiang are expected as a result of these impacts. The variation in sediment discharge is of great importance to the geomorphology of the Changjiang Delta. Furthermore, considering that the delivery of water and sediment from the Changjiang affects the chemistry of the entire northwest Pacific Ocean, the variation in water and sediment discharge also has a major impact on the biochemistry and ecology of the estuary and adjacent continental shelf (Chaps. 5 and 6 of this volume).

4.2 Temporal Variation in Riverine Water and Sediment Discharge

4.2.1 Water Discharge

Key hydrological stations in the Changjiang basin are shown in Fig. 4.1a. Measurements of water discharge in the Changjiang began in 1865 at Hankou station, ~1200 km upstream from the East China Sea, with an interruption in data collection in 1944 and 1945. Discharge measurements at Yichang station, where water from the upper catchment enters the river, began in 1878, with an interruption in data collection in 1880 and 1881. Water discharge measurements at Datong station, the tidal limit of the river, began in 1923, but there are several data gaps before 1950 (Yang et al. 2005a). However, because water discharge at Hankou station is highly correlated with water discharge at Yichang and Datong stations, there was no common interruption in data collection, enabling the reconstruction of a continuous time series of annual water discharge and monthly water discharge since 1865 (Yang et al. 2005a, 2010).

4.2.1.1 Long-term Variation in Water Discharge

Periodicity of Water Discharge

The time series of annual water discharge at Datong station exhibits frequent fluctuations (Fig. 4.4) (Yang et al. 2005a) that are in good agreement with temporal fluctuations in catchment-wide precipitation (Yang et al. 2005a). Based on spectral analysis of a 44-year water discharge dataset from Datong station, Shen et al. (2000) identified periodicities of 2–3 and ~7 years. Spectral analysis also shows that discharge and precipitation have common periodicities of ~3, 7–8, 14, 21–23, and 35–39 years, although with different significance levels (Yang et al. 2010; Table 4.1). This suggests that periodic changes in discharge are governed by climate (precipitation) variability. As indicated by earlier studies, periodic variations in precipitation of 2–3, 6–14, ~22, and ~36 years correspond to variations in solar activity (Parthasarathy and Mooley 1978; Liu et al. 1996, 2008).

Temporal Trends in Water Discharge

Linear regression analysis and the Mann–Kendall test (Khaliq et al. 2008) both reveal significant decreasing trends in annual water discharge, as well as discharge for the months of August to November, for the period 1865–2008 (Table 4.2). In addition, significant increasing trends in water discharge occur in January and February for the same period (Table 4.2; Yang et al. 2010).

As shown in Fig. 4.4, the temporal trends in water discharge are monotonic overall. They reflect the cumulative anthropogenic impact of water consumption and dam construction and are not attributable to climate change. The

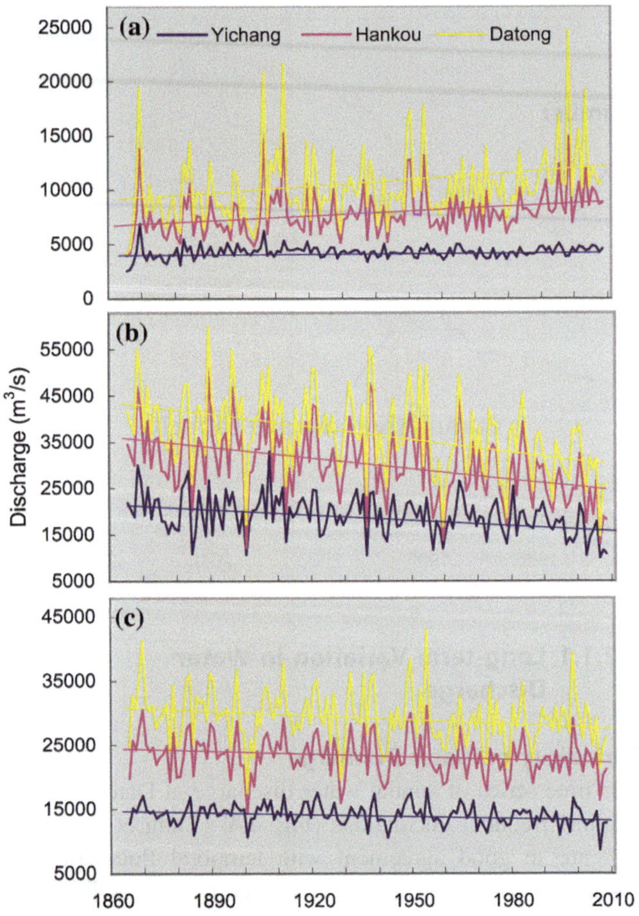

Fig. 4.4 Time series and linear regression trend lines of discharge at Yichang, Hankou, and Datong as typical examples of significant temporal trend: **a** January discharge. **b** October discharge. **c** Annual discharge (after Yang et al. 2010)

changes in discharge at Datong station reported in Table 4.2 which reveals that over the period 1865–2008, the annual water discharge decreased by ~11 %. Furthermore, the discharge in November decreased by ~47 %, whereas the discharge in January increased by ~30 %.

Given that there is no evidence of a decrease in precipitation, or an increase in evapotranspiration, it is reasonable to attribute the decreasing trend in water discharge from the Changjiang to human activities (Yang et al. 2010). The primary cause of decreasing discharge is thought to be increased water consumption (Yang et al. 2005a). However, because the water and land area available to each person are, on average, much higher in the Changjiang drainage basin (~2000 m^3/person and 500 × 10^3 m^3/km^2, respectively) than those in the adjacent Huanghe drainage basin (~300 m^3/person and 40 × 10^3 m^3/km^2, respectively), the decreasing discharge rate has been lower in the Changjiang than that in the Huanghe River (Liu and Zhang 2004; Wang et al. 2006). The mean annual water discharge of the Huanghe River decreased by more than 60 % from 1952–1961 (44 km^3/year) to 2002–2011 (17 km^3/year). Half of this decrease is attributed to human activities, mainly increased water consumption (Wang et al. 2006). In other words, the human-induced decrease in water discharge was 0.75 % per year within the Huanghe drainage basin. As shown above, the annual water discharge in the Changjiang decreased by only 11 % from 1865 to 2008, which represents an annual decrease of 0.077 % per year.

The construction of dams has also contributed to reduced discharge from the Changjiang. As indicated above, the total storage capacity of reservoirs constructed within the Changjiang drainage basin has reached ~300 km^3 or one-third of the annual water discharge of the Changjiang. It is estimated that about 10 % of the long-term decrease in water

Table 4.1 Periodicity and significance level of precipitation and discharge in the Changjiang based on wavelet analysis (cf. Yang et al. 2010)

	Time series	Maximum power		Second maximum power		Other maximum power	
		Period (years)	Significance level (p)	Period (years)	Significance level (p)	Period (years)	Significance level (p)
Precipitation (3 stations)	1883–2008	20	<0.1, >0.05	39	>0.1	2.9, 7.0	>0.1
Precipitation (22 stations)	1951–2008	22	>0.1	2.9	>0.1	8.1	>0.1
Precipitation (33 stations)	1957–2008	8.2	>0.1	2.9	>0.1	22	>0.1
Discharge at Datong	1883–2008	14	>0.1	6.2	>0.1	21, 35	>0.1
	1951–2008	7	>0.1	3	>0.1	14, 21	>0.1
	1957–2008	7.1	>0.1	3	>0.1	14, 23	>0.1
	1865–2008	14	0.05	7.2	>0.1	3.1, 38	>0.1
Discharge at Hankou	1865–2008	14	<0.05	6.8	0.1	38	>0.1
	1865–2008	14	<0.05	6.9	0.1	21	>0.1

The three stations used for the 1883–2008 calculations are located along the main branch of the river and include Chongqing station (upper reaches), Yichang station (transition between the upper and middle reaches), and Hankou station (middle reaches). The remaining stations are scattered throughout the Changjiang basin

Table 4.2 Changes in water discharge (%) at Yichang, Hankou, and Datong stations from 1865 to 2008 (cf. Yang et al. 2010)

	Linear regression method			MK method		
	Yichang	Hankou	Datong	Yichang	Hankou	Datong
January		26	30		29	30
February		22	22		23	21
August	−16	−16	−14	−15	−19	−18
September	−15	−18	−16		−17	−17
October	−26	−36	−33	−27	−37	−34
November	−20	−48	−49	−15	−43	−44
December	−9.2	−19				
Annual	−9.1	−9.4	−10	−9.3	−10	−12

discharge into the Changjiang Estuary is due to the increase in reservoir storage capacity, not including the reservoir-generated evaporation and water leakage (Yang et al. 2005a).

The decreasing water discharge during the late wet season and the increasing water discharge during the dry season are mainly attributed to reservoir operation. Reservoirs within the Changjiang drainage basin typically impound water from September to November and release water in January and February. The Three Gorges Reservoir (TGR) is a good example of typical reservoir operation in the Changjiang basin. During the water-impoundment season, the TGR can reduce the mean daily discharge by more than 60 %, and the mean monthly discharge by more than 20 %. On the other hand, during the water-release season, the TGR can increase the mean daily discharge by more than 130 %, and the mean monthly discharge by more than 30 % (Yang et al. 2010).

4.2.1.2 Seasonal Variation in Precipitation and Water Discharge

The East Asian monsoon modulates the seasonal variation in precipitation and water discharge from the Changjiang. The mean monthly precipitation in the Changjiang basin is lowest in December (25 mm) and highest in June (184 mm) (Fig. 4.5). The water discharge into the Changjiang Estuary increases from minimum values in January (11×10^3 m^3/s) to maximum values in July (50×10^3 m^3/s), and then decreases for the remainder of the year (Fig. 4.5). The seasonal patterns of precipitation and discharge in the Changjiang basin reflect the control of both the East Asian and Indian monsoons (Zhu et al. 1986; Tao and Cheng 1987; Zhao et al. 2000).

Figure 4.6 shows catchment precipitation plotted against water discharge and has a counter clockwise configuration. However, when the one-month lag is taken into account (i.e., monthly precipitation data from January to December were plotted against monthly discharge from February to January), the correlation between precipitation and discharge is very strong (Fig. 4.6). This one-month time lag likely reflects the average residence time of precipitation in the catchment before it is discharged to the sea (Yang et al. 2010).

The ratio of maximum monthly precipitation in June to minimum monthly precipitation in December is 7.5, whereas the ratio of maximum monthly discharge in July to minimum monthly discharge in January is 4.5. The smaller seasonal change in discharge than in precipitation likely reflects the dampening effect of the catchment on water transport.

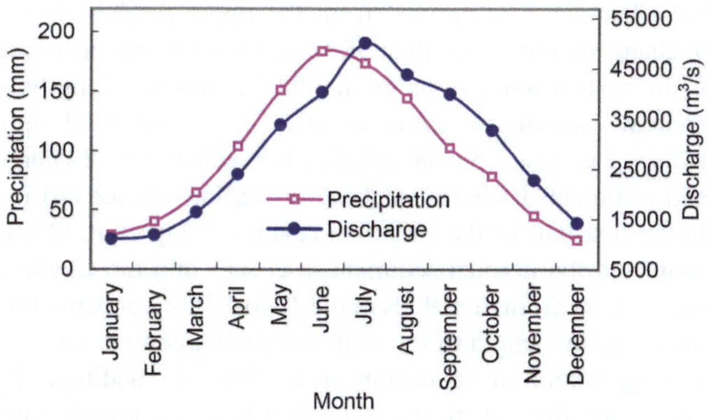

Fig. 4.5 Mean monthly catchment-wide precipitation and water discharge into the Changjiang Estuary (1951–2010)

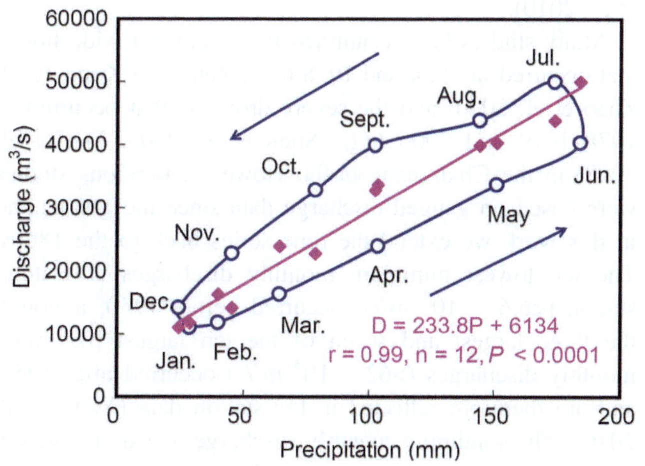

Fig. 4.6 Plot of monthly precipitation and monthly discharge

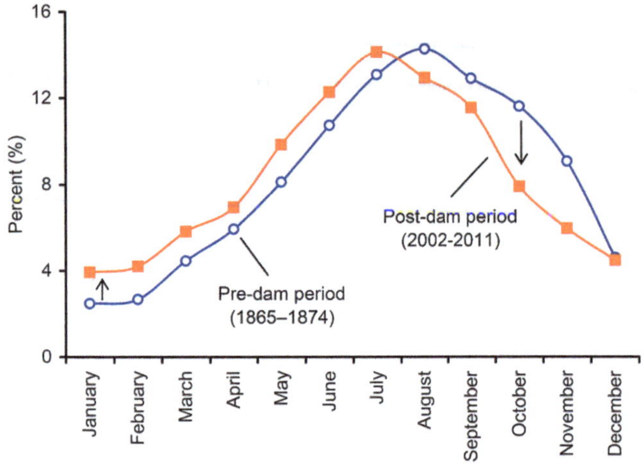

Fig. 4.7 Changes in the seasonal distribution of water discharge into the Changjiang Estuary in pre- and post-dam periods

The seasonal distribution of discharge into the Changjiang Estuary has changed during the past 150 years. Most significantly, the monthly water discharge has increased by 60 % in January and February, and decreased by 30 % in October and November (Fig. 4.7). These long-term changes in the seasonal distribution of water discharge are mainly attributed to the construction and operation of reservoirs (Yang et al. 2002).

4.2.1.3 Extreme Water Discharge Events

Based on the reconstructed time series of water discharge from 1865 to 2008 (Yang et al. 2005a, 2010), the maximum monthly discharge at Datong station occurred in August 1954 (84×10^3 m^3/s) and the minimum monthly discharge occurred in January 1865 (4.1×10^3 m^3/s). At Yichang and Hankou stations, the maximum monthly discharge occurred in August 1998 (52×10^3 and 67×10^3 m^3/s, respectively) and the minimum monthly discharge occurred in January 1865 (2.6×10^3 and 3.9×10^3 m^3/s, respectively) (Yang et al. 2010).

Many studies have examined the catchment-wide floods that occurred in 1954 and 1998 (e.g., Zeng and Kong 1999; Zhao et al. 2000) and the severe droughts that occurred in 1978–1979 and 2006 (e.g., Shen et al. 2003; Dai et al. 2008) in the Changjiang basin. However, previous studies were based on gauged discharge data since the 1950s, and in this work we extend the time series back to the 1860s. The ten lowest minimum monthly discharges at Datong station ($<6.6 \times 10^3$ m^3/s) occurred before 1950, although the three largest and seven of the ten largest maximum monthly discharges ($>62 \times 10^3$ m^3/s) occurred after 1950, and are therefore reflected in the station data (Yang et al. 2010). The minimum monthly discharge at Datong station shows a significant increasing trend, whereas the maximum monthly discharge did not change significantly between 1865 and 2008.

4.2.1.4 Predictions of Future Water Discharge

It is likely that water discharge into the Changjiang Estuary will continue to decrease as a result of anthropogenic effects. When complete, the South-to-North Water Diversion (SNWD) project will divert water from the Changjiang to north China at a rate of 45 km^3/year, thereby reducing the annual water discharge into the estuary by another 5 %. The East Line Water Diversion (with a volume of 15 km^3/year) and the Middle Line Water Diversion (with a volume of 13 km^3/year) are both phases of the SNWD that will begin operation at the end of 2013 and 2014, respectively. The West Line Water Diversion (with a volume of 17 km^3/year) will be in operation by 2050 (http://www.nsbd.gov.cn).

In addition, it is expected that water consumption will continue to increase along with a growing population and economy. It is predicted that the Chinese population will steadily increase to 1.44 billion by 2030 (from ~1.31 billion in 2005), although it will likely decrease after this time (Chen 2006). By 2020, China's GDP will have doubled in comparison with 2010 (http://www.taoguba.com.cn/Article/702854/1). In addition, it is likely that many new reservoirs will be constructed in the Changjiang basin in future decades (http://www.chinawater.net.cn/Journal/Three_Gorges/199903/990314.htm). These reservoirs will further reduce river discharge. For these reasons, it is expected that water discharge from the Changjiang will decrease by another 10 % by 2100 (Yang et al. 2010). Meanwhile, the seasonal change in water discharge from the Changjiang will probably be enhanced in future decades as a result of the construction of new reservoirs.

4.2.2 Sediment Discharge

Measurements of sediment discharge in the Changjiang basin began in the 1950s. Most of the sediment carried by the Changjiang is derived from the upper reaches, above Yichang station. More than 98 % of the total sediment load is in suspension (Yang et al. 2002), and the long-term median suspended sediment size is around 0.01 mm (Chap. 3). Based on the relation between water discharge and sediment discharge, and the increasing surface erosion in the first half of the twentieth century, Yang et al. (2004) estimated the decadal sediment discharge into the Changjiang Estuary from the 1900s to the 1940s. This reconstruction increased the length of the sediment discharge time series at Datong station to a century (Fig. 4.8). In addition, by employing the statistical relationship between annual water and sediment discharge, Wang et al. (2007a, b) estimated the

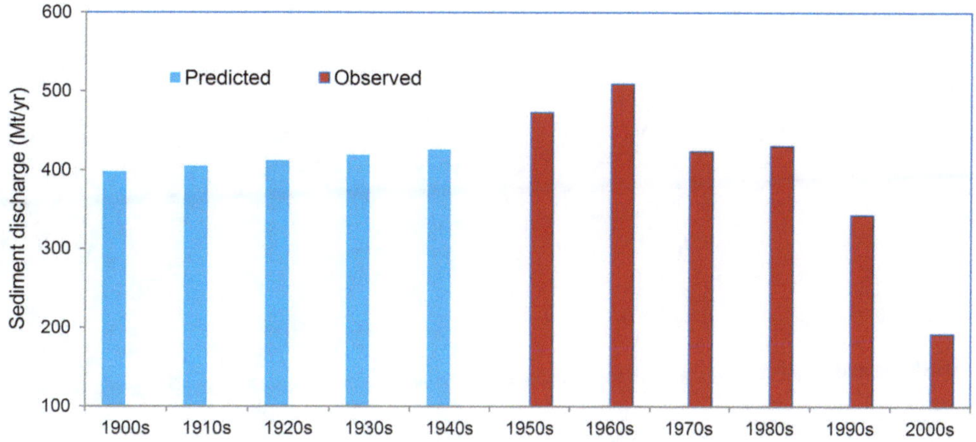

Fig. 4.8 Mean decadal suspended sediment discharge into the Changjiang Estuary in the twentieth century (modified from Yang et al. 2004)

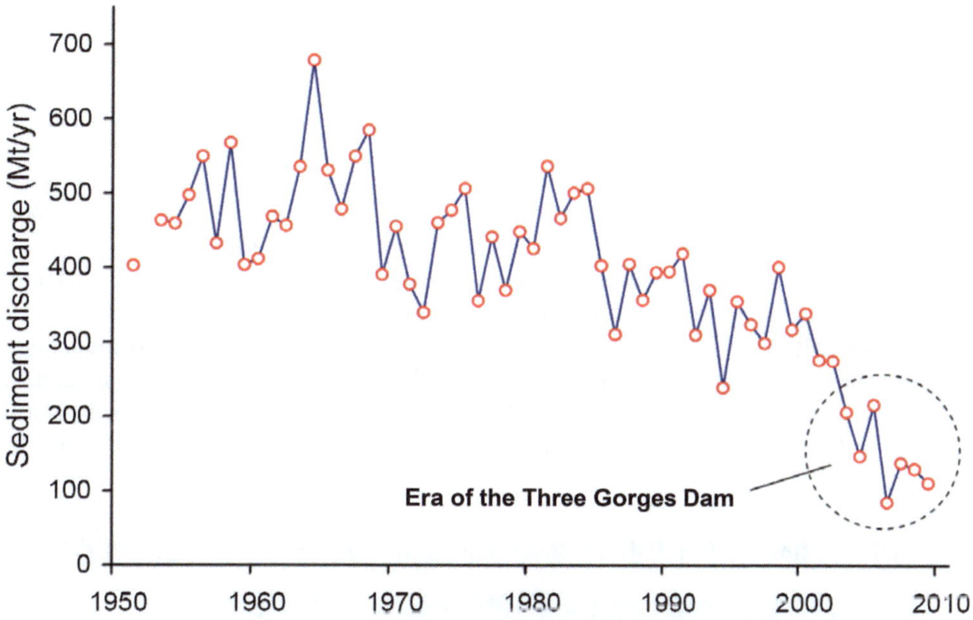

Fig. 4.9 Annual sediment discharge into the Changjiang Estuary using data from Datong station

missing annual sediment discharge data at Datong station for 1865–1950.

4.2.2.1 Long-term Variation

Increase in Sediment Discharge Before the 1960s

The surface erosion area in the Changjiang basin increased from 364×10^3 km^2 in the 1950s (Shi 1999; Zhang and Zhu 2001) to 711×10^3 km^2 in 2002 (CRWCCWCMC 2003) due to deforestation spurred by a population boom (Yang et al. 2004). The decadal suspended sediment discharge (SSD) was estimated by assuming that the SSD increased progressively from 1900 to 1949 (Fig. 4.8). The sediment discharge into the Changjiang Estuary increased by 25 % during the seven decades from the 1900s to the 1960s. Based on population trends and the relationship between the area of surface erosion and sediment yield within the Changjiang catchment, we estimate that the suspended sediment flux into the Changjiang Estuary was 395 Mt/year in 1900 and 429 Mt/year in 1949 (Fig. 4.9). The population boom after the Second World War and the Civil War (which ended 1949) likely caused the sharp increase in annual sediment discharge in the 1950s and 1960s.

Decrease in Sediment Discharge Since the 1960s

Around 50,000 dams have been constructed in the Changjiang drainage basin, of which more than 150 have a reservoir storage capacity exceeding 0.1 km^3 (Fig. 4.10). As the reservoir storage capacity increased in the Changjiang basin, more riverine sediment was trapped. For example, in the upper reaches of the Changjiang (i.e., upstream from Yichang station), the total deposition in reservoirs was 2.9 Mt/year in the 1950s, 19 Mt/year in the 1960s, and 240 Mt/year in the 1990s (Yang et al. 2005b).

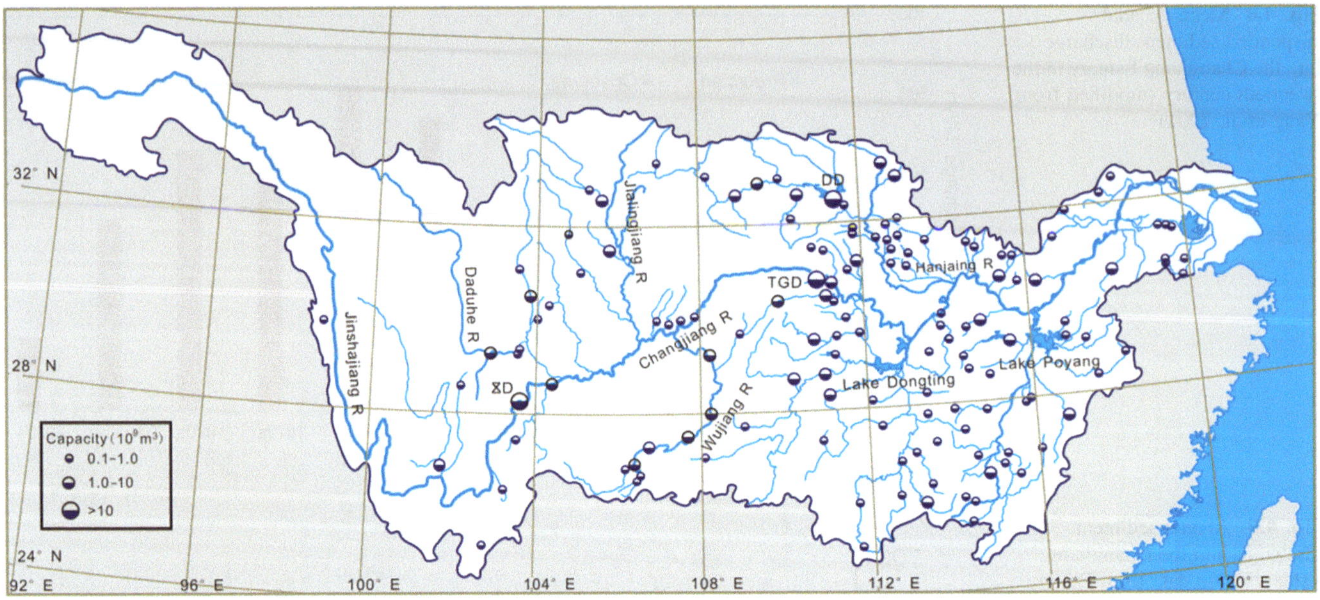

Fig. 4.10 Distribution of large reservoirs (i.e., >10^8 m^3 in storage capacity) in the Changjiang basin. DD, TGD, and XD represent Danjiangkou Dam, Three Gorges Dam, and Xiluodu Dam, respectively

It has been estimated that the total deposition in reservoirs in the Changjiang basin was 740 Mt in 2002 (Yang et al. 2005b). It is reasonable to assume, therefore, that the total deposition in reservoirs in the Changjiang basin currently exceeds 900 Mt.

The Danjiangkou Reservoir on the Hanjiang River was the largest reservoir to become operational in the twentieth century and began to impound water in 1968. Since it became operational, more than 90 % of the sediment entering the reservoir has been retained (Yang et al. 2005b). By 1994, a total 1.41 km^3 of sediment was trapped in the Danjiangkou Reservoir (www.cjh.com.cn, 2003), resulting in a mean depositional rate of 68 Mt/year. In 2003, the TGR became operational and, since then, the Three Gorges Dam (TGD) has trapped most of the sediment supplied from the upper catchment (Yang and Zhang 2007; Box 4.1).

Box 4.1 Impact of the TGD on sediment discharge to the Changjiang Estuary

Dam construction has changed the world's river sediment discharge from having a long-term gradually increasing trend (due to deforestation) to a short-term rapid decreasing trend. The TGD is the world's largest dam (Nilsson et al. 2005) and was constructed across the Changjiang River at a site 4500 km downstream from the headwaters and 1800 km upstream from the river mouth. Nearly 90 % of the Changjiang River sediment load originates from the catchment upstream of the TGD (Yang et al. 2002). Since its closure in June 2003, nearly 80 % of this sediment has been retained behind the TGD (Yang et al. 2014). In spite of significant downstream channel erosion in response to this sediment retention (Yang et al. 2011), the sediment discharge into the Changjiang Estuary has experienced a significant decrease (of 150 Mt/year) (Yang et al. 2014). Prior to the construction of the TGD, nearly 50,000 dams had been constructed in the tributaries of the Changjiang River, and the sediment discharge into the estuary had shown a decreasing trend (Yang et al. 2002, 2005a, b). The TGD has further decreased the sediment discharge into the estuary by 30 % Yang and Zhang (2007), Yang et al. (2014) and has changed the long-term evolution of the delta front from a state of progradation to one of the erosion (Yang et al. 2011).

As a result of dam construction, the SSD at Datong station has shown a decreasing trend since the late 1960s when the sediment discharge was >500 Mt/year (Fig. 4.9). After the completion of the TGD, the sediment discharge at Datong station decreased to an average of <150 Mt/year (Yang et al. 2005b; Fig. 4.9). In other words, dam construction has greatly exceeded land use changes (e.g., deforestation) in influencing the rate of SSD into the estuary.

The decrease in sediment discharge into the Changjiang Estuary can be divided into three phases (Yang et al. 2006b; Fig. 4.9). In the first phase, sediment discharge at Datong station decreased from 490 Mt/year during 1951–1968 to 440 Mt/year during 1969–1985, mainly due to the operation of the Danjiangkou Reservoir on the Hanjiang tributary. In the second phase, from 1985 to 2002, sediment discharge at Datong station decreased to 340 Mt/year, mainly due to the installation of numerous dams and water–soil conservation works in the tributaries above Yichang station. After the TGD became operational in 2003, the Changjiang entered a third phase of sediment reduction, with the annual sediment load at Datong station decreasing to 143 Mt/year from 2003 to 2011. Alongside the impact of the TGD on sediment supply in this time period, the construction of other new dams, soil conservation projects, and a decrease in rainfall contributed to the decrease in sediment discharge into the Changjiang Estuary (Yang et al. 2007). Precipitation in the Changjiang drainage basin, for example, was 4 % lower in 2003–2011 than in 1951–2002, which may have resulted in ~5 % decrease in sediment discharge (Yang 2013).

4.2.2.2 Seasonal Variation

As for water discharge, the sediment discharge at Datong station shows a significant seasonal pattern (Fig. 4.11). The highest monthly sediment discharge generally occurs between July and August, whereas the lowest generally occurs between January and February. In the period from 1951 to 2009, the highest sediment discharge in July was 33 times higher than the lowest sediment discharge in January.

Along with the temporal decrease in annual sediment discharge (Fig. 4.9), the seasonal difference in monthly sediment discharge decreased after the construction of the TGD (Fig. 4.11). For example, the ratio of the maximum monthly sediment discharge in summer to the minimum monthly sediment discharge in winter was 35 for 1951–1990 and 15 for 2000–2009 (Xu and Milliman 2009).

4.2.2.3 Predictions of Future Sediment Discharge

Sediment discharge into the Changjiang Estuary will most likely decrease further in the near future. Although old reservoirs are being silted and eventually abandoned, new reservoirs will continue to trap sediment and reduce the supply to the estuary. The total capacity of reservoirs in the upper Changjiang, either currently being constructed or in the planning stages, exceeds 130 km^3 (Zhao et al. 2000), or 3.3 times greater than the capacity of the TGR.

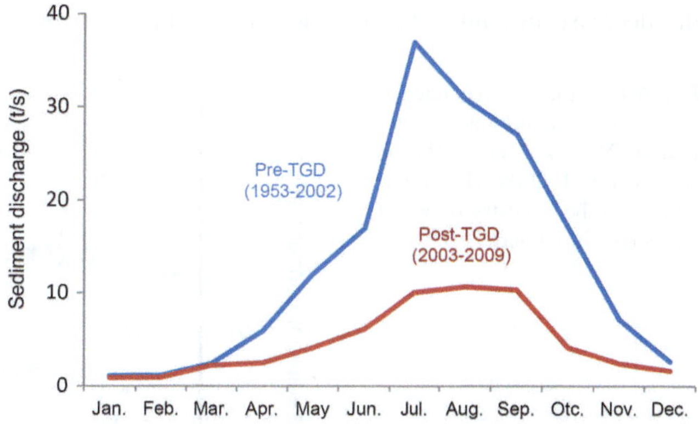

Fig. 4.11 Average multi-year monthly suspended sediment discharge into the Changjiang Estuary using data from Datong station in the pre-TGD and post-TGD periods

Furthermore, as shown above, water discharge into the Changjiang Estuary is expected to continue to decrease due to human activities, such as the SWND, which will further decrease sediment transport to the estuary. Thirdly, because of the increase in crop yields due to advances in agriculture, more cropland is being developed, which will also reduce the sediment yield in the catchment. Under the combined influence of these factors, the sediment discharge into the Changjiang Estuary is predicted to decrease to ~120 Mt/year in the following decades (Yang et al. 2014).

4.2.3 Suspended Sediment Concentration

4.2.3.1 Long-term Variation

From 1951 to 1968, the annual suspended sediment concentration (SSC) in the Changjiang showed an increasing trend ($p = 0.01$; see Fig. 4.12 for original data) alongside the increase in the annual SSD ($p = 0.04$; see Fig. 4.9 for original data), based on linear regressions using SPSS statistics. The annual water discharge over the same period, by contrast, did not show any significant trends ($p = 0.3$; see Fig. 4.4c for original data). The increasing trend in SSC in 1951 1968 has been attributed primarily to enhanced deforestation (Yang et al. 2002).

In 1969, the SSC and sediment discharge from the Changjiang decreased abruptly, due mainly to the construction of Danjiangkou Dam. Since the mid-1980s, the SSC has experienced a rapid and continuous decrease due to the construction of many new dams and soil conservation works. After the construction of the TGD in 2003, the SSC decreased to 0.18 g/L, which is only 32 % of the level in the 1950s and 1960s (0.56 g/L) (see Fig. 4.12 for original data).

Unsurprisingly, the long-term variation in the annual SSC in the Changjiang is in agreement with the long-term variation in the annual suspended sediment load (Fig. 4.9 and Fig. 4.12). This is in contrast to the Huanghe River, where the decrease in annual SS was slower than the decrease in sediment discharge, because the annual water discharge has also shown a significant decreasing trend (Wang et al. 2007a; Dai et al. 2009). Nevertheless, specific changes in the annual SSC in the Changjiang can be quite different to the changes in the annual suspended sediment load. At Datong station in 1954, for example, the SSC was very low while the sediment discharge was around the multi-year average. In fact, at Yichang station, 40 km downstream from the TGD and 1200 km upstream from Datong station, the sediment discharge in 1954 was the highest on record (754 Mt) (Wang et al. 2011). This specific example is due to a dike bursting during the 1954 flood, causing a large amount of sediment to be deposited on the alluvial plain in the middle reaches of the Changjiang (Yang 2006b). As a result, the sediment discharge at Datong station was only 460 Mt or ~300 Mt lower than that at Yichang station. On the other hand, the water discharge at Yichang and Datong stations in 1954 was the highest on record for both (Fig. 4.4).

4.2.3.2 Seasonal Variation

On average, the seasonal pattern of SSC is similar to that of the SSD (Fig. 4.13); i.e., on average, the highest monthly

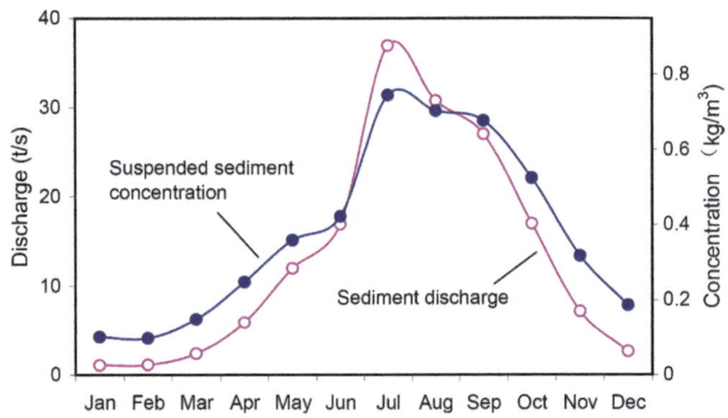

Fig. 4.13 Multi-year averaged monthly suspended sediment concentration and suspended sediment discharge into the Yangtze Estuary (Datong station) (based on data in 1951–2009 from the Ministry of Water Resources in China)

Fig. 4.12 Annual suspended sediment concentration of water into the Yangtze Estuary (Datong station) (1951–2009) (based on data from the Ministry of Water Resources in China)

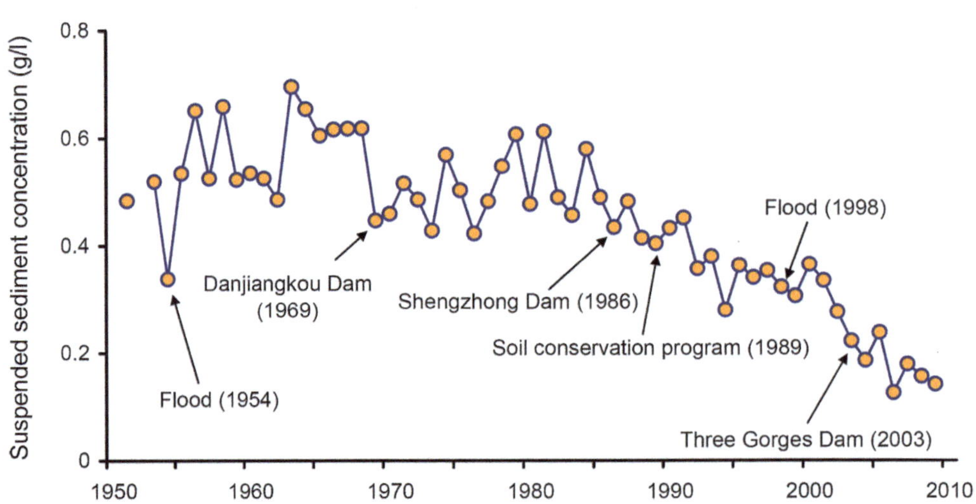

value occurs in July and the lowest in February. However, due to the seasonal variation in water discharge (Fig. 4.4), the annual range in SSC (the difference between the highest and lowest monthly values) is 7.8 times smaller than the annual sediment load range.

Along with the long-term decrease in annual SSC, the seasonal range in the SSC has been reduced. For example, the ratio of the highest to lowest monthly SSC was 7.9 in the 1950s and was only 3.9 since the construction of the TGD.

4.3 Delta Response to Variations in Riverine Sediment Supply

4.3.1 Intertidal Wetland Response

In the history, seawalls tended to be constructed along the high tide line of intertidal wetlands during reclamation in the Changjiang Delta. For this reason, the sequential seawalls well reflect the history of coastal progradation (Yang et al. 2001). The progradation of the modern Changjiang Delta (i.e., the delta development since Holocene sea level rise to its present position at 6500 years BP) has exhibited a long-term gradual increase until recent decades. For example, the progradation rate of the Eastern Nanhui Coast increased from 2 m/year in 6500–2000 BP to 18 m/year in 2000 BP–1950 AD, and to 35 m/year in 1950–1995 (Yang et al. 2001 and references therein). In Eastern Chongming, the progradation rate increased from 2 m/year in 825–1762 AD to 7 m/year in 1762–1955 AD, and to 230 m/year in 1955–1990 AD (Yang et al. 2001 and references therein; Fig. 4.14).

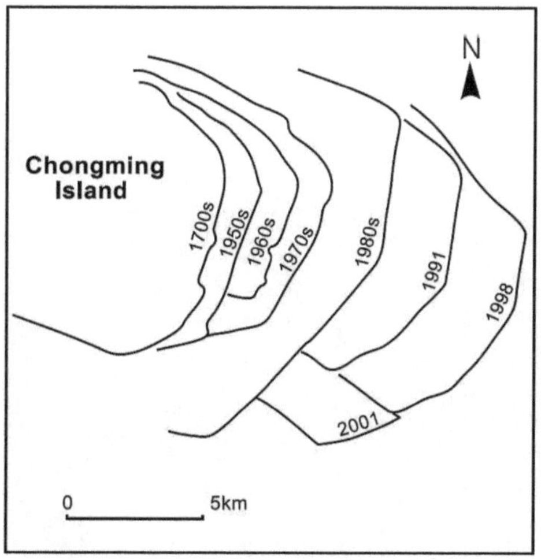

Fig. 4.14 Advancement of the coastline due to seawall construction on eastern Chongming Island (location is shown in Fig. 4.1) (cf. Yang et al. 2005b and references therein)

The total progradation rate of the main intertidal wetlands in the delta front based on the changes in low tide line in bathymetric maps (i.e., Eastern Chongming, Eastern Hengsha, Jiuduansha, and Eastern Nanhui) has recently exhibited a rapid decrease. Analysis of systematic data suggests that the total progradation rate in the late 1990s was less than 30 % of that in the early 1970s. Furthermore, the retreat of the 0-m isobath has even been observed in Eastern Chongming and Eastern Hengsha since the early 1990s (Table 4.3). The only localized exceptions to this are Jiuduansha and Eastern Nanhui, which show increases in the progradation rate since the 1970s (Table 4.3). The low margin of the intertidal wetlands seems having been influenced by the decline of riverine sediment supply earlier than the high margin.

A comparison of cross-shore profiles in the intertidal wetland in Eastern Chongming shows a total progradation rate of ~350 m/year in the 1980s. The progradation rate in the upper intertidal wetland (i.e., salt marsh) decreased to ~60 m/year for 2006–2010, and retreat was also observed in the lower intertidal wetland (i.e., mudflat) over this time period (Fig. 4.15).

The progradation rates of coastal wetlands in the Changjiang Delta reflect a rapid response of the sedimentary system to sediment supply from the Changjiang. The total sediment discharge of the five major rivers in East and Southeast Asia (i.e., the Huanghe, Changjiang, Zhujiang, Red, and Mekong rivers) was ~600 Mt/year before 2 ka BP (Wang et al. 2011). Between 2 and 1 ka BP, sediment supply from these rivers began to increase due to catchment deforestation, and in the 1950s and 1960s, the total sediment discharge of these rivers reached a peak of more than 2000 Mt/year (Wang et al. 2011). Since the 1970s, however, the sediment discharge of these rivers has decreased abruptly, mainly due to dam construction and soil conservation. Between 2000 and 2008, the total sediment discharge of the five major rivers in East and Southeast Asia decreased to ~600 Mt/year, or the level before 2 ka BP (Wang et al. 2011).

As shown above, the sediment load of the Changjiang exhibited an increasing trend before the 1960s, reached its historical peak in the 1960s, and abruptly decreased after this time (Fig. 4.8). The concurrent decrease in the progradation rate of the intertidal wetlands in the delta front suggests that this has a close association with sediment supply from the Changjiang (Fig. 4.16). The estuarine SSC has also decreased in response to the decline in riverine sediment supply (discussed in detail below). The decrease in estuarine SSC reduces the deposition flux during the inundation of intertidal wetlands, and net erosion occurs when the deposition flux decreases below the erosion flux (Box 4.2).

Table 4.3 Growth rates (km^2/year) of the intertidal wetlands at the delta front of the Changjiang, based on the 0-m isobath (low tide line) (modified from Yang et al. 2005b)

Period	Eastern Chongming[a]	Eastern Hengsha[a]	Jiuduansha[a]	Eastern Nanhui[a]	Total
1971–1975	11.1	1.1	1.4	−1.8	11.8
1975–1979	8.8	0.8	2.2	−1	10.8
1979–1983	6.6	0.5	2.4	−0.3	9.2
1983–1987	4.5	0.4	2.7	0.5	8.1
1987–1991	1.9	0.2	2.9	1.2	6.2
1991–1995	−0.5	0	3.1	2	4.6
1995–1998	−2.5	−0.1	3.3	2.6	3.3

[a]Locations are shown in Fig. 4.1

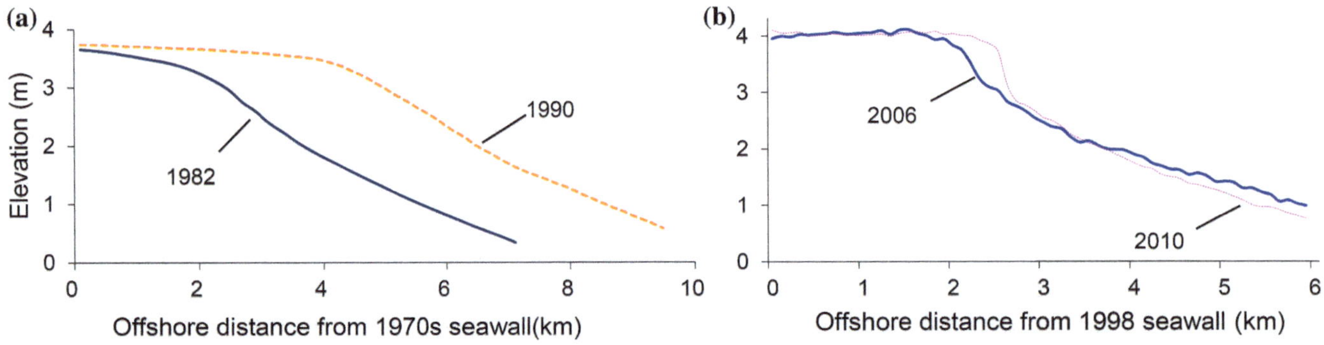

Fig. 4.15 Changes in eastern Chongming Island topographic profiles (location is shown in Fig. 4.1). Profile A is for 1982–1990 (cf. Yang et al. 2001) and Profile B is for 2006–2010 (cf. Yang et al. 2011)

Box 4.2 Intertidal wetland sediment dynamics under a combined current–wave action at the delta front of the Changjiang River

The Changjiang delta front is meso–macrotidal and exposed to wave action, making the intertidal wetlands very dynamic. On the mudflat, the bed shear stress due to waves (τ_w) is generally as large as the bed shear stress due to currents (τ_w). During the tidal cycle, the combined current–wave action (τ_{cw}) is sometimes greater and sometimes smaller than the critical bed shear stress for the erosion of sediments (τ_{ce}), suggesting frequent alternations between erosion and deposition (Shi et al. 2012). On the salt marsh, however, due to the attenuation of waves and currents by vegetation (Yang et al. 2008, 2012), τ_{cw} values are generally smaller than τ_{ce} values and the critical bed shear stress for deposition (τ_{cd}) dominates over erosion (Shi et al. 2012).

Bed-level change on the intertidal wetlands is a result of the balance between erosion and deposition fluxes (Shi et al. 2014; Zhu et al. 2014). Erosion is assumed to occur if $\tau_{cw} > \tau_{ce}$, and the erosion flux is proportional to the difference between these two values (Whitehouse et al. 2000). On the other hand, deposition is assumed to occur if $\tau_{cw} < \tau_{cd}$, and the deposition flux is proportional to the SSC (Whitehouse et al. 2000). In other words, the SSC value is an important factor in bed-level changes. Thus, the drastic decrease in suspended sediment load from the Changjiang River and the decline of SSC in the estuary (Li et al. 2012) would have affected the sediment dynamics of the intertidal wetlands at the Changjiang delta front. In addition, if the sediment supply from the Changjiang River had not decreased, the recently observed mudflat erosion would probably not have occurred (Shi et al. 2012; Zhu et al. 2014).

4.3.2 Subaqueous Delta Response

Along with the decline in sediment supply from the Changjiang, the progradation rate of the subaqueous decreased, and this was followed by recession of the 10- and 20-m isobaths in the subaqueous delta (Table 4.4). In the subaqueous delta off the Changjiang mouth, the shift of the 10- and 20-m isobaths from 1958 to 1997 was characterized by: (a) progradation from 1958 to 1978, which was followed by a minor recession from 1978 to 1997 in the southern delta; (b) a decrease in the progradation rate (between 31.14°N and 31.33°N, at the 10-m isobath; between 31.02°N and 31.15°N; and 31.39°N and 31.46°N, at the 20-m isobath); (c) continual recession between 31.39°N and 31.48°N at the 10-m isobath; and (d) transitional situations, with variable progradation and erosion (Yang et al. 2003). The responses of the 10- and 20-m isobaths are thought to reflect spatial differences in hydrodynamics (see Chap. 2) and sediment supply (Yang et al. 2003).

Based on the correlation between SSD and the progradation rate shown in Table 4.4, the interception of SSD is 390×10^6 t/year for the 10-m isobath and 400×10^6 t/year for the 20-m isobath, or approximately 395×10^6 t/year on average. This value could be considered a sediment-supply threshold, beyond which isobaths switch from progradation to recession (Yang et al. 2003).

However, it is also necessary to examine the net accretion rate across the entire subaqueous delta and, if bathymetric data are not available for the whole delta, the study area should be as large as possible. Yang et al. (2003) studied an area of 6000 km^2 and found that the accretion rate has decreased dramatically from 1978 to 1997, along with the decline in sediment supply. Compared with the preceding two decades, the accretion area decreased from 74 to 60 % of the study area, and the mean accretion rate decreased from 3.8 to 0.8 cm/year (Yang et al. 2003).

To examine recent erosion/accretion trends in the Changjiang subaqueous delta, Yang et al. (2011) compared bathymetric data from 1958, 1977, 2000, 2004, and 2007 within an area of 1825 km^2 in the South Branch through which more than 98 % of the Changjiang's water and sediment are discharged (Chen et al. 1985). Bathymetric

Table 4.4 Progradation rates of the 10- and 20-m isobaths and sediment discharge of the Yangtze River during three periods, and the correlation between them (cf. Yang et al. 2003)

	10-m isobath			20-m isobath		
	1958–1965	1965–1978	1978–1997	1958–1965	1965–1978	1958–1965
Time interval (years)	7	13	20	7	13	20
Progradation (m)	1385	1855	−232	1376	1804	−425
Progradation rate (m/year)	198	143	−12	197	139	−22
Riverine SSD (10^6 t/year)	498	449	394	498	449	394
Correlation	$R = 2.036$ SSD = 800; $r = 0.97$; $n = 3$			$R = 2.12$ SSD = 844; $r = 0.97$; $n = 3$		

Negative values indicate recession; *SSD* suspended sediment discharge; *R* progradation rate

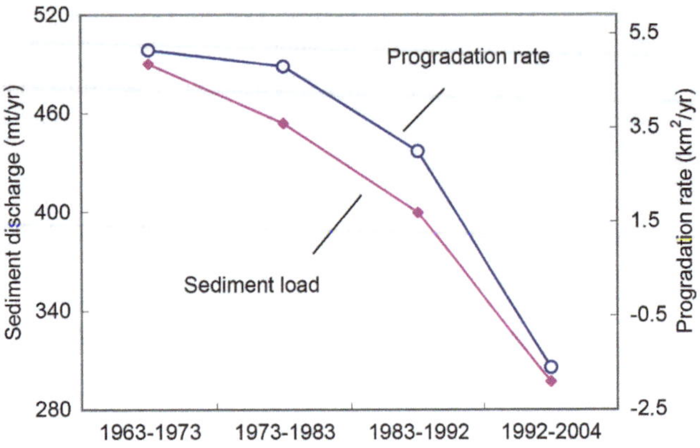

Fig. 4.16 Temporal changes in the progradation rate of the intertidal wetland on eastern Chongming Island (location is shown in Fig. 4.1) in comparison with sediment supply from the Changjiang (at Datong station). Data for the progradation rate are from Yang et al. (2006a, b)

extensions of Chongming, Hengsha, and Jiuduansha define the western limits of the study area, and 15- to 20-m-depth contours define the eastern limits. Area 1 (1280 km^2; Fig. 4.1B) was surveyed in all five years, and Area 2 (545 km^2; Fig. 4.1b), directly off the North Channel of the South Branch, was not surveyed in 2007 (Yang et al. 2011).

Delineating bathymetric changes was somewhat complicated in 1997 by construction of the Twin Jetty–Groyne Complex (TJ–GC), which was created to facilitate navigation and covered an area of 103 km^2 within Area 1 (Du and Yang 2007). By 2004, the TJ–GC extended into the southwestern portion of the study area (Fig. 4.1b). In conjunction with the TJ–GC, a channel with a width of 350 m and an area of 8.5 km^2 (Fig. 4.1B) was dredged regularly to 10 m below lowest tide level, with most spoil being dumped beyond the 10–m-depth contour (Yang et al. 2011).

Historically, most of the Changjiang's annual sediment load was deposited on the subaqueous delta during the flood season. Much of this material was eroded during winter storms (DeMaster et al. 1985; Milliman et al. 1985) and transported south to form an extensive nearshore mud wedge extending into Taiwan Strait (Liu et al. 2007). Isotopic data from several cores suggest that sediment accumulation in the subaqueous delta between the 1950s and 1980s averaged approximately 4.5 cm/year (DeMaster et al. 1985; Zhang et al. 2008).

Between 1958 and 1977, when the Changjiang discharged an average of 470 Mt/year of sediment, bathymetric data indicate that the 5-m contour east of Hengsha Island prograded as much as 5 km (Fig. 4.17). The 10-m isobath prograded more than 12 km overall, but retreated ~2 km in the northern delta between 1958 and 1977. The 20-m contour, by contrast, showed little overall change (Fig. 4.17). During this 19-year period, 55 % of the area shoaled by >5 cm/year, whereas 8 % deepened by more than 5 cm/year, mostly off the North Channel and the North Passage of the South Channel (Fig. 4.18a). Average shoaling within the study area was 6.8 cm/year (Yang et al. 2011).

Between 1977 and 2000, most of the Changjiang subaqueous delta continued to prograde, but presumably due to declining sediment discharge, the accretion rate was generally <5 cm/year. Although some of the north–central delta and areas off Jiuduansha shoaled by as much as 15 cm/year, there was little net change in the southern delta (Yang et al. 2003). Calculated net accretion for the entire study area was 3.2 cm/year (Yang et al. 2011).

Between 2000 and 2004, sediment discharge from the Changjiang fell from 340 to 150 Mt/year. Areas below the 5-m contour east of Hengsha Island continued to prograde as much as 2 km, but there was little change east of Chongming or Jiuduansha islands. The 10-m contour, by contrast, retreated between 0.5 and 1.5 km throughout much of the study area, particularly east of Hengsha (Fig. 4.17). Overall, 70 % of the study area experienced erosion, and the average vertical erosion rate was −3.8 cm/year. There was a notable difference in vertical erosion between Area 1 (−0.7 cm/year) and Area 2 (−11 cm/year), with much of Area 2 deepening by as much as 25 cm/year (Fig. 4.18a).

Between 2004 and 2007, after the TGD was closed, the 5-m contour off both Henhsha and Chongming islands continued to prograde by as much as 1–2 km (Fig. 4.17). In contrast, the 10-m isobath retreated on average by ~1 km and locally by nearly 2 km. Erosion seems to have been concentrated between depths of 5 and 8 m, and this depth range decreased in Area 1 by 30 % (~150 km^2) (Yang et al. 2011). In contrast, the 8 m to 11 m depth interval increased in area by 120 km^2. Depths greater than 11 m showed no significant change in total area (Yang et al. 2011), which may in part reflect dumping of the dredge spoil removed from the TJ–GC channel. The calculated net erosion rate for Area 1 was −4.5 cm/year (Yang et al. 2011).

Since 1958, bathymetric changes within the study area have shown a strong linear correlation with the Changjiang sediment load measured at Datong station (Fig. 4.18b, after Yang et al. 2011). Between 1958 and 1977, when the average SSD was 470 Mt/year (i.e., 362 Mm^3/year, with a bulk density of 1.3 t/m^3; Zhu 2000), the subaqueous delta shoaled by ~125 Mm^3/year. In contrast, between 2000 and 2004, an average discharge of 245 Mt/year (i.e., 188 Mm^3/year) resulted in 70 Mm^3/year of delta-front erosion (Fig. 4.18b). These data and trends suggest that the Changjiang delta will continue to erode as long as the river discharges less than ~270 Mt/year of sediment annually (after Yang et al. 2011), but whether the erosion is confined primarily to depths of between 5 and 8 m remains to be seen.

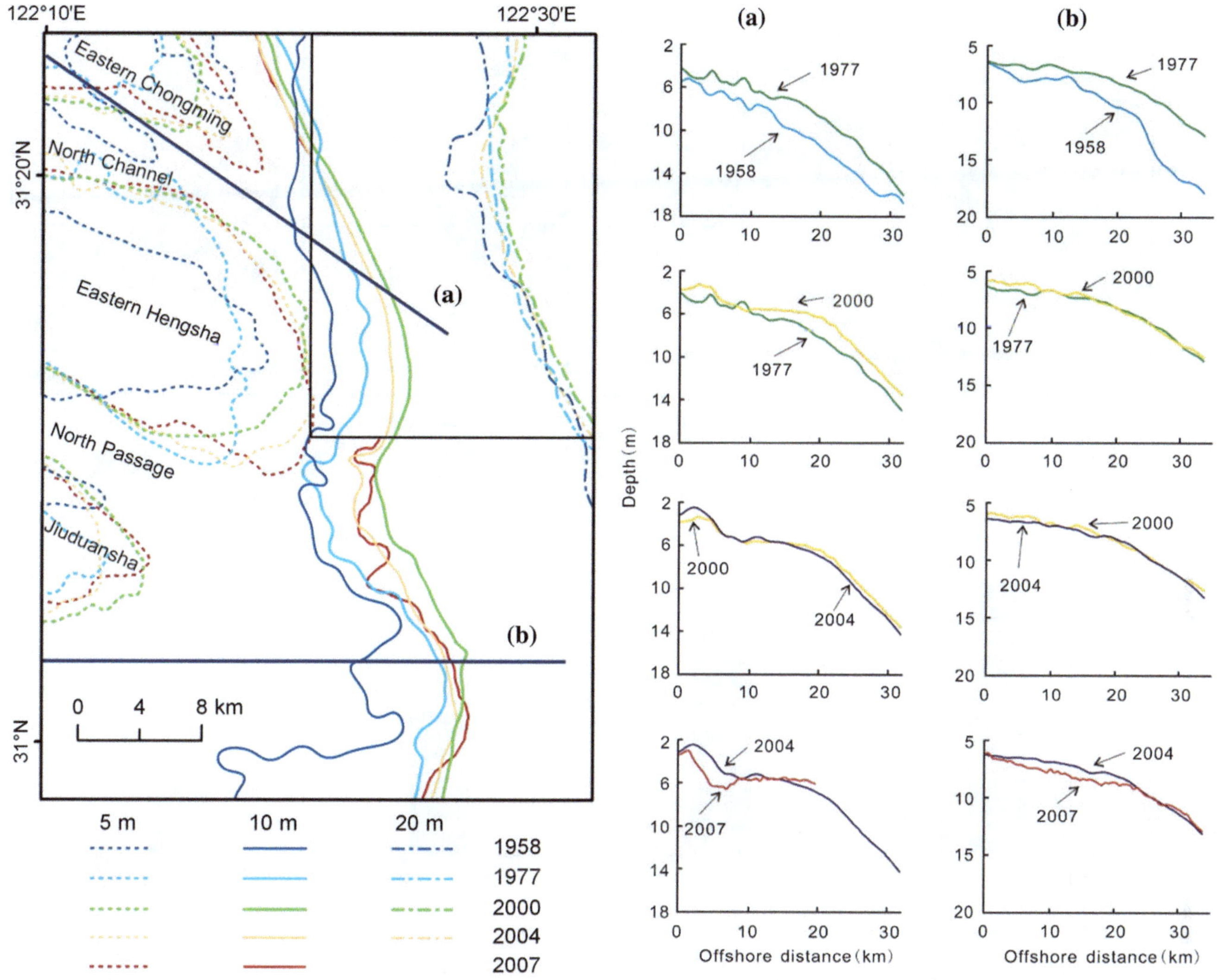

Fig. 4.17 *Left* 5-, 10-, and 20-m bathymetric contours on the Changjiang delta for the years 1958, 1977, 2000, 2004, and 2007 (the location of the study area is shown in Fig. 4.1). *Right* changes in bathymetric profiles **a** and **b** (cf. Yang et al. 2011)

4.3.3 Complexity of Delta Response

The morphodynamic behavior of the subaqueous area of the Changjiang Delta is influenced by many factors, and not simply sediment supply; this complicates the analysis of the delta's response to riverine sediment supply (Yang et al. 2003). The subaqueous delta of the Changjiang is sensitive to the decline in the river sediment load. However, the accretion rate of the subaqueous delta is not simply proportional to sediment discharge from the river. For instance, the mean accretion rate of the subaqueous delta decreased by 79 %, from 3.8 cm/year in 1958–1978 to 0.8 cm/year in 1978–1997, while the riverine sediment load of the Changjiang decreased by 15 %, from 466 Mt/year to 394 Mt/year during the same period. Comparing the time periods 1958–1965 and 1978–1997, the riverine sediment load decreased from 500 to 390 Mt/year (29 %) but ~200 m/year progradation rate did not decrease proportionally. Instead, erosion began to occur at the 10- and 20-m isobaths.

In addition to sediment load, the response of the delta may also be related to SSC. Without a change in sediment quantity, an increase in water discharge will enhance the dispersion of riverine sediments into the East China Sea shelf (see Chap. 2). In river-dominated situations, effluent diffusion and sediment dispersion depend upon the relative importance of turbulent diffusion, turbulent friction, and outflow buoyancy (Coleman and Wright 1975; Coleman 1981; Wright 1985). Although the Changjiang delta is not included in the ternary classification diagram of Wright (1985), it can be treated as a transitional form between the three basic types, which are river-, wave-, and tide-dominated systems.

The distribution of riverine water and sediment supply among the branches in the Yangtze Estuary are frequently

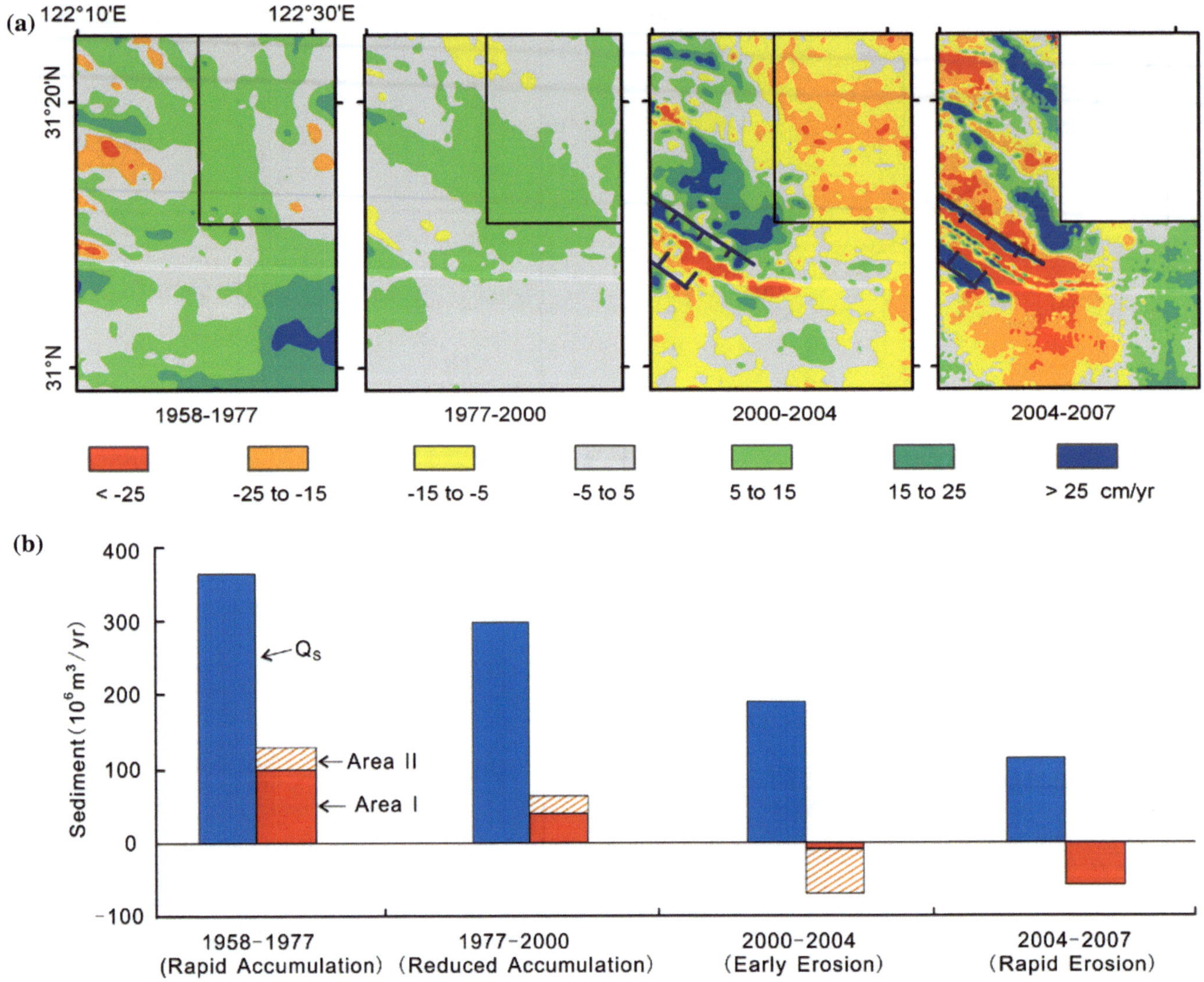

Fig. 4.18 **a** Accumulation/erosion in study areas 1 and 2 (the locations of these areas are shown in Fig. 4.1) for the periods 1958–1977, 1977–2000, 2000–2004, and 2004–2007. By 2004, the TJ–GC and dredged channel extended into the study area. **b** Comparison of the sediment load at Datong station and sediment accumulation/erosion in study areas 1 and 2. The sediment load was converted to Mm^3/year assuming a bulk density of 1.25 t/m^3 (cf. Yang et al. 2011)

modified due to morphological evolution at the bifurcations (Xie and Yang 1999). The shift of the main source of riverine water and sediment from one channel to another often results in sediment progradation downstream of the new outlet and recession of the delta front adjacent to the abandoned outlet (Chernicoff and Venkatakrishnan 1995; Trenhail 1997). In this case, spatial changes in accretion/erosion can be observed without changes in sediment supply from the watersheds.

To examine the different accretion/erosion responses to the decrease in riverine sediment supply between the estuary and its adjacent offshore slope, Xu et al. (2008) used ArcGIS to calculate the accretion/erosion rates of the South Passage and its offshore extension. The results showed that the accretion rates during 1990–1996, 1996–2000, and 2000–2004 were 2.1, 1.5, and 12 cm/year in the South Passage, and 6.1, 4.3 and −6.6 cm/year in the offshore area, respectively. They also showed that, compared with its offshore extension, the South Passage exhibited a more complicated spatial distribution of accretion/erosion. It was concluded that the accretion/erosion in the South Passage was caused mainly by the DWP in the North Passage and the movement of sand bars, rather than by the decrease in riverine sediment supply. The offshore slope, by contrast, is more sensitive to the decrease in riverine sediment supply, although it is also influenced by the bed-level changes in the South Passage (Xu et al. 2008).

In the same area where Yang et al. (2011) did their study for the period 1958–2007, the comparison of bathymetric data between 2007 and 2011 suggests a slight net accretion

(~0.2 cm/year). Further studies are required to understand why erosion of the subaqueous delta front has ceased and been replaced by accretion, and whether or not this trend will continue to occur.

4.4 Variations in Surface Suspended Sediment Concentration in the Estuary and Adjacent Coastal Waters

In the Changjiang Estuary and adjacent coastal environment, surface SSC is highly variable through time and space. For instance, the surface SSC observed twice-daily at nine gauging stations in the Changjiang Estuary and Hangzhou Bay in 2009, differed by three orders of magnitude and ranged from 0.001 to 4 g/L (Li et al. 2012).

4.4.1 Spatial Pattern of Suspended Sediment Concentration

The mean annual surface SSC along the southern channels of the Changjiang Estuary in 2009 increased from 0.06 g/L at the most landward station to 0.6 g/L at the river mouth, and then decreased farther offshore again to a minimum of 0.06 g/L at the most seaward station. Along the shore, the mean SSC increased from 0.4 g/L at Sheshan station, off the Changjiang mouth, to 1.3 g/L at Tanhu station in Hangzhou Bay (Fig. 4.19).

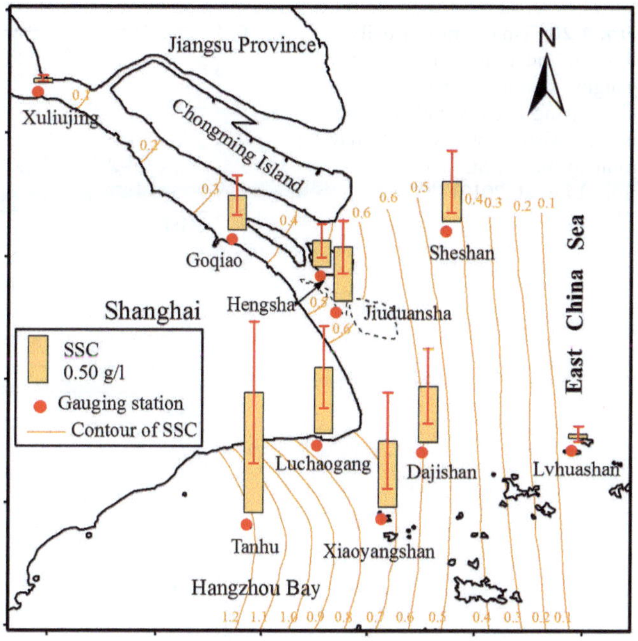

Fig. 4.19 Averages and standard deviations of surface SSC in 2009 based on daily observations at Luchaogang station, and twice-daily observations at other stations. Contours show the average interpolated SSC in areas between the gauging stations (modified from Li et al. 2012)

4.4.2 Temporal Variation in Suspended Sediment Concentration

4.4.2.1 Neap–Spring Tide Variation

The SSC in Changjiang Estuary and adjacent Hangzhou Bay is typically correlated with tidal range, with maximum values during spring tides and minimum values during neap tides (Fig. 4.20). Generally, SSC during spring tides is two to three times greater than those during neap tides. The SSC at all nine gauging stations shows a statistically significant ($p < 0.001$) spring–neap signal of 14–15 days, although there are differences among the stations (Li et al. 2012). The high regularity of spring–neap SSC cycles at Xiaoyangshan station (Fig. 4.20g) and Tanhu station (Fig. 4.20h), which are both located within Hangzhou Bay, likely reflects high tidal ranges and greater water depths (i.e., >5 m). This presumably results in greater tidal control and less wave control than at the other stations. In contrast, the spring–neap SSC cycles are least apparent in the upper part of Changjiang Estuary; e.g., Xuliujing station (Fig. 4.20a). This most likely reflects the dominating influence of river discharge over a relatively small tidal range.

4.4.2.2 Seasonal Variation

Fifteen-day SSC running averages were used to filter out spring–neap cycles, so that the seasonal variability could be identified. The annual cyclicity in SSC is apparent in waters off the Changjiang mouth, particularly at Sheshan, Lvhuashan, Dajishan, Xiaoyangshan, and Tanhu stations (Fig. 4.20). At these stations, wintertime SSC was three to four times greater than that in summer. Elsewhere, SSC was also higher in winter than that in summer, although the annual cyclicity was less pronounced, with the mean SSC being 1.5–2.0 times greater in winter than that in summer (Fig. 4.20).

Because of local resuspension due to tides and waves, SSC in the Changjiang Estuary and Hangzhou Bay (Fig. 4.19) is greater overall than that in the Changjiang. Thus, the river discharge acts to dilute the SSC of the estuary and adjacent waters (Gao et al. 2008).

The seasonal cyclicity in SSC is attributed to the regulating effect of the monsoon on local climate. The southwest monsoon results in higher precipitation in the river basin and higher water discharge into the estuary (Fig. 4.4), and thereby causes lower SSC in the estuary and adjacent waters. Increased river discharge in summer can also result in

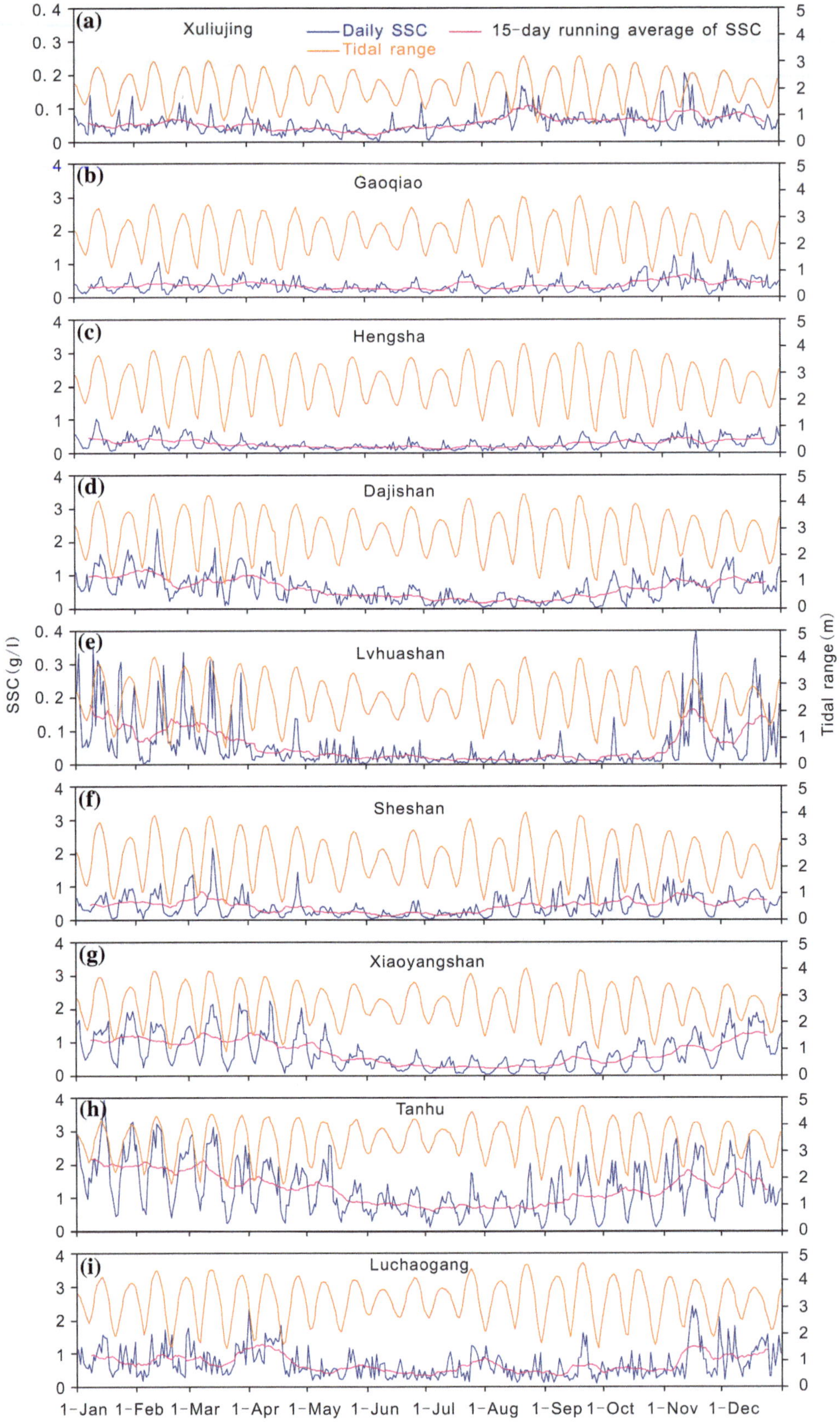

Fig. 4.20 Time series of daily SSC and tidal range in 2009 at gauging stations in the Changjiang Estuary and adjacent waters. Also shown is the 15-day running mean SSC (modified from Li et al. 2012)

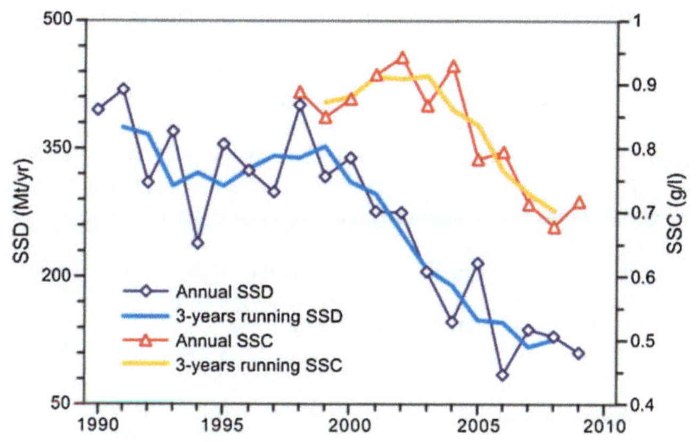

Fig. 4.21 Time series of average annual surface SSC at Xiaoyangshan station and suspended sediment discharge (SSD) at Datong station (modified from Li et al. 2012)

enhanced stratification and reduced vertical mixing of the water column (Li et al. 2012). In addition, stronger winds and waves may enhance sediment resuspension and maintain higher SSC in winter (Chen et al. 2006), especially in shallow waters.

4.4.2.3 Long-term Trends

Over the past 10–20 years, the annual SSD and SSC at Datong hydrological station have decreased by 63 and 55 %, respectively (Li et al. 2012). In the Changjiang Estuary and adjacent coastal environment, average annual surface SSC at all gauging stations has decreased by between 5 and 56 % (26 % on average). The highest decrease was at Xuliujing station in the upper estuary, and the lowest decrease was at Sheshan station at the delta front. Although the decrease in SSC at Sheshan station was not statistically significant (rank-sum test, $p > 0.05$), the decreases at the other six stations were significant ($p < 0.01$) (Li et al. 2012). At Xiaoyangshan station, where observations of SSC have been conducted for more than 10 years, the average annual SSC has decreased significantly (Fig. 4.21).

4.5 Summary and Conclusions

Due to the East Asian monsoon climate, the discharge from the Changjiang is much larger in summer than that in winter, but reservoir operation has modified the seasonal distribution of annual discharge. Interannual changes in river discharge have also occurred due to climate variability. Water consumption and reservoir construction have decreased the annual water discharge by ~10 % in the past 150 years, and human impacts on long-term variation in the Changjiang sediment load have been even greater. Deforestation increased the sediment discharge of the river from less than 400 Mt/year in the 1900s to more than 500 Mt/year in the 1960s. After this time, dam construction and soil preservation decreased the sediment discharge to less than 150 Mt/year.

In the Changjiang Estuary and the adjacent coastal environment, sedimentary responses to changing river discharge and sedimentary load are significant. The subaqueous delta front exhibited accretion of several centimeters per year in the 1950s to 1970s, but began to erode by the same amount after the closure of the TGD. The erosion-generated resuspension of fine-grained sediments slowed the decline of SSC in the delta front. However, overall SSC in the Changjiang Estuary and adjacent coastal waters has shown a significant decrease.

This study suggests that estuarine and coastal sedimentary processes are highly variable and susceptible to human activities and climate fluctuations. This variability has great ecological, environmental, and engineering implications and should be considered in river–ocean management.

Acknowledgments This work was funded by the Natural Science Foundation of China (grant #: 41130856 and 41021064). We thank Mr. P. Li and Mrs. X.Y. Zhang for their help with the production of figures. Data of water and sediment discharge from Changjiang were collected by the Changjiang Water Resources Commission, Ministry of Water Resources in China.

References

Chen JY, Zhu HF, Dong YF, Sun JM (1985) Development of the Changjiang estuary and its submerged delta. Cont Shelf Res 4(1/2):47–56

Chen SL, Zhang GA, Yang SL, Shi JZ (2006) Temporal variations of fine suspended sediment concentration in the Changjiang River estuary and adjacent coastal waters China. J Hydrol 331:137–145

Chen W (2006) Future trend in Chinese populations: 2005–2050. Popul Res 30(4):93–95 (in Chinese)

Chernicoff S, Venkatakrishnan R (1995) An introduction to physical geology. Worth Publishers Inc., New York, p 593

Coleman JM, Wright LD (1975) Modern river deltas: variability of processes and sand bodies. In: Broussard ML (ed) Deltas—models for exploration. Houston Geological Society, Houston, pp 99–149

Coleman JM (1981) Deltas: Processes of deposition and models for exploration. MN: Burgess, Minneapolis, (2nd ed.) pp 124

CRWCCWCMC (2003) Changjiang River Almanac. Changjiang River Almanac Press, Wuhan

Dai SB, Yang SL, Li M (2009) Sharp decrease in suspended sediment supply from China's rivers to the sea: anthropogenic and natural causes. Hydrolog Sci J 54(1):135–146

Dai Z, Du J, Li J, Li W, Chen J (2008) Runoff characteristics of the Changjiang River during 2006: effect of extreme drought and the impounding of the Three Gorges Dam. Geophys Res Lett 35: L07406. doi:10.1029/2008GL033456

DeMaster DJ, McKee BA, Nittrouer CA, Qian JC, Cheng GD (1985) Rates of sediment accumulation and particle reworking based on radiochemical measurements from continental shelf deposits in the East China Sea. Cont Shelf Res 4:143–158

Du JL, Yang SL (2007) Morphological impact of the north passage deeper waterway project on the surrounding areas, Changjiang Estuary. Sci Geogr Sinica. 27:390–394 (in Chinese with English abstr.)

Gao A, Zhao HY, Yang SL, Dai SB, Chen SL, Li P (2008) Seasonal and tidal variations in suspended sediment concentration under the influence of river runoff, tidal current and wind waves: taking the Nanhui Headland, the joint area between Changjiang Estuary and Hangzhou Bay as an example. Adv Mar Sci 26(1):44–50 (in Chinese with English abstract)

Khaliq MN, Ouarda TBMJ, Gachon P, Sushama L (2008) Temporal evolution of low-flow regimes in Canadian rivers. Water Resour Res 44:W08436. doi:10.1029/2007WR006132

Li P, Yang SL, Milliman JD, Xu KH, Qin WH, Wu CS, Chen YP (2012) Spatial, temporal, and human-induced variations in suspended sediment concentration in the surface waters of the Changjiang Estuary and adjacent coastal areas. Estuar Coast 35:1316–1327

Liu G, Mi J, Qi C, Lin X, Yang C (1996) Relationship between the solar forcing and periodic variations of meteoric water in the northeastern China. J Changchun Uni Earth Sci 26:422–427 (in Chinese with English abstract)

Liu C, Zhang X (2004) Causal analysis on actual water flow reduction in the mainstream of the Yellow River. Acta Geogr Sinica 59:323–330 (in Chinese with English abstract)

Liu G, Liu J, Zhao R, Zhang Q (2008) Analysis of periodical variation of Lancang River based on wavelet transform. Comput Eng Appl 44:236–237 (in Chinese with English abstract)

Liu JP, Xu KH, Li AC, Milliman JD, Velozzi DM, Xiao SB, Yang ZS (2007) Flux and fate of Changjiang River sediment to the East China Sea. Geomorphology 85:208–224

Milliman JD, Shen HT, Yang ZS, Meade RH (1985) Transport and deposition of river sediment in the Changjiang estuary and adjacent continental shelf. Cont Shelf Res 4:37–45

Milliman JD, Farnsworth KL (2011) River discharge to the coastal ocean: a global synthesis. Cambridge University Press, New York, p 384

Nilsson C, Reidy CA, Dynesius M, Revenga C (2005) Fragmentation and flow regulation of the world's large river systems. Science 308:405–408

Parthasarathy B, Mooley DA (1978) Some features of a long homogeneous series of Indian summer monsoon rainfall. Mon Weather Rev 106:771–781

Shen HT, Mao ZC, Zhu JR (2003) Saltwater Intrusion in the Changjiang Estuary. Ocean Press, Beijing, p 175 (in Chinese)

Shen H, Wang X, Yang Q (2000) The spectrum analysis of discharge and salinity in the Changjiang estuary. Acta Oceanol Sinica 22 (4):17–23 (in Chinese with English abstract)

Shi DM (1999) Analysis of relationship between soil and water loss and flood disasters in Changjiang River basin. J Soil Erosion Soil Water Conserv 5(1):1–7 (in Chinese with an English abstract)

Shi BW, Yang SL, Wang YP, Bouma TJ, Zhu Q (2012) Relating accretion and erosion at an exposed tidal wetland to the bottom shear stress of combined current–wave action. Geomorphology 138:380–389

Shi BW, Yang SL, Wang YP (2014) Intertidal erosion and deposition interpreted from field observations of hydrodynamic and sedimentary processes: A case study from an exposed, meso–macrotidal and highly turbid mudflat–saltmarsh transition. Cont. Shelf Res. 90: 109–116

Tao S, Cheng L (1987) A review of recent research on East Asian monsoon in China. Monsoon Meteorol 60–92

Trenhaile AS (1997) Coastal dynamics and landforms. Oxford University Press, Oxford, p 366

Vörösmarty CJ, Meybeck M, Fekete B, Sharma K, Green P, Syvitski JPM (2003) Anthropogenic sediment retention: major global impact from registered river impoundments. Glob Planet Change 39:169–190

Walling DE (2006) Human impact on land-ocean sediment transfer by the world's rivers. Geomorphology 79:192–216

Wiegel RL (1996) Nile delta erosion. Science 272:338–340

Wang H, Yang ZS, Saito Y, Liu JP, Sun X (2006) Interannual and seasonal variation of the Huanghe (Yellow River) water discharge over the past 50 years: connections to impacts from ENSO events and dams. Glob Planet Change 50:212–225

Wang H, Yang Z, Saito Y, Liu JP, SunX Wang Y (2007a) Stepwise decreases of the Huanghe (Yellow River) sediment load (1950–2005), Impacts of climate change and human activities. Glob Planet Change 57:331–354

Wang H, Yang Z, Wang Y, Saito Y, Liu JP (2007b) Reconstruction of sediment flux from the Changjiang (Changjiang River) into estuary since the 1860s. J Hydrol 349:318–332

Wang H, Saito Y, Zhang Y, Bi N, Sun X, Yang Z (2011) Recent changes of sediment flux to the western Pacific Ocean from major rivers in East and Southeast Asia. Earth Sci Rev 108:80–100

Whitehouse R, Soulsby R, Roberts W, Mitchener H (2000) Dynamics of Estuarine Muds: A Manual for Practical Applications. Tomas Telford Limited, London

Wright LD (1985) River deltas. In: Davis RA Jr (ed) Coastal Sedimentary environments, 2nd edn. Springer, New York, pp 1–76

Xie WH, Yang SL (1999) Evolution of the Jiuduansha Shoaland its influence on adjacent channels in the Changjiang estuary. China Ocean Eng 13(2):185–195

Xu K, Milliman JD (2009) Seasonal variations of sediment discharge from the Changjiang River before and after impoundment of the Three Gorges Dam. Geomorphology 104:276–283

Xu JX (2003) Sediment flux into estuary as influenced by changing human activities and precipitation: example of the Yellow River, China. Environ Manag 31:328–341

Xu X, Yang SL, Li P (2008) Different bed responses to the decrease in riverine sediment supply between the river mouth channel and the open sea slope on its seaward side: a case study from the Changjiang Estuary. Mar Sci Bull 27(5):100–104

Yang SL, Ding PX, Chen SL (2001) Changes in progradation rate of the tidal flats at the mouth of the Changejiang River, China. Geomorphology 38:167–180

Yang SL, Zhao QY, Belkin IM (2002) Temporal variation in the sediment load of the Yangtze River and the influences of the human activities. J Hydrol 263:56–71

Yang SL, Belkin IM, Belkina AI, Zhao QY, Zhu J, Ding XD (2003) Delta response to decline in sediment supply from the Yangtze River: evidence of the recent four decades and expectations for the next half-century. Estuar Coast Shelf Sci 57:589–599

Yang SL, Shi Z, Zhao H, Li P, Dai S, Gao A (2004) Effects of human activities on the Yangtze River suspended sediment flux into the estuary in the last century. Hydrol Earth Syst Sci 8(6):1210–1216

Yang SL, Cao A, Hotz HJ, Zhu J, Dai SB, Li P (2005a) Trends in annual discharge from the Yangtze River into estuary (1865–2004). Hydrolog Sci J 50(5):825–836

Yang SL, Zhang J, Zhu J, Smith JP, Dai SB, Gao A (2005b) Impact of dams on Yangtze River sediment supply into estuary and delta wetland response. J Geophys Res 110:F03006. doi:10.1029/2004JF000271

Yang SL, Li M, Dai SB, Liu Z, Zhang J, Ding PX (2006a) Drastic decrease in sediment supply from the Yangtze River and its challenge to coastal wetland management. Geophys Res Lett 33: L06408. doi:10.1029/2005GL025507

Yang ZS, Wang H, Saito Y, Milliman JD, Xu K, Qiao S, Shi G (2006b) Dam impacts on the Yangtze (Changjiang) River sediment load to the sea: the past 55 years and after the Three Gorges Dam. Water Resour Res 42:W04407. doi:10.1029/2005WR003970

Yang SL, Zhang J, Xu XJ (2007) Influence of the Three Gorges Dam on downstream delivery of sediment and its environmental implications, Yangtze River. Geophys Res Lett 34:L10401. doi:10.1029/2007GL029472

Yang SL, Li H, Ysebaert T, Bouma TJ, Zhang WX, Wang YY, Li P, Li M, Ding PX (2008) Spatial and temporal variations in sediment grain size in tidal wetlands, Yangtze delta: on the role of physical and biotic controls. Estuar Coast Shelf Sci 77:657–671

Yang SL, Liu Z, Dai SB, Gao ZX, Zhang J, Wang HJ, Luo XX, Wu CS, Zhang Z (2010) Temporal variations in water resources in the Yangtze River (Changjiang) over the Industrial Period, based on reconstruction of missing monthly discharges. Water Resour Res 46:W10516. doi:10.1029/2009WR008589

Yang SL, Milliman JD, Li P, Xu K (2011) 50,000 dams later: erosion of the Yangtze River and its delta. Glob Planet Change 75:14–20. doi:10.1016/j.gloplacha.2010.09.006

Yang SL, Shi BW, Ysebaert T, Luo XX (2012) Wave attenuation at a salt marsh margin: a case study of an exposed coast on the Yangtze Estuary. Estuar Coast 35:169–182

Yang SL (2013) Morphological evolution of the Changjiang Delta and causes. In: Ding PX, Wang HJ, Meng XW, Zhu JR (eds) Evaluation and cause analysis of typical coastal zones in China during the last 50 years. Science Press, Beijing, pp 22–61 (in Chinese)

Yang SL, Milliman JD, Xu KH, Deng B, Zhang XY, Luo XX (2014) The impact of sedimentation in the three Gorges Reservoir and downstream sediment delivery. Earth Sci Rev 138:469–483

Zhang R, Wang YP, Gao JH, Pan SM, Zhan SL (2008) The vertical sedimentary structure and its implications for environmental evolutions in the Changjiang Estuary in China. Acta Oceanologia Sinica 30(2):80–91

Zhang S, Zhu C (2001) Soil loss and its effect on flooding catastrophe in Changjiang drainage basin. J Soil Water Conserv 15(6):9–16

Zhao CH, Zhu ZH, Zhou DZ (2000) Worldwide Rivers and dams. China Water Conservancy and Hydroelectric Press, Beijing, p 1059 (in Chinese)

Zeng G, Kong X (1999) Preliminary study on the causes and countermeasures of the 1954 and 1998 extraordinary Changjiang River floods. J Catastrophol 14:1–3 (in Chinese with English abstract)

Zhu JY (2000) Variation of sediment transportation in the Changjiang River and the way for its reduction. J Hydroelectr Eng 70:38–48 (Chinese with English abstract)

Zhu Q, He J, Wang P (1986) A study of circulation difference between East Asia and Indian summer monsoon with their interaction. Adv Atmos Sci 3:446–477

Zhu Q, Yang SL, Ma YX (2014) Intra-tidal sedimentary processes associated with combined current–wave action on an exposed, erosional mudflat, southeastern Yangtze River Delta, China. Mar Geol 347:95–106 (in revision)

Plant Nutrients and Trace Elements from the Changjiang Watersheds to the East China Sea

5

Jing Zhang, Ying Wu, and Ying Ying Zhang

Abstract

Over the last several decades, the Changjiang (Yangtze River) watersheds has suffered from an acceleration of economic activities and changing land-use patterns. Monitoring programs at downstream hydrological stations and data from field campaigns have shown that these changes have caused an increase in the loss of chemical elements (e.g., plant nutrients such as N, P, and Si) from catchment areas and changes in pollutant discharges (e.g., trace elements). Changes in plant nutrients and trace elements in the river have altered the partitioning of chemical species between water and solid phases (e.g., total suspended matter) in aquatic systems, in both freshwater and marine environments. Irreversible element partitioning in rivers and coastal waters can affect the bioavailability of chemical species in adjacent marine environments, which can in turn impact ecosystem structure and functioning. In the East China Sea, the continual increase in plant nutrient fluxes from land sources, such as the Changjiang, has caused eutrophication in surface waters of the coastal environment. The degradation of organic matter, either allochthonous or autochthonous in origin or both, has fueled heterotrophic processes that induce the seasonal depletion of dissolved oxygen (DO) and even hypoxia in the Changjiang Estuary and the inner shelf of the East China Sea. Based on observational hydrography and chemistry data, a box model approach is used in this study to describe the plant nutrient and trace element budgets. The results illustrate that incursion across the shelf break of the western boundary current system (i.e., Kuroshio) has a significant influence on the inventory and residence time of chemical elements such as plant nutrients on the East China Sea Shelf.

Keywords

Changjiang • Watersheds • Estuary • East China Sea • Plant nutrients • Trace elements

5.1 Introduction

Estuaries and coasts play an important role in the exchange of energy and materials between the land and ocean. They regulate the fluxes of weathering products and pollutants from land sources and their delivery to the ocean (LOICZ 2005). Coastal circulation, which is driven by climate (e.g., the East Asian monsoon), controls the delivery of chemical elements in dissolved and particulate phases from both terrestrial and marine sectors, which in turn modifies the transfer of elements to the food web (IMBER 2005).

J. Zhang (✉) · Y. Wu
State Key Laboratory of Estuarine and Coastal Research, East China Normal University, 3663 Zhongshan Road North, Putuo, 200062 Shanghai, China
e-mail: jzhang@sklec.ecnu.edu.cn

Y. Wu
e-mail: wuying@sklec.ecnu.edu.cn

Y.Y. Zhang
School of Environmental Science and Engineering, Yancheng Institute of Technology, 9 Xiwang Road, Yancheng, 224051 Jiangsu, China
e-mail: yyzhang@ycit.edu.cn

J. Zhang (ed.), *Ecological Continuum from the Changjiang (Yangtze River) Watersheds to the East China Sea Continental Margin*, Estuaries of the World,
DOI 10.1007/978-3-319-16339-0_5,

Large aquatic systems, such as the Changjiang (Yangtze River) (note: here "jiang" means river in Chinese), are the site of mixing of weathering products (the production of which is related to climate) and pollutants (the production of which is related to anthropogenic activities) that have drained from watersheds, and disperse land-sourced materials to marine waters over hundreds to thousands of square kilometers, and nourish ecosystems (e.g., food web structure and biodiversity) in adjacent coastal oceans. The effluent plume from large rivers can be tracked hundreds of kilometers off the river mouth, reaching the continental margin and even beyond. Hence, by the mechanisms of shelf-break dynamics and continental pumping, terrestrial materials from rivers and coasts can be actively carried to the open ocean.

In this chapter, we examine and compare plant nutrient data (e.g., N, P, and Si) and trace elements (e.g., Al, As, and Se) along a continuum from the Changjiang watersheds to the shelf break in the East China Sea. We then synthesize the major biogeochemical issues that affect the element cycles in the East China Sea and assess their impact on marine ecosystems. An overview of previous studies of the Changjiang Estuary and the East China Sea is also provided, based on statistical analyses of data and budgetary approaches.

5.2 Materials, Methods, and Data Collection

5.2.1 Field and Oceanic Observations, Sample Collection, and Analysis

This study incorporates field data and laboratory simulation experiments for the region from the upper part of the watershed of the Changjiang, at an elevation of ca. 5000 m above sea level (asl), to the Okinawa Trough and the Kuroshio off the continental margin of the East China Sea at water depths of 1500–2000 m. The study considers the last 10–20 years and includes a comparison with data in national and international scientific literature.

A field campaign was conducted in April–May 1997, covering an area from the river mouth to the upper Changjiang watershed. It included ca. 4500 km of the main river channel, together with 15 important tributaries (Fig. 5.1). This survey was repeated in April–May 2003 and September–October 2009 to explore different water regimes of the Changjiang. Field observations were also made in 2006, 2008, and 2010 in the region between the river mouth and Yichang, which is the location of the Three Gorges Dam (TGD), during various discharge conditions (i.e., the dry and wet seasons). In the meantime, monitoring programs were conducted at Xuliujing, a station 150 km upstream from the river mouth, where samples were collected for chemical analysis, and hydrographic profiles of variables such as temperature, conductivity, total suspended matter (TSM), and pH were made on a monthly basis (Table 5.1 and Fig. 5.1).

Observational cruises were made between 1998 and 2010 from the river mouth to the Kuroshio waters with depths of ca. 2000–2500 m in the East China Sea Shelf (Table 5.1 and Fig. 5.1). During the cruises, conductivity–temperature–depth (CTD) hydrographic profiles were made of parameters such as temperature, salinity, epi-fluorescence, and pH. Optical backscatter sensor (OBS) measurements of turbidity and the amount of total suspended matter were undertaken at grid and anchor stations (Note: TSM was estimated based on turbidity data and calibrated using an empirical relationship). Discrete water samples were collected for chemical analysis based on CTD profiles and turbidity measurements from the surface to the near-bottom (ca. 2 m above the seabed). Drifters were deployed in the Changjiang Estuary and the East China Sea Shelf to track the mixing of water masses with different hydrographic properties. Other onboard measurements included dissolved oxygen (DO) by titration, pH by chemical sensor, amount of total suspended matter by filtration, optical properties by spectrophotometer, and chlorophyll a (*Chl-a*) by fluorescence measurements.

In situ incubation experiments were also carried out to understand the response of the phytoplankton community to external nutrient inputs and the effect of mixing between coastal and open-boundary waters on the community structure of micro- and pico-phytoplankton groups. Bottom sediments were collected by a box corer, multi-corer, and/or gravity corer and used for chemical and sedimentary measurements.

In the laboratory, plant nutrient species [i.e., NO_3^-, NO_2^-, NH_4^+, PO_4^{3-}, SiO_3^{2-}, and organic forms of nutrients such as dissolved organic nitrogen (DON) and dissolved organic phosphorous (DOP)] were measured by photometry using a continuous-flow spectrophotometer according to Grasshoff et al. (1999). Dissolved concentrations of As and Se and their chemical speciation [i.e., As (III/IV) and Se (IV/VI)] were determined by hydride generation and atomic absorption spectrometry (HG-AAS) and/or atomic fluorescence spectrometry (HG-AFS) (Yao 2005; Ren et al. 2010). Dissolved Al was determined by fluorescence measurements after complexion with lumogallion and extraction into *n*-hexanol (Zhang et al. 2000).

Particulate samples, including TSM and bottom sediments, were digested in $HF\text{-}HNO_3\text{-}HClO_4$ and then analyzed using an electrothermal AAS (ET-AAS; Model: AA-800) and inductively coupled plasma mass spectrometer (ICP-MS; Model: Element 2) for major and trace elements. Stable isotopes of carbon ($\delta^{13}C$) and nitrogen ($\delta^{15}N$) in organic matter of TSM and bottom sediments were measured using an isotope ratio mass spectrometer (Model: Finnigan Delta Plus XP), after removing carbonates using HCl before analysis for $\delta^{13}C$. Radioisotopes such as radium (i.e., ^{223}Ra, ^{224}Ra, ^{226}Ra, and ^{228}Ra), ^{7}Be, ^{137}Cs, and ^{210}Pb were

(a)

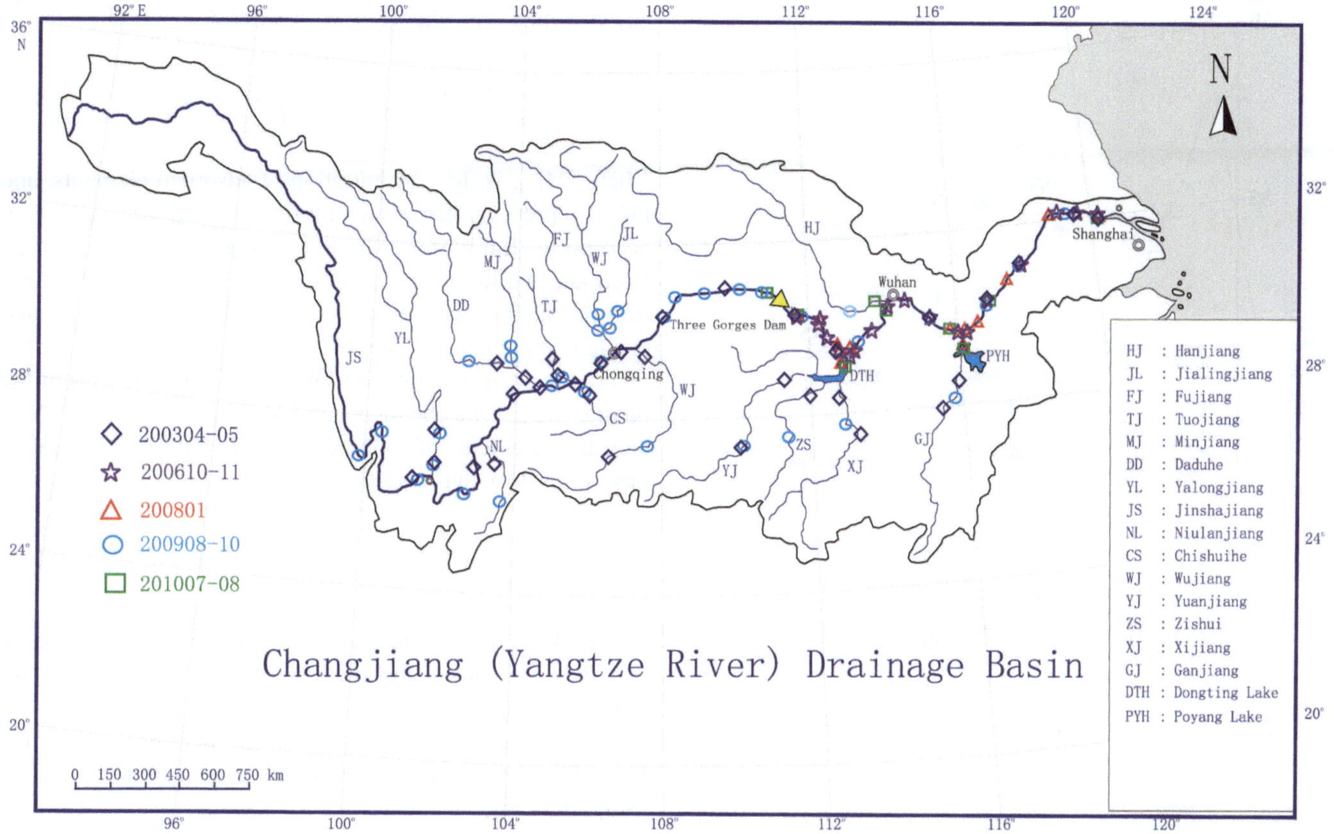

Fig. 5.1 Key geographical features of the study, including **a** the Changjiang watershed; **b** Changjiang Estuary; and **c** the East China Sea. In the subfigures, the locations of field and oceanic observations are shown with different symbols, and the isobaths in the marine environment are shown to differentiate the coast from the open shelf. Note that the dates in the legend are given in the order of year and month; for example, "200304–05" means that the observations were made in April–May 2003

determined using γ-ray and α-ray counting facilities (Zhang et al. 2007b; Du et al. 2011).

All experimental wares used in the field and on board of research ships, as well as in the laboratory for the analysis of nutrients and trace elements, were washed with hot detergent and rinsed in distilled water. They were then soaked in HCl (v/v: 1:5) for 5–10 days and rinsed with Milli-Q water before use. Quality assurance and quality control (QA/QC) was conducted by measuring national and international standards and/or reference materials, and intercomparisons were made with different laboratories.

5.2.2 Simulation of Element Partitioning Between Water and Particulate Phases

Laboratory simulation experiments were conducted to understand the partitioning of trace metals (i.e., Cd, Co, Cu, Ni, and Zn) and plant nutrient species (i.e., NH_4^+, NO_2^-, NO_3^-, PO_4^{3-}, and SiO_3^{2-}) along with the hydrographic gradients in the estuary and coastal environments (e.g., normoxic and hypoxic conditions). The variables and/or parameters considered in the simulation experiments included temperature, salinity, TSM, pH, and redox potential (cf. Zhang 2007; Zhang et al. 2008b). Briefly, the following results were reported in the study:

- Kinetic processes involving solid- and water-phase partitioning of chemical elements at different salinities and TSMs.
- The isotherms of element partitioning between water and solid phases at different pH and temperature levels.
- The effect of redox conditions on the adsorption and remobilization of elements of interest in coastal environments and the change in chemical speciation.

In the experimental batches, trace metals were analyzed by inductively coupled plasma atomic emission spectrometry (ICP-AES; Model: Perkin Elmer Plasma 2000) and plant nutrients by spectrophotometry (Model: Skalar SAN^{plus}).

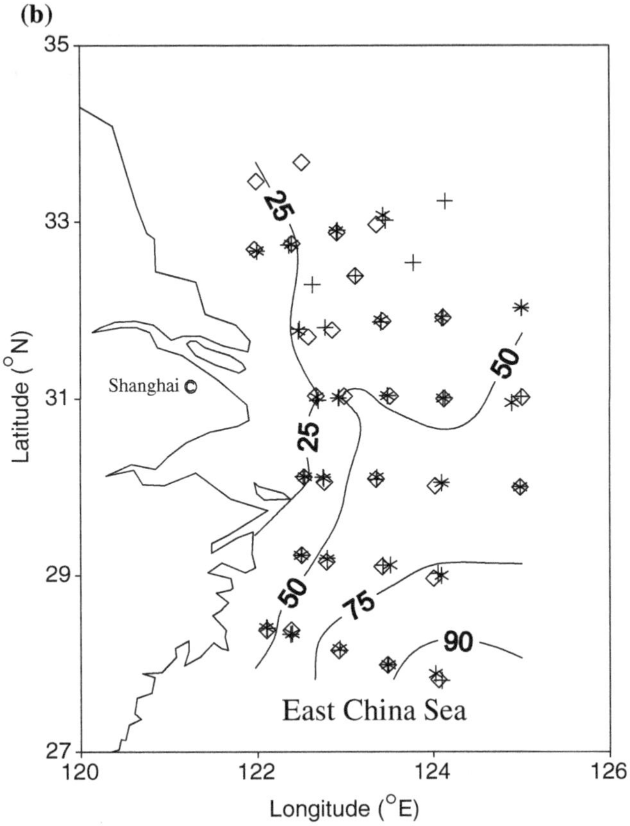

Fig. 5.1 (continued)

In this chapter, the percentage of dissolved elements (E) absorbed onto the TSM is expressed as

$$E(\%) = \frac{[M]_{\text{ads}}}{[M]_{\text{ads}} + [M]_{\text{dis}}} \times 100\,\% \quad (5.1)$$

where $[M]_{\text{ads}}$ is the concentration of adsorbed elements, and $[M]_{\text{dis}}$ represents the concentration of dissolved elements at adsorption equilibrium. In the aquatic environment (i.e., fresh and marine waters), the partitioning of a given chemical species between dissolved and solid phases (e.g., suspended particles in the water column) can be either reversible or irreversible, depending on the properties of the media, such as the chemical composition of a solution, TSM, and the element of interest. Hence, the percentage of desorption (D) can be defined as

$$D(\%) = \frac{[M]_{\text{des}}}{[M]_{\text{ads}}} \times 100\,\% \quad (5.2)$$

where $[M]_{\text{des}}$ is the concentration of the element of interest that was previously absorbed onto solid phases but is remobilized into solution.

The adsorption capacity can be determined by the Langmuir isotherm model (cf. Lopez et al. 1996; Stumm and Morgan 1996); that is,

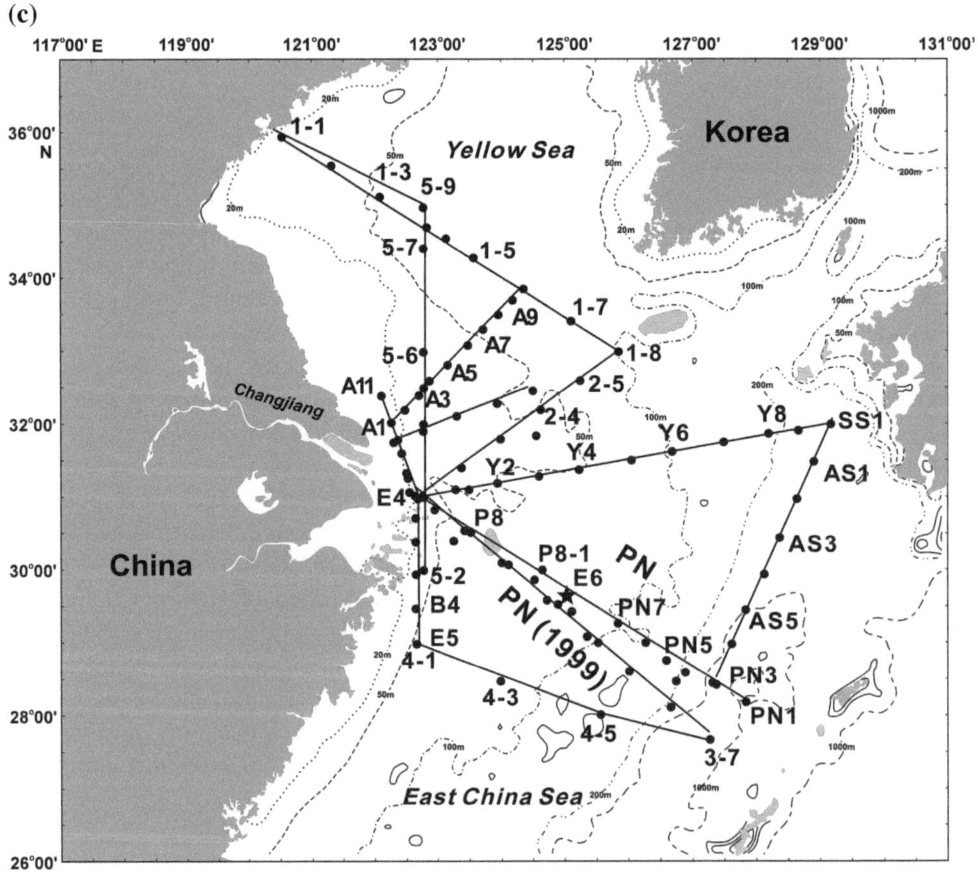

Fig. 5.1 (continued)

Table 5.1 Description of field observations in the Changjiang watershed, estuary, and the East China Sea

Research periods	Details of field observations and cruises
Changjiang catchment areas	
1997–2010	Field expeditions were made from the river mouth of the Changjiang upstream over a distance of 4000–4500 km and into 15 major tributaries in 1997, 2003, and 2009. Five other field campaigns were also conducted in the middle and lower reaches (i.e., from the TGD to the river mouth) over a distance of 2000 km in 2006–2010. In situ observations include hydrography (e.g., temperature and conductivity), turbidity, and TSM, and samples were also collected for chemical analysis
Changjiang Estuary	
2001–2010	Field observations were made from Xuliujing Station (i.e., 150 km upstream from the river mouth) down to the estuary until salinities of 30 or higher. In situ measurements included the turbidity and CTD profiles, epi-fluorescence, and *Chl-a*, and samples were also collected for chemical and biological analysis (e.g., pigments and taxonomy of phytoplankton)
East China Sea	
1998–2010	Cruises covered a region from the river mouth of the Changjiang at a water depth of 20 m until the Kuroshio at a water depth of 2500 m. This region encompasses the entire East China Sea Shelf and the Okinawa Trough. Onboard measurements included profiles for CDT, turbidity, DO, and epi-fluorescence. Discrete water samples were collected for chemical analysis, and sediment samples were collected by grab sampler, box sampler, multi-corer, and gravity corer

$$\Gamma = \Gamma_{\max} K \cdot C_{\text{ep}} / (1 + K \cdot C_{\text{ep}}) \tag{5.3}$$

where C_{ep} is the adsorption equilibrium concentration (mg l^{-1}) for chemical elements (e.g., phosphorus and trace metals), Γ represents the adsorption capacity (μg g^{-1} dry weight) at equilibrium, $\Gamma_{\max}$ stands for the maximum adsorption capacity (μg g^{-1} dry weight), and K is the equilibrium constant.

The distribution coefficient (K_{d}) is a useful parameter to describe the solid–solution partitioning of chemical elements. The K_{d} is calculated using

$$K_{\text{d}} = \frac{10^6 \cdot [M]_{\text{ads}}}{[\text{SPM}] \cdot [M]_{\text{res}}} = \frac{10^3 \cdot C_p}{C_w} \tag{5.4}$$

where $[M]_{\text{ads}}$ and $[M]_{\text{res}}$ are the concentrations of adsorbed chemical elements in suspended particulate matter (SPM) and in solution at adsorption equilibrium (mg l^{-1}), C_p is the concentration of absorbed elements (mg g^{-1}), and C_w is the dissolved element concentration (mg l^{-1}). In aquatic environments, K_{d} is a conditional constant that is readily implemented within a modeling framework (Hatje et al. 2003). Combining Eqs. 5.3 and 5.4, K_{d} can be rewritten as

$$K_{\text{d}} = \frac{10^3 \cdot \Gamma}{C_{\text{ep}}} = \frac{10^3 \cdot K \cdot \Gamma_{\max}}{1 + K \cdot C_{\text{ep}}} \tag{5.5}$$

where C_{ep} is the equilibrium concentration of chemical elements, and the K_{d} for a given elements depends on the concentration at equilibrium in adsorption experiments. It was reported that the K_{d} value for phosphorus is the same magnitude as for trace metals such as Cu, Ni, and Zn (Balls 1989), suggesting that phosphorus has a higher affinity with particulate matter compared to other plant nutrients (e.g., nitrogen) in the aquatic environment.

Details of laboratory manipulation and experimental procedures are described elsewhere (cf. Zhang 2003, 2007). Laboratory simulation experiments show that water samples from the Changjiang have a pH close to 8.0 and a salinity of 0 at Xuliujing station. At $C_{\text{ep}} = 0.285$ mg l^{-1}, the distribution coefficient on a log scale (i.e., $\log K_{\text{d}}$) ranged from 3.26 to 5.72, as follows.

Element	Cd	Co	Cu	Ni	P	Zn
$\log K_d$	4.13	3.85	4.85	3.26	5.72	4.68

5.3 Plant Nutrients and Trace Elements from the Watershed to the River Mouth of the Changjiang

5.3.1 Longitudinal Profiles of Plant Nutrient Concentrations

The examination of longitudinal profiles of plant nutrients from the upper reaches to the river mouth reveals that the Changjiang has a number of distinct characteristics compared with other large rivers in the world (Fig. 5.2). Generally, concentrations of plant nutrients are comparable between surface and near-bottom waters in the main channel from upstream areas to the river mouth over a distance of 4500 km. In the upper watershed, at an elevation of ca. 3000 m asl, the concentration of dissolved phosphorus is rather low, with values of 0.1–0.2 μM or even less.

The concentration of phosphate increases downstream in the main channel and reaches values of 2–2.5 μM when the river flows through Sichuan Province, where tributaries such as the Minjiang (MJ), Tuojiang (TJ), and Jialingjiang (JL) empty into the main stream (Fig. 5.2). This is most likely due to the elevated phosphate concentrations in the tributaries of this region (Zhang et al. 1999).

After passing through the Three Gorges Reservoir (TGR), which is about 600 km long in the deep Three Gorges Valley, concentrations of phosphate in the Changjiang tend to decrease and stay in the range of 0.5–1.0 μM, based on data from 1997 to 2003 (Fig. 5.2). As shown in Fig. 5.2, observations made in November 2006 show similar phosphate concentrations in the river downstream from the TGD over a distance of ca. 1850 km to the river mouth. However, the concentration of phosphate in 2006 is systematically higher than in 1997 and 2003, with values of 1.5–2.0 μM close to the river mouth. This is partly due to the fact that 2006 was an extremely dry period with a considerable reduction in water flow (see Chap. 3).

In the Changjiang, nitrate accounts for ≥90 % of dissolved inorganic nitrogen (i.e., DIN = NO_3^- + NO_2^- + NH_4^+) from upstream to the river mouth. Nitrate in the water column is also relatively low in the upper reaches of the Changjiang, at ca. 3000 m asl, with concentrations of ca. 1–5 μM in 1997 and 10–20 μM in 2003. Levels of nitrate increase to ca. 100–120 μM in the TGR region. Downstream from the TGD, concentrations of nitrate either fall to ca. 80 μM, as in 1997, or remain at 100–120 μM, as in 2003 (Fig. 5.2). Observations in November 2006 show that the concentration of nitrate increases downstream of the TGD, but the levels of nitrate are comparable between 2003 and 2006 at the river mouth.

Compared to phosphate and nitrate, dissolved silicate has a different pattern of distribution along the river course (Fig. 5.2). High concentrations (i.e., 100–125 μM) of dissolved silicate can be found in the high-elevation area, i.e., 3000–4500 km upstream from the river mouth, illustrating the extensive loss of chemical elements in the catchment areas (Box 5.1). Relatively low concentrations of dissolved silicate at 80–100 μM are found downriver from the TGD, presumably because of the uptake of silicate by diatoms during photosynthesis in the reservoir and the dilution by water from tributaries with low SiO_3^{2-} and high carbonates. About 1500–2000 km further downstream and until the river mouth, the concentration of dissolved silicate tends to increase again, most likely because of the collective contribution of tributaries in the southern watershed (e.g., the Ganjiang and Xiangjiang) with elevated SiO_3^{2-} concentrations and high discharges (Fig. 5.2).

Box 5.1 Loss of Chemical Elements from the Changjiang Watershed

Surface and riverine waters in the Changjiang catchment area receive chemical elements from natural and anthropogenic sources. The natural sources of chemical elements in the river are mainly related to the weathering and erosion of minerals and rocks in earth crust. Chemical elements can also enter the river through leaching of soil profiles and groundwater recharge. The watershed also receives chemical elements through atmospheric wet (e.g., rainfall and snow) and dry (i.e., aerosols and gas species) depositions, and in some places, atmospheric deposition can be an important factor in regulating riverine chemistry.

The anthropogenic sources of chemical elements are related to human activities in the watersheds, such as the application of synthesized fertilizers and pesticides in agriculture, and pollution discharge. Hence, the contribution from human activities can be categorized as nonpoint (e.g., release in agriculture) and point (e.g., pollutant drainage from urban sites) sources.

Chemical elements can be fixed through the accumulation of biomass (e.g., forestation) and removed from the watershed through the harvest of agricultural products (e.g., food and legumes). Thus, estimating the loss of chemical elements from watersheds involves a mass balance calculation, taking into account the difference between removal and gain terms on the annual average.

The loss of chemical elements (or the "chemical yield") from the watershed as the seaward flux in the river can be estimated by calculating the product of the element concentration and the water flux at the river mouth divided by the catchment areas. For instance, based on extensive sample collection and measurements, Liu et al. (2003a) give an estimate of plant nutrient loss from 10 to 15 major tributaries of the Changjiang as 25×10^3 to 125×10^3 mol km^{-2} year^{-1} for SiO_3^{2-}, 1.5×10^3 to 220×10^3 mol km^{-2} year^{-1} for NO_3^-, and 0.01×10^3 to 10×10^3 mol km^{-2} year^{-1} for PO_4^{3-}. Similarly, Zhang et al. (2003) estimated the chemical yield to be 50–75 mol km^{-2} year^{-1} for dissolved Al in the Changjiang watersheds. The loss of chemical elements does show some geographic pattern in the watersheds. For example, the estimated chemical yield of dissolved silicate in the tributaries in the southern part of the Changjiang watershed is statistically higher than that from the northern tributaries. This indicates that rivers in the subtropical climate

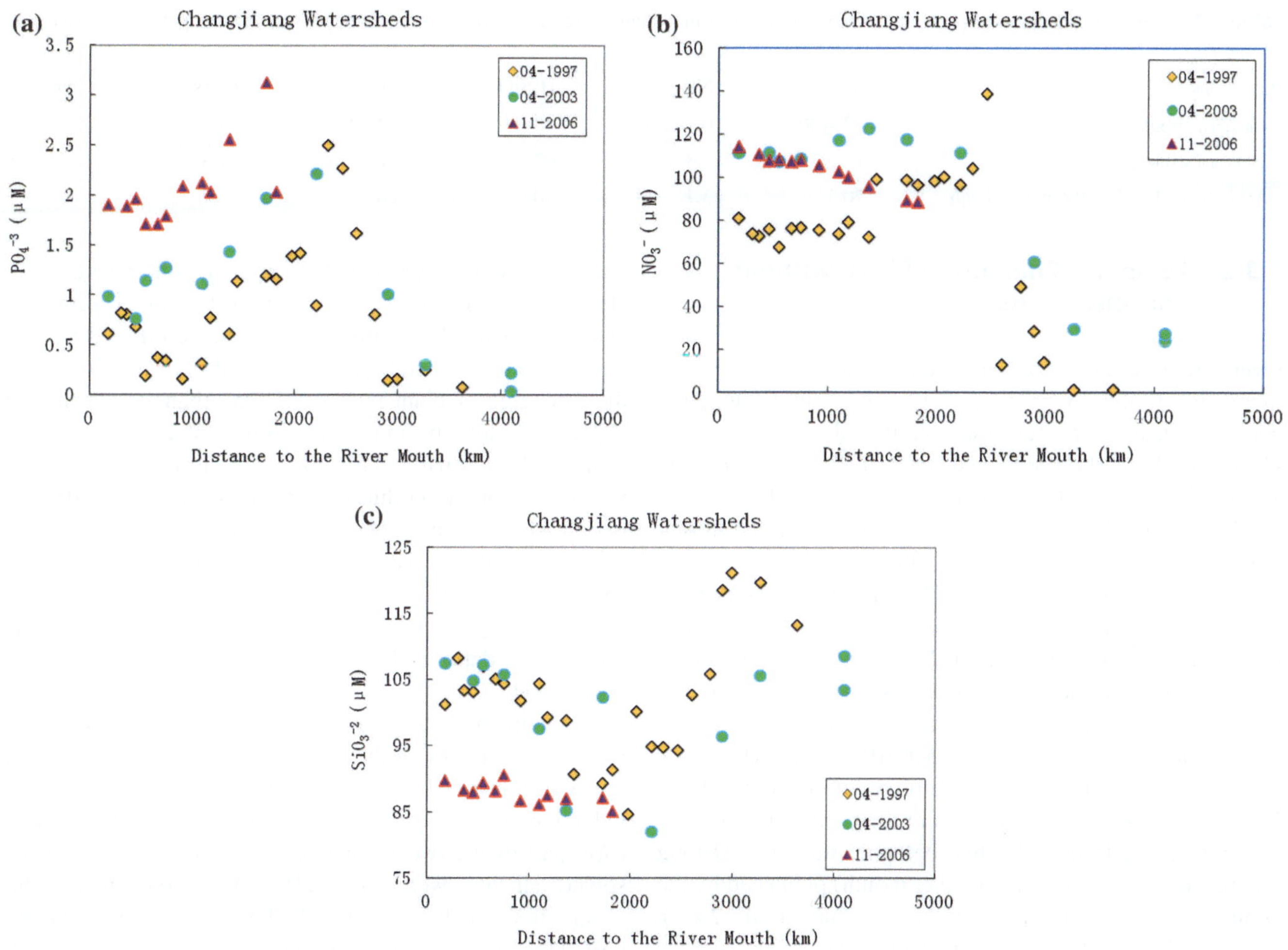

Fig. 5.2 Profiles of major plant nutrient species over a distance of ca. 4500 km along the main Changjiang channel from different expeditions, with **a** PO_4^{3-}, **b** NO_3^-, and **c** SiO_3^{2-}, showing the longitudinal distributions in April–May 1997, April–May 2003, and November 2006, respectively. Note that the TGD is located ca. 1850 km upstream from the river mouth with a 600-km-long reservoir situated in a deep valley

zone have higher SiO_3^{2-} than in temperate regions, because the weathering rate of silicates increases with higher temperatures and is affected by vegetation coverage.

In the present study, we find that the downstream levels of dissolved silicate are comparable between 1997 and 2003, whereas silicate concentrations in 2006 were up to 20 % lower, most likely because of the reduced water discharge in the dry season, and the limited contribution of water from tributaries in the southern watershed (Fig. 5.2).

It should be noted that the longitudinal profiles of major plant nutrient species described above are reflected in other field studies conducted during different water discharge conditions in the Changjiang (Shen et al. 2003; Shen 2006).

Plant nutrients in the Changjiang show considerable variation in speciation (i.e., chemical forms) from the upper to the lower reaches of the water course. For instance, data in Table 5.2 show that major tributaries in the watersheds differ in plant nutrient concentrations, and in their organic-to-inorganic molar ratios. Briefly, organic matter can be an important component of total dissolved nitrogen, and DOP can be higher than DIP when the concentration of phosphorus is low in the water column (Liu et al. 2003a). Tributaries in the highly populated northern parts of the watershed (e.g., Tuojiang) have elevated DIN and DIP concentrations compared with tributaries draining through the southern part of the watershed. However, high concentrations of organic forms are not observed in the lower reaches of the river and/or at the river mouth, presumably due to dilution and mixing in the main channel (Liu et al. 2003a).

Table 5.2 Chemical speciation of plant nutrients (Unit: μM) in the Changjiang based on the measurements in April–May 1997 (cf. Liu et al. 2003a)

Locations	NO_3^-	NH_4^+	H_2SiO_3	DIP	DON	DOP
Northern tributaries	13–400	0.7–171	48–114	0.1–19	4.8–163	0.1–3.9
Southern tributaries	14–97	0.8–11	62–125	0.02–6.1	7.8–68	0.1–0.8
Main channel and at the river mouth	70.3	3.7	102	0.83	10.9	0.56

5.3.2 Temporal Changes in Plant Nutrient Concentrations

Over the past 50 years, the concentration of DIN has increased in the lower section of the main Changjiang channel. Data show, for instance, that the DIN was ca. 40 μM in the early 1980s, and reached 90 μM in the late 1990s (Zhang et al. 1999). Nitrate levels in the 1960s were 15–20 μM, and then increased to 65 μM in the early 1980s at the river mouth (Gu et al. 1981, 1982).

Between 2000 and 2009, the DIN concentration at the river mouth of the Changjiang increased to 140 μM, which is higher than levels measured in the 1990s. The data show, however, that the increasing trend has slowed down and/or eventually will be phased out in the near future, with a recent stabilization of plant nutrient levels at 100–140 μM (Fig. 5.3).

Other studies have shown similar temporal trends in inorganic nitrogen concentrations in the Changjiang, mostly based on the data from the hydrographic station at Datong (ca. 650 km upstream from the river mouth) published by the Ministry of Hydraulics of China (cf. Yan et al. 2003). It should be noted, however, that until the 1990s riverine chemistry data in national reports and/or yearbooks were generated using different methods, and without strict QA/QC, and/or comparisons with international/national reference materials. For example, water samples were not properly filtered through 0.2-μm and/or 0.4-μm acid-cleaned filters, and sometimes, the sample was left decanted before chemical analysis. Furthermore, differences in the protocols for chemical analysis of plant nutrients used before the 1990s and those used today cause an unknown bias that is not addressed in most publications. Recent studies have shown that the deviation and waste drainage of water resources in the lower reaches (i.e., from Datong to the river mouth over a distance of ca. 650 km) of Changjiang could amount to 4×10^9 m^3 $month^{-1}$ to 5×10^9 m^3 $month^{-1}$ on average, which add another uncertainty to the estimation of seaward flux of chemical elements (cf. Zhang et al. 2012).

It has been suggested that the increase of nitrate and DIN in the lower reaches of the Changjiang between the 1960s and the 1990s could be related to the application of synthetic fertilizers in the watersheds, along with change in land-use patterns, such as deforestation and the conversion of land for agriculture. Indeed, there is a positive correlation between the amount of chemical fertilizer applied over the watershed and the DIN concentration at the river mouth (Zhang 2002). However, taking into consideration the rapid economic expansion and urbanization of China since the 1980s, other pathways such as atmospheric wet and dry depositions and the discharge of plant nutrients from urban settlements are expected to also be important sources of DIN in the river channel. This is particularly likely during the winter season when atmospheric pollution is high, and in the dry/drought period when the discharge of groundwater becomes a major component of the river flux. These pathways (i.e., atmospheric deposition and groundwater discharge) have not been properly evaluated in previous studies due to the scarcity of data, and hence, these processes are poorly understood.

In the main channel of the Changjiang, dissolved silicate was ca. 70–80 μM in the late 1950s, but increased to high levels in the mid-1960s (i.e., 150–155 μM), followed by a reduction of ca. 30 % in 1990s (cf. Zhang et al. 2006). In the early part of the twenty-first century, concentrations of dissolved silicate were 100–150 μM, based on monthly observations at the river mouth (Fig. 5.3). It has been suggested that the concentration of dissolved silicate in the main channel of the Changjiang has decreased over the past several decades due to intensive dam construction and the biological uptake of nutrients in reservoirs (cf. Li et al. 2007).

Approximately 50×10^3 large and small dams have been constructed in the Changjiang watershed since 1950, and the consequences of these engineering works are evident in data from the 1950s and 2000s regarding to sediment loads (see Chap. 4). The lower levels of dissolved silicates in the 1950s (ca. 75 μM) presumably give a baseline concentration for the watershed prior to the large-scale changes in land use (cf. Zhang et al. 2006). Higher levels of dissolved silicates in the Changjiang in the 1960s and 1970s correspond to deforestation and the expansion of agriculture in the watershed. During this period, dissolved silicate concentrations were positively correlated with the amount of TSM in the river (cf. Zhang et al. 2006). While a certain amount of dissolved silicate, as well as other plant nutrients, can be depleted because of dam construction and the uptake by phytoplankton in reservoirs, the loss of dissolved silicate can be mitigated by the contribution of large tributaries with high concentrations of dissolved silicate (e.g., Hanjiang, Xiangjiang–Dongtinghu, and Ganjiang–Poyanghu) in the

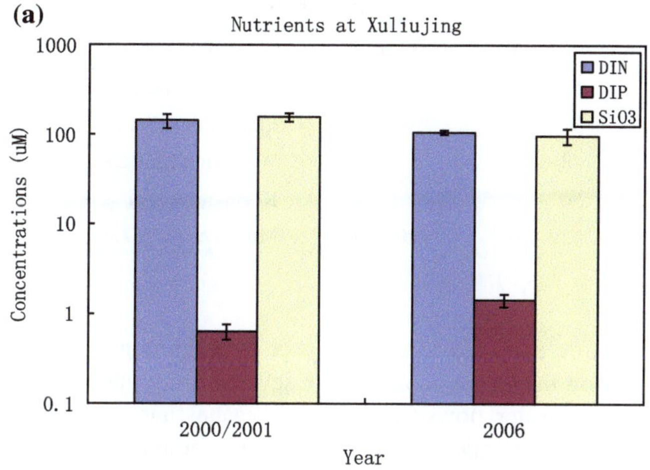

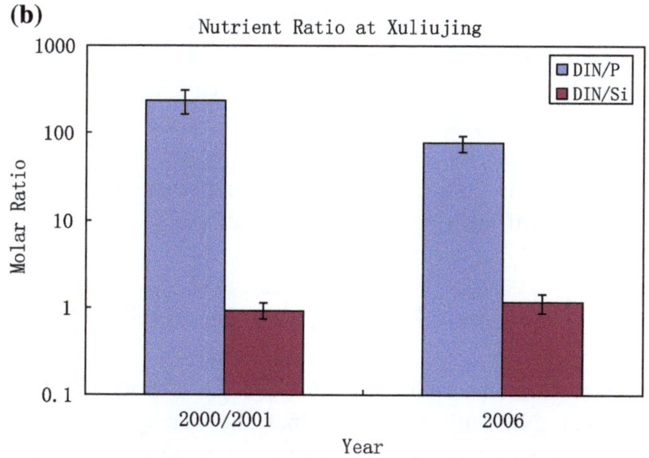

Fig. 5.3 Change in **a** plant nutrient concentrations, and **b** plant nutrient molar ratios in the lower reaches of the Changjiang based on monthly sample collection and analysis at Xuliujing Station, which is ca. 150 km upstream from the river mouth

lower reaches of the Changjiang (note: here "hu" means lake in Chinese) (Figs. 5.1 and 5.2). Therefore, the whole data sets of the past 60 years (i.e., 1950s–2010s) do not show the depletion of dissolved silicate relative to other plant nutrients in the Changjiang, while the so-called silicate limitation that is attributed in the literature to dam construction can be misleading and/or result of misinterpretations of data.

Phosphate data from Chinese rivers are rather limited relative to other plant nutrients (e.g., nitrogen and dissolved silicate). Early observations in the 1980s showed that dissolved inorganic phosphorus (DIP) or phosphate (PO_4^{3-}) in the lower reaches of the Changjiang was 0.5–1.0 μM, higher than that in the Huanghe (Yellow River), but lower than that in the Zhujiang (Pearl River) (Zhang 2002). Moreover, the observed DIN/DIP molar ratio at the river mouth of the Changjiang has tended to increase since the early 1980s. The DIN/DIP ratio in the water column was ca. 80 in 1980, but reached to 120 by the end of the twentieth century. This reflects the different characters of phosphate in the Changjiang compared with the rapid increase in nitrogen that began in the 1980s and continued to the beginning of this century (Zhang 2002).

Compared to inorganic nitrogen, phosphorus is less soluble in natural waters and combines with calcium, aluminum, and iron, which tend to be in particulate form. For this reason, phosphorus can be retained in watersheds during weathering. Inorganic nitrogen, by contrast, is readily leached out of soil profiles because it does not readily form its own minerals in nature (i.e., convert to mineral form) and because nitrogen compounds are highly soluble in the aquatic environment.

5.3.3 Concentrations of Trace Elements in the Changjiang

With regard to trace species in the water column, the range of dissolved Al is 0.1–1.5 μM in the northern tributaries and 0.1–1.8 μM in the southern tributaries, with concentrations of 0.3–1.2 μM in the main channel of the Changjiang at a distance of ca. 4000–4500 km upstream of the river mouth (Table 5.3). Overall, dissolved Al in the Changjiang has a stable longitudinal profile from the upper part of the water course to the river mouth, with 65 % of samples in the range of 0.4–0.8 μM and a mean value of 0.7 μM. These values are comparable with river systems in undeveloped regions of the world (Zhang et al. 2003). Moreover, based on data from monthly samples taken at the river mouth, there are no simple relationships between dissolved Al and water flow and/or TSM levels.

In the Changjiang, the dissolved-phase Al/Si molar ratio increases with higher $1/w$ values (w = Si_d/Si_t, where the subscript d represents dissolved and t stands for total

Table 5.3 Characteristics of dissolved Al, As, and Se in the Changjiang watersheds from tributaries and the main channel

Locations	Al (μM)	As (nM)	Se (nM)	References
Northern tributaries	0.1–1.5	7–68	1.5–3.3	Zhang et al. (2003) and Yao et al. (2007)
Southern tributaries	0.1–1.8	10–57	1.5–9.9	Zhang et al. (2003) and Yao et al. (2007)
Main channel	0.3–1.2	12–65	1.2–5.9	Zhang et al. (2003) and Yao et al. (2007)

Table 5.4 Dissolved trace element data reported in the literature for the lower reaches of the Changjiang after 1980

Period	Fe (nM)	Mn (nM)	Cu (nM)	Ni (nM)	Cd (pM)	Pb (pM)	References
1981	325–672	10–24	18–21	2.8–3.1	14–25		Edmond et al. (1985)
1986			20–35		31–54	237–271	Elbaz–Poulichet et al. (1990)
1998	109–642		21–25		28–61		Wang and Liu (2003)
2001		8.6	19	3.5	50		Koshikawa et al. (2007)
2006	127–165		26–41	5.6–14.4	129–249	290–531	Müller et al. (2008)

concentrations of silicate). This shows that there is lower Al relative to Si in the water column when the contribution of dissolved (*d*) to total (*t*) silicate is high, and hence, Al and Si are separated during weathering processes in the watershed (Zhang et al. 2003).

Dissolved arsenic and selenium in the Changjiang have considerable variability in the main channel and tend to increase from the upper reaches to the river mouth over a distance of ca. 4000–4500 km (Table 5.3). In 1997, dissolved inorganic arsenic (DIAs) was 21–65 nM, and dissolved inorganic selenium (DISe) was 1.3–3.1 nM. In 2003, DIAs in the main Changjiang channel was 12–36 nM, with an As (III)-to-DIAs ratio of 0.2–0.4, and DISe was 1.2–5.9 nM along the river course (Yao et al. 2007). Also, in 2003, DIAs was 10–68 nM and DISe was 1.5–10 nM in the major tributaries in the Changjiang watershed.

Measurements of other dissolved trace elements from the Changjiang are rather limited. In Table 5.4, we compare the available information for trace elements from the river mouth and/or the lower reaches of the Changjiang over a period of 30–40 years (i.e., 1980–2010). From Table 5.4, it appears that, although dissolved trace element data from the Changjiang show considerable variability owing to different sampling times and locations, as well as different methods for sample collection and analysis, concentrations are comparable over the 1981–2001 period. By contrast, in 2006, the values of dissolved trace elements (e.g., Cd, Ni, and Pb) were 2–4 times higher than concentrations reported in earlier studies.

Based on data collected in November 2006, Müller et al. (2008) found that elements such as dissolved As, Cu, Ni, and Se tended to increase by up to a factor of two over a distance of ca. 1500 km in the lower reaches of the Changjiang. The authors also illustrated that the 2006 concentrations of particulate trace elements from the main channel were higher than those of the 1980s and 1990s, presumably due to the enormous increase in anthropogenic pollutants discharged into the river. It has been noted that November 2006 was an extremely dry period, and the water discharge at the hydrographic station of Datong was 11.1×10^3 m^3 s^{-1} (Müller et al. 2008), which is only 40–50 % of the average long-term monthly water flow (ca. 20×10^3 m^3 s^{-1}). It is therefore expected that the impact of anthropogenic pollutants on the Changjiang is more severe during the period of reduced water flows in the lower reaches and at the river mouth, where higher trace element concentrations are found in the water column.

Moreover, a comparison with trace element concentrations in TSM enables estimation of the distribution coefficient (K_d) in the Changjiang. On a log scale, K_d is estimated to be 5.17 for Cd, 4.42 for Cu, and 5.89 for Pb, which is similar to that for other large rivers in the world (Zhang 1995), but the difference has been found when compared to the results of laboratory simulations (cf. Sect. 5.2.2).

5.4 Results of Laboratory Simulations

The adsorption kinetic curves of trace elements (e.g., Cd, Co, Cu, Ni, and Zn) show that the adsorption density increases asymptotically and then reaches stable values (i.e., equilibrium) in 5 h for Cu; 20 h for Zn, Co, and Ni; and up to 40 h for Cd (Zhang et al. 2008b). In the experiments, the adsorption edges for trace elements onto SPM illustrates the sigmoid nature and that the *E* (%) increased considerably with higher pH values, in the range of pH = 6–9, but decreased as the concentration of SPM increased (Zhang 2003). Moreover, the simulation data showed that the adsorption of trace elements onto SPM was considerably reduced as salinity increased (Fig. 5.4). It has been observed that for a given percentage of adsorption (e.g., *E* = 50 %), the equilibrium pH increased considerably with higher salinities (e.g., ΔpH = 0.5–1.0 for a salinity range of 0–25).

The Kurbatov adsorption model can be used to diagnose the characteristics of the exchange of chemical species between aqueous solution and active groups of SPM, and the change in adsorption density can be simulated by the Langmuir equation (Zhang et al. 2008b). The adsorption of trace elements increases sharply from 10–20 % to 80–90 % within a rather narrow pH range of 6.5–8.5. Because the pH value of the Changjiang is ca. 8.0–8.1, trace elements tend to be combined onto SPM in aquatic solutions. An increase in SPM concentration also drives the adsorption of trace elements, but the adsorption density (i.e., mass of the

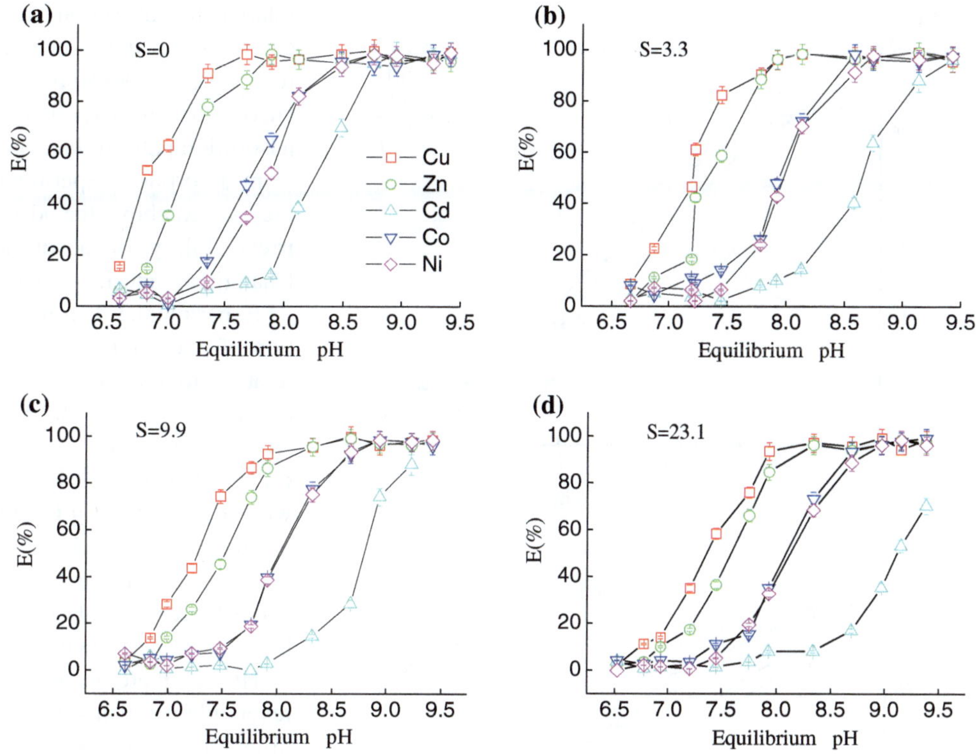

Fig. 5.4 Simulation of adsorption edges (i.e., E % as in Eq. 5.1) for trace elements (i.e., Cd, Co, Cu, Ni, and Zn) onto SPM in the Changjiang Estuary at different salinities (S) and pH for **a** $S = 0$; **b** $S = 3.3$; **c** $S = 9.9$; and **d** $S = 23.1$. In the experiment, SPM = 300 ± 9 mg l^{-1}, and the total (*T*) concentration of the simulation system is TCu = 1.47 mg l^{-1}, TZn = 1.48 mg l^{-1}, TCd = 1.51 mg l^{-1}, TCo = 1.42 mg l^{-1}, and TNi = 1.53 mg l^{-1}, respectively. Data are from Zhang (2003)

adsorbate per unit of SPM) decreases with higher SPM loadings in the river.

Our experimental data reveal that the K_d on a log scale increases with higher pH levels, but is relatively independent of the amount of SPM in the range of 0.1–1.5 g l^{-1}. When compared with other studies, it was found that K_d values in the world's large rivers are relatively stable, such as the Huanghe, Mississippi River, and St. Lawrence River, indicating again that concentrations of trace elements are regulated by the level of SPM in the aquatic environment (Zhang et al. 1994; Zhang 2002).

It can be summarized that the ability of Cu, Zn, Cd, Co, and Ni to adhere to suspended particles is considerably reduced as salinity increases, and the extent of adsorption is smaller in seawater compared to freshwater environments. Hence, terrestrial trace elements combined with SPM tend to be remobilized in the marine environment. The equilibrium pH value also increases with higher salinities, which leads to the higher bioavailability of pollutants (e.g., trace elements) in seawater.

With regard to the plant nutrient species, phosphate adsorption kinetics indicate that equilibrium can be reached within ca. 15 h, which is comparable to the flood-ebb tide cycle in the Changjiang Estuary. Similar to the trace elements, the adsorption behavior of phosphorus in the Changjiang Estuary can be described using Langmuir isotherm algorithms (Zhang 2007). The adsorption edges for phosphorus onto the SPM have an upended sigmoid character, and the *E* (%) of phosphorus decreases with higher pH (Fig. 5.5).

Moreover, the adsorption of phosphorus in the Changjiang Estuary is inversely related to phosphate concentrations in water samples, but is directly related to suspended matter loadings, and hence increases with higher SPM concentrations (Fig. 5.5). Salinity in the Changjiang Estuary has a dramatic effect on the partitioning of phosphorus between the solid phase and solution. For instance, unlike trace elements at pH = 8.0 and SPM level of 1 g l^{-1}, the adsorption percentage can be increased by 50 % from freshwater to samples with a salinity of 30; the adsorption percentage can be 100 % higher when the SPM loading is 10 g l^{-1} (Fig. 5.5).

In this study, the adsorption of phosphorus is endothermic and, for a given amount of SPM, the percentage adsorption at pH = 8.0 increases by twofold when temperature increases from 280 to 303 °K (Zhang 2007). Furthermore, an increase in concentration of dissolved organic matter (DOM) in samples from the Changjiang results in a dramatically higher proportion of phosphorus being attached onto the SPM, indicating that the discharge of organic pollutants may considerably modify the partitioning of phosphate between solid and liquid phases, and hence its availability to the biota (e.g., photosynthesis). The percentage adsorption of phosphorus also increased with higher levels of dissolved oxygen

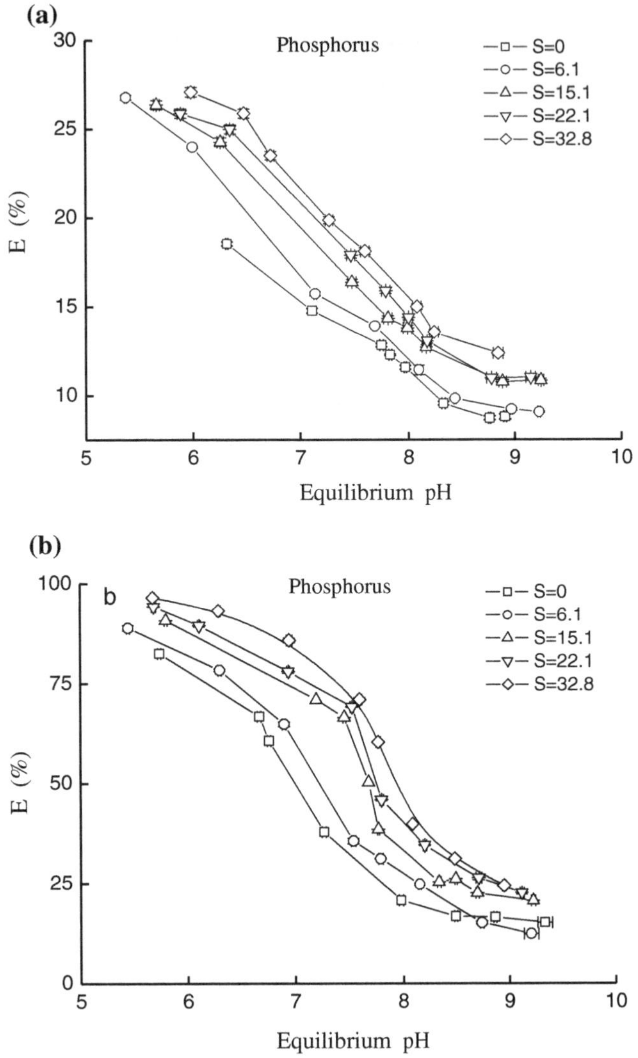

Fig. 5.5 Simulation of adsorption edges for phosphorus onto SPM in the Changjiang Estuary at different salinities for: **a** SPM = 1 g l^{-1}, and **b** SPM = 10 g l^{-1}, respectively. Data are from Zhang (2007)

(DO) in solution. At pH = 8.0, the observed *E* (%) can be doubled when DO increases from 1.4 to 8.3 mg l^{-1} (cf. Zhang 2007). The implication of this result is that adsorbed phosphorus and/or phosphorus combined with particulate matter can be regenerated when hypoxic conditions occur in coastal waters, which usually happens off the river mouth of the Changjiang in summer.

It should be noted, however, that the partitioning of chemical species (e.g., trace elements and plant nutrients) between solution and solid phases (i.e., SPM) in the aquatic environment may not necessarily be reversible (Han et al. 2013). Figure 5.6 shows a Langmuir sorption isotherm curve of phosphorus and selected trace elements for SPM in the Changjiang Estuary. The *y*-axis of Fig. 5.6 shows the degree of saturation (i.e., the percentage of elements bound to SPM relative to adsorption saturation) of phosphorus and trace elements. If the processes of adsorption–desorption are reversible, the desorption will follow the adsorption isotherm curve; that is, experimental data will be placed on the adsorption path. In fact, as can be seen from Fig. 5.6, results from desorption experiments deviate positively from and/or are located above the adsorption curve, which illustrates the irreversibility of adsorption–desorption processes in the Changjiang and its estuary.

Experimental data in Fig. 5.6 also show that the desorption of phosphorus is 3.18 % (i.e., point "*B*" in Fig. 5.6b), relative to the adsorption saturation at the "equilibrium" concentration, C_1 = 0.055 mg l^{-1} in solution. In Fig. 5.6b, "*A*" is in the adsorption equilibrium state when C_0 = 0.285 mg l^{-1}. If the adsorption process is reversible, when C_1 = 0.055 mg l^{-1}, the desorption of phosphorus could again be described by the Langmuir sorption isotherm equation, but in the reverse direction. The new equilibrium could be reached at "*C*" with the desorption of phosphorus by 7.66 %, as shown in Fig. 5.6b.

Hence, there is a deviation between the "ideal" (i.e., reversible mechanism) and actual desorption processes, and difference (Δ) in desorption relative to adsorption saturation for phosphorus is 4.88 % (i.e., "*B*–*C*" in Fig. 5.6b); that is, 4.88 % of phosphorus still remains attached onto SPM relative to the initial situation at "*A*" (Fig. 5.6b).

Experimental results show that desorption relative to the adsorption saturation for Cu, Zn, Cd, Co, and Ni are 0.63, 0.72, 1.53, 3.68, and 1.15 %, respectively, at the equilibrium concentration, C_1 = 0.1 mg l^{-1} in solution (Fig. 5.6c). Again, if the adsorption–desorption processes were reversible for trace elements, when C_1 = 0.1 mg l^{-1}, the observed desorption relative to the adsorption saturation for these heavy metals would follow the Langmuir isotherm equation, and the theoretical desorption would be 4.19, 4.64, 7.64, 9.36, and 12.67 % for Cu, Zn, Cd, Co, and Ni, respectively. Therefore, there is also an important deviation between the "ideal" (i.e., reversible mechanism) and actual desorption processes for trace elements. The difference (Δ) in desorption relative to the adsorption saturation for Cu, Zn, Cd, Co, and Ni is 3.56, 3.92, 6.11, 5.67, and 11.52 %, respectively (Fig. 5.6).

The observed irreversibility of the partitioning of plant nutrients and trace elements between liquid and solid phases (e.g., SPM) in this study can help explain a number of characteristics of element distributions in the Changjiang and its estuary reported in previous studies. For instance, although pollution of the aquatic environment by plant nutrients and trace elements has been reported in the Changjiang watershed, the observed concentrations of trace metals are still relatively low compared with other large rivers in Europe and North America, where the water has lower TSM concentrations. This is in part related to the high affinity of elements with SPM and the irreversibility of element partitioning

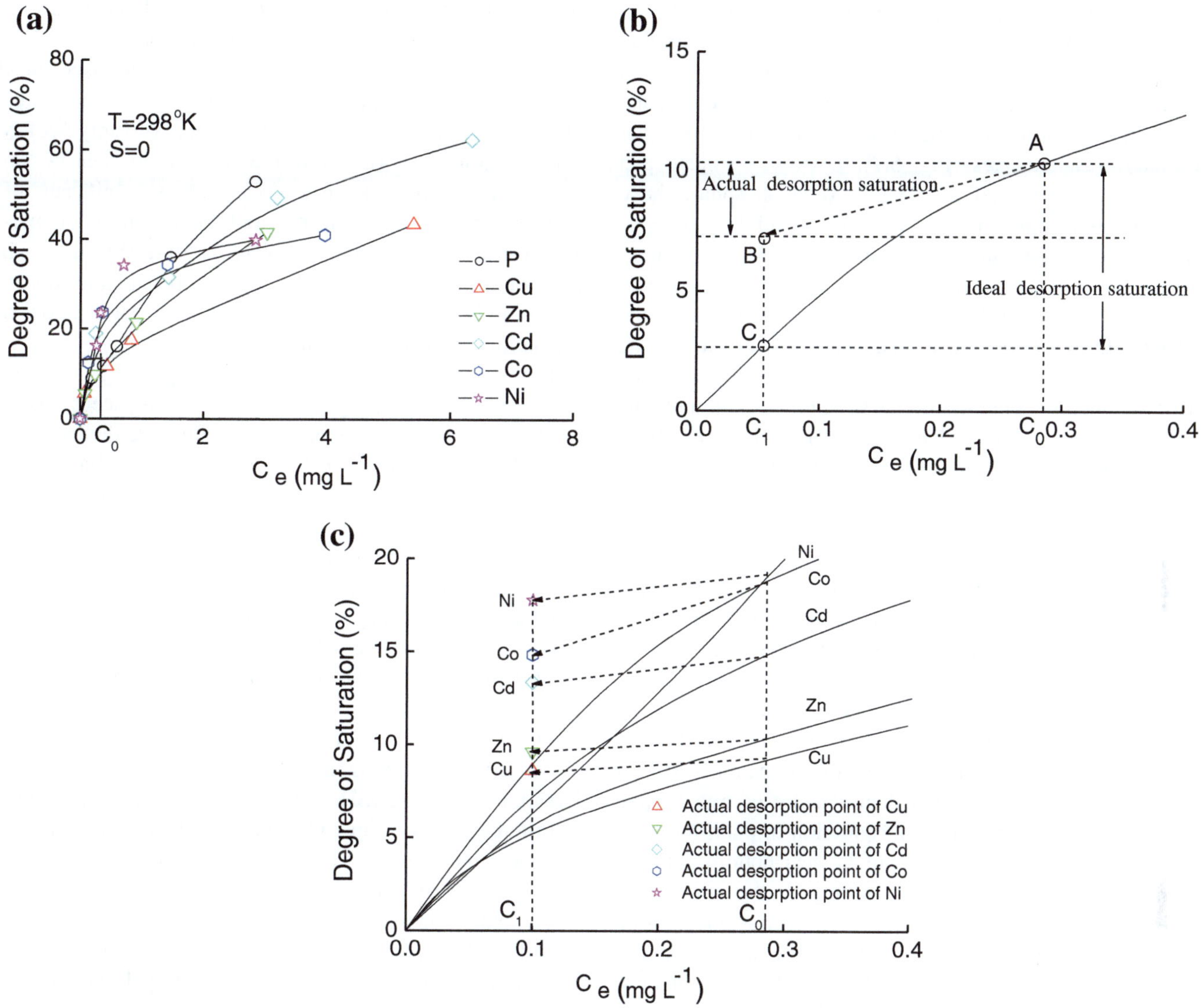

Fig. 5.6 Irreversible partitioning model of chemical elements between solution and solid phases in the aquatic environment, with **a** Langmuir adsorption isotherms of phosphate and trace elements (i.e., Cu, Zn, Cd, Co, and Ni) onto the SPM of the Changjiang Estuary ($T = 298$ °K and $S = 0$); **b** an enlargement to show the case of phosphorus; and **c** an enlargement to show trace elements (Cu, Zn, Cd, Co, and Ni). In the figure, "*A*" is the thermodynamic equilibrium state of adsorption–desorption when the $C_e = C_0$, "*B*" represents the kinetic steady state of the desorption process, that is, the actual desorption point when $C_e = C_1$ starting from "*A*", and "*C*" represents the thermodynamic equilibrium state of desorption, that is, the "ideal" desorption situation when $C_e = C_1$ from "*A*" following the Langmuir isotherms

between solid phases and solution (cf. Müller et al. 2008). The high capacity of SPM to combine with solutes and the elevated turbidity of the Changjiang likely play an important role in regulating trace element levels in the aquatic environment, which can induce the scavenging of chemical species onto the TSM and transport downstream with the solid phases. Moreover, the observed irreversibility of element partitioning between solid phases and solution can result in the lack of significant regeneration in the early stages of mixing between fresh and marine waters for some of the trace metals (e.g., Cd, Cu, and Ni) in the Changjiang Estuary, seen in previous studies (Edmond et al. 1985; Elbaz-Poulichet et al. 1990; Koshikawa et al. 2007).

5.5 Modification of Chemical Species in the Changjiang Estuary and Adjacent Coastal Environment

In the Changjiang Estuary, brackish water dispersal plumes, with salinities <25–30, have a high seasonal variability, regulated by riverine discharge and monsoon-induced coastal circulation (see Chap. 2). Plant nutrients (i.e., NO_3^-, NO_2^-, NH_4^+, PO_4^{3-}, and SiO_3^{2-}) demonstrate concentration gradients due to mixing between fresh and marine waters in the Changjiang Estuary. For nitrate and dissolved silicate, gradients are particularly high due to the significant input from terrestrial sources (Fig. 5.7). Dissolved silicate and inorganic nitrogen, and to a lesser extent phosphate, have linear relationships with salinity in the mixing zone of Changjiang Estuary until a salinity of ca. 30. Laboratory simulations using bulk river samples and filtered seawater produced similar results (Zhang 1996). Furthermore, dissolved silicate, nitrate, and phosphate in surface waters decrease with distance from the river mouth, following the dispersal of effluent plumes off the Changjiang, and nutrients such as nitrate can be exhausted beyond the plume front (Zhang 2002). In contrast, ammonium in surface waters can be higher than riverine concentrations in the area beyond the effluent plumes.

Because of the estuarine environment and the buoyancy effect, which is related to gradients in hydrographic properties (e.g., temperature and salinity), samples collected from near-bottom waters have different characteristics off the Changjiang. For example, although plant nutrient samples from different observational cruises have various freshwater end-member compositions, the difference in near-bottom concentrations between the river mouth and the mid-shelf off the plume region can be less than a factor of two (Zhang et al. 2007a). Moreover, the concentration of plant nutrients such as ammonia and phosphate in near-bottom samples from offshore regions can be even higher than in surface waters due to the regeneration of chemical species, along with heterotrophic degradation of organic matter, and/or exchange at the sediment–water interface. Shelf water masses in the deep part of the water column can also be enriched in plant nutrients and contribute to coastal areas because of dynamic mixing between the Changjiang Diluted Water (i.e., effluent plumes from the Changjiang), Shelf Mixed Water (SMW), Taiwan Strait Warm Water (TSWW), as well as water incursions from Kuroshio to the East China Sea Shelf, known as the Kuroshio Surface Water (KSW) and Kuroshio Subsurface Water (KSSW) (Zhang et al. 2007a).

Observations from mooring stations in the lower estuary and adjacent marine areas off the Changjiang show that near-bottom offshore water with salinities ≥30 can reach the coast at water depths of ca. 20–30 m. In the coastal environment, the Changjiang effluent plumes cap the upper 0–5 m of the water column with salinities of 20–25. Mid-depth and near-bottom waters have salinities of 25–30 or higher (Zhang et al. 2007a).

The oscillation of salinity in surface waters at coastal monitoring sites is likely due to the tidal regulation of effluent plumes off the river mouth of the Changjiang (see Chap. 2). The water column in the adjacent coasts off the river mouth of the Changjiang can be well mixed during spring tides, but tends to be stratified in neap tides.

Further offshore and in areas affected by the dispersal of Changjiang effluent plumes and in the broad shelf area, nitrate and dissolved silicate show an approximately twofold concentration decrease from the surface to near-bottom waters; phosphate and nitrite do not show significant differences between the surface and near-bottom samples. However, the concentration of NH_4^+ is 50–100 % higher in near-bottom waters than at the surface, presumably due to heterotrophic processes that modulate the degradation of organic matter (e.g., ammonification).

The limited published information regarding trace elements in the Changjiang Estuary and the adjacent coastal environment (Edmond et al. 1985; Elbaz-Poulichet et al. 1990; Abe et al. 2003; Koshikawa et al. 2007; Jiann et al. 2009; Jiann and Wen 2012) is summarized as follows:

- Dissolved Fe is rapidly removed from the water column at salinities of 5–10, most likely because of the scavenging effect (e.g., flocculation). Further down the Changjiang Estuary, the concentration of dissolved Fe decreases slightly to 100 nM or less. The concentration of dissolved Fe in the shelf waters of the East China Sea is 1–2 nM.
- Dissolved Cu and Ni concentrations are dominated by conservative mixing in the Changjiang Estuary up to salinities of ca. 30. However, these two elements have a different relationship with salinity in the estuarine mixing zone: Cu decreases in the water column by a factor of 4–5 at a salinity of 30 relative to riverine concentrations, while Ni tends to increase by twofold down the estuary.
- In the Changjiang Estuary, dissolved Cd increases in the estuarine mixing zone and reaches a broad maximum in the salinity range of 15–25, presumably owing to desorption from the particulate phase and the formation of complexes with ligands (e.g., Cl^-). Further offshore, dissolved Cd drops to 0.05–0.10 nM in the surface waters of the East China Sea.

The available dissolved Al data in the Changjiang Estuary and adjacent coastal waters show that it ranges widely in concentration, with values up to 200 nM at very low salinities, reflecting the variability of freshwater end-member compositions. In high salinity samples, such as oceanic surface waters (i.e., Kuroshio), Al is ca. 20–30 nM, which is comparable with other coastal environments in the world (Ren et al. 2006). There is no strong evidence from field and laboratory studies that dissolved Al concentrations either increase through regeneration or decrease via removal with salinity gradients in the mixing zone between freshwater and seawater end members in the Changjiang Estuary (Ren et al. 2006). Moreover, in the East China Sea, vertical profiles of dissolved Al show surface enrichment, and then, concentrations decrease with water depth, compared with plant

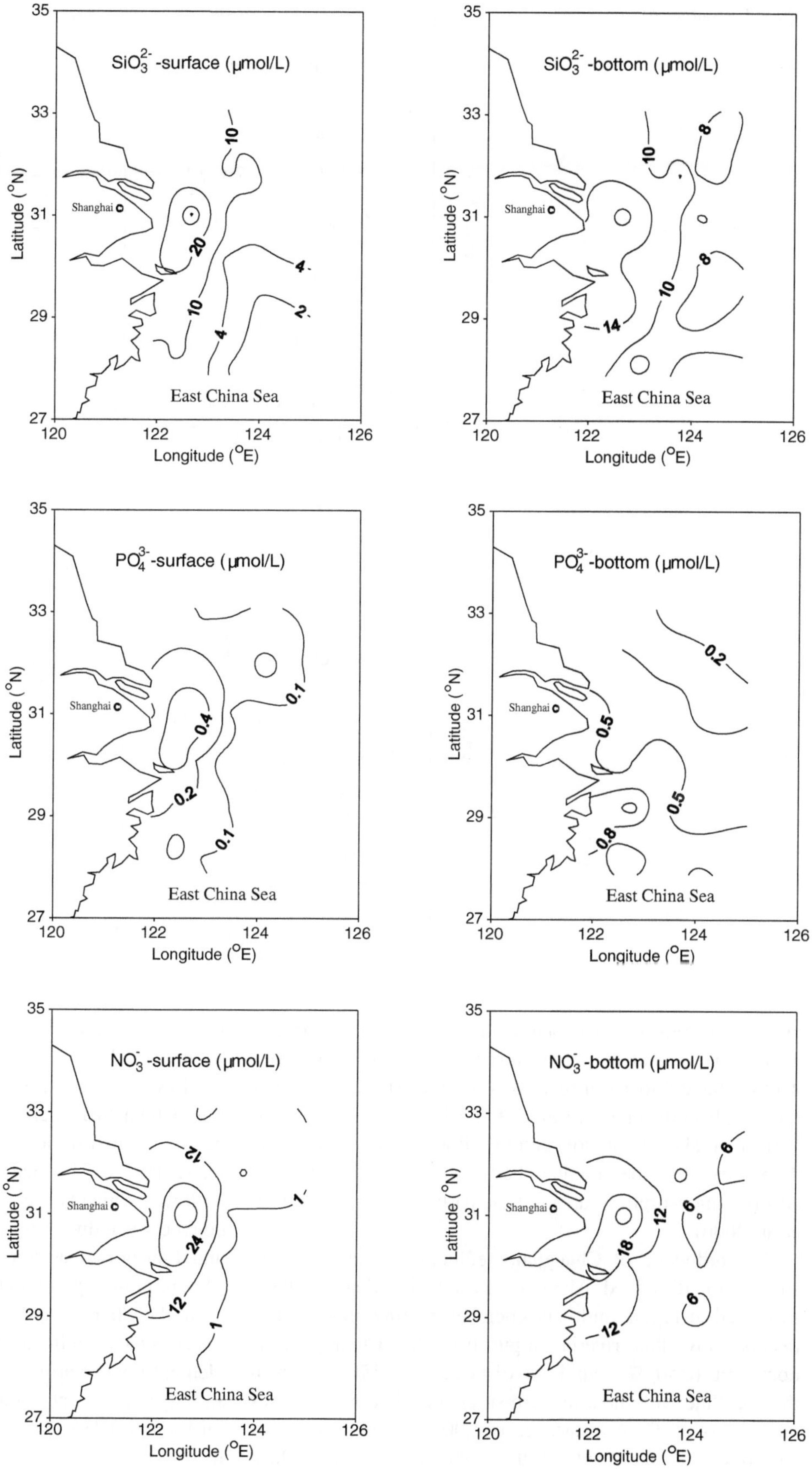

Fig. 5.7 Distribution of plant nutrient species in the Changjiang Estuary and adjacent coastal waters, showing the dispersal of terrestrial materials in marine recipients and the potential impact on the ecosystem

Fig. 5.7 (continued)

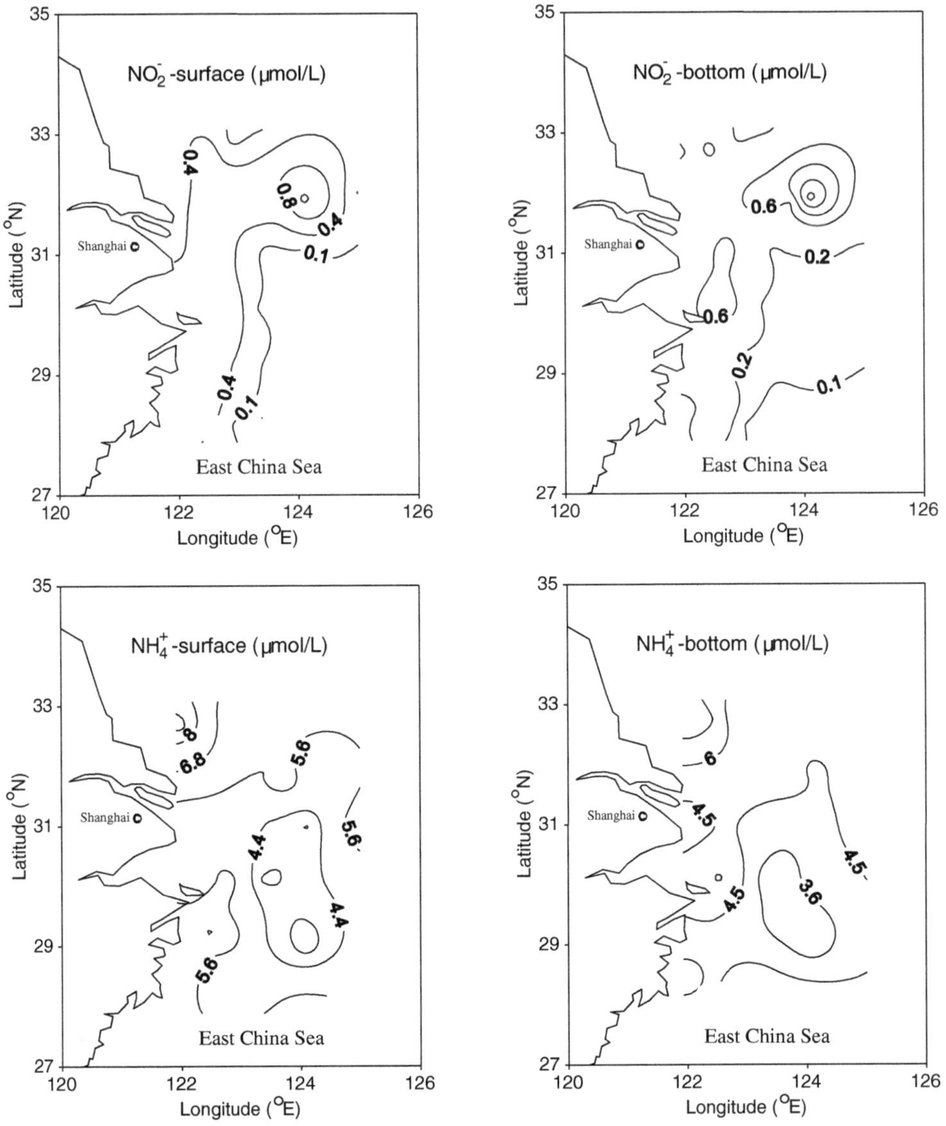

nutrient profiles (e.g., dissolved silicate) that show depletion at the surface and then concentrations increase with water depth. This comparison of vertical profiles shows that Al is not significantly involved in biological processes (e.g., biological uptake by phytoplankton). Rather, dissolved Al is dominated by the mixing between water masses and adsorption/desorption at solution–particle interfaces (Ren et al. 2006).

The influence of Changjiang effluent (i.e., a terrestrial source) on dissolved Al in the East China Sea has been evaluated using a multi-end-member mixing model. The results show that riverine input from the Changjiang is dominant (≥50 %) within a distance of ≤150 km from the coastline, and then, the terrestrial contribution decreases quickly to <5 % at a distance of 200–250 km from coast and beyond. Dissolved Al then remains low in surface waters until the shelf break area of the East China Sea (Ren et al. 2006).

However, terrestrial contributions to the deep water of the East China Sea Shelf as well as the shelf break have also been identified through measurements of particulate organic matter (POC). This material increases in concentration when bottom sediments are resuspended during periods of deep convection induced by northerly monsoon winds in winter and transported across the shelf under the mechanism of Ekman pumping (Zhu et al. 2006).

In the Changjiang Estuary, dissolved As and Se illustrate the dominance of linear dilution of the riverine influx by trace-element-poor marine waters. Although concentration and salinity data may be variable between different cruises, typical DIAs and DISe concentrations are 25–40 and 3–5 nM, respectively, for the freshwater end member and 10–20 nM and ca. 1–2 nM, respectively, for open-shelf seawater (cf. Yao 2005).

Further offshore, profiles of dissolved arsenic in the water column of the East China Sea show that it is regulated by a combination of metabolism and changes in boundary influxes (Ren et al. 2010). Briefly, the influxes from Kuroshio waters and land sources (e.g., the Changjiang) have a high As (V)/As (III) ratio relative to the shelf waters (i.e., SMW), although As concentrations from the Changjiang (20–40 nM) can be a factor of two higher than KSW (i.e., 10–15 nM) and KSSW (15–20 nM) (Yao et al. 2007; Ren et al. 2010) (Fig. 5.8). For instance, data from 1996 to 2003 show that total dissolved inorganic arsenic [TDIAs = As (III) + As (V)] concentration in the lower part of the Changjiang is 18.6 ± 7.2 nM, and the arsenite concentration is 1.26 ± 0.8 nM.

Cruise data give total average dissolved arsenic concentrations of 14.4 nM for KSW and 18.0 nM for KSSW, with As (III) being highly depleted relative to As (V), with the ratio of arsenite to total dissolved inorganic arsenic being ≤0.05–0.1. Furthermore, surface waters in the shelf area of the East China Sea can have As (III)-to-TDIAs ratios that are twofold–fourfold higher than those of near-bottom samples (cf. Ren et al. 2010). In the East China Sea, the ratio of As (III) to TDIAs in water samples has a positive correlation with chlorophyll a (γ^2 = 0.5–0.8). Given that land sources

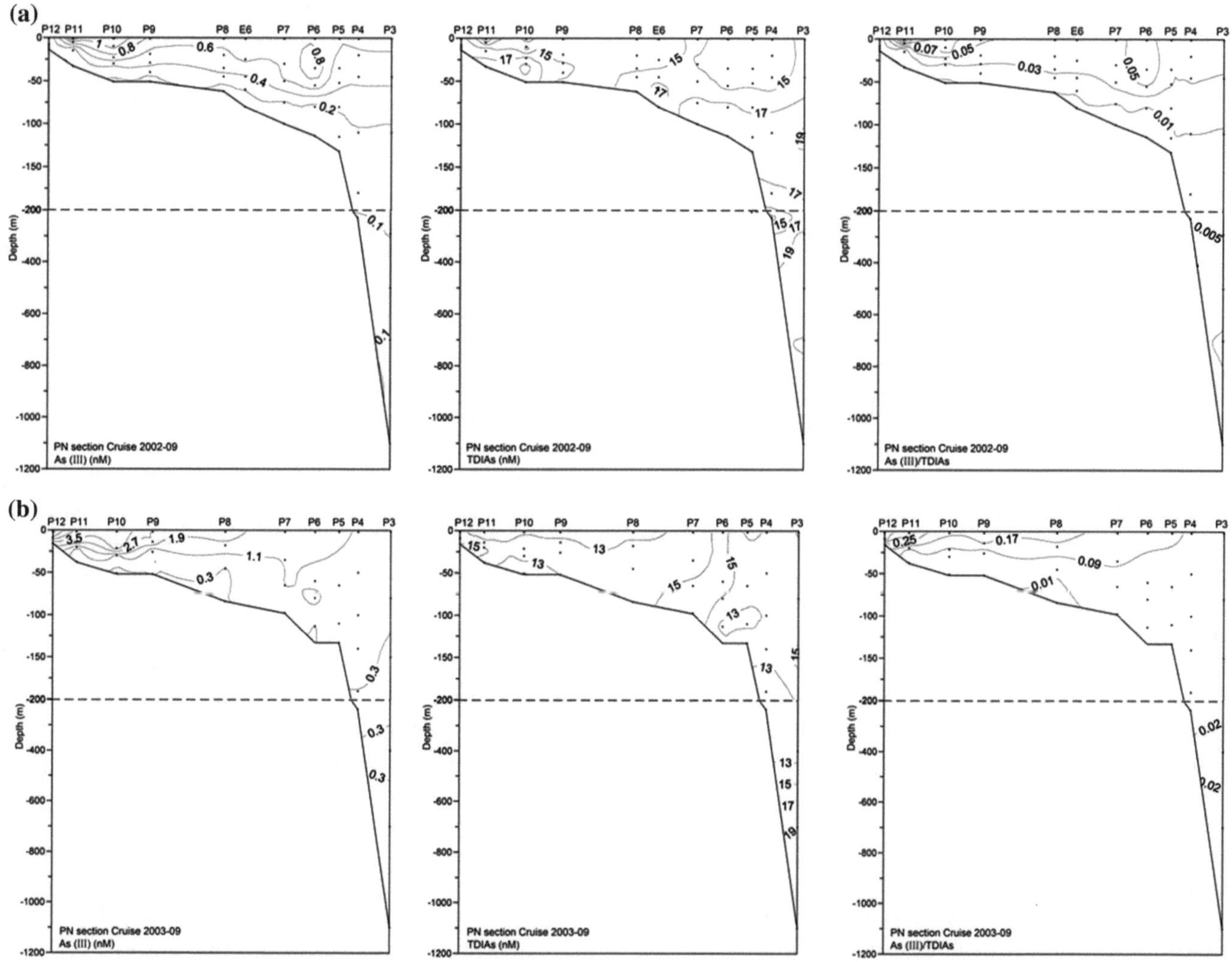

Fig. 5.8 Profiles of concentrations (nM) for As (III), total dissolved inorganic arsenic (i.e., TDIAs), and As (III)/TDIAs across the East China Sea Shelf from the river mouth of the Changjiang to the Okinawa Trough in **a** September 2002 and **b** September 2003. Data are from Ren et al. (2010)

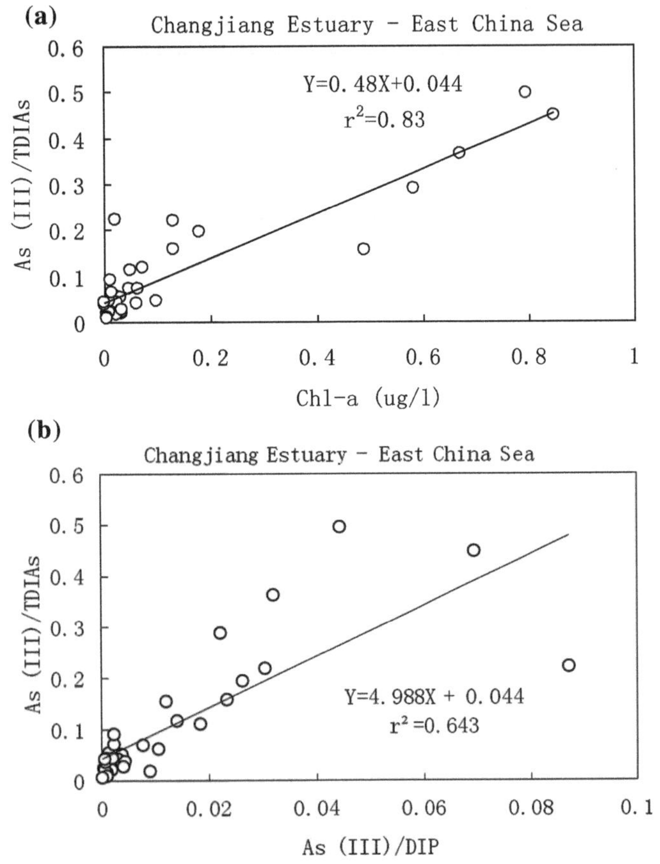

Fig. 5.9 As (III)/TDIAs plotted against **a** chlorophyll a (μg l^{-1}) and **b** phosphate (DIP) concentrations (μM) in a reciprocal mode (i.e., As (III)/[PO_4^{3-}]) in the East China Sea. Data are from Ren et al. (2010)

and/or open-boundary samples (e.g., Kuroshio waters) have very low levels of As (III) in TDIAs, it is likely that As (V) is transformed to As (III) by synthetic metabolic processes (e.g., photosynthesis) in the photic zone (Fig. 5.9). Also, it has been found that the proportion of As (III) in the TDIAs increases with depletion of DIP in the water column, thereby raising the possibility that the reduction of As (V) to As (III) is related to the biogeochemical cycle of plant nutrients.

Indeed, the ratio of As (III) to TDIAs increases exponentially when the concentration of PO_4^{3-} is smaller than 0.1–0.2 μM in the water column, with an apparent As-to-P stoichiometric molar ratio of 2.0×10^{-3} to 2.5×10^{-3}, which is similar to values from other coastal waters (cf. Ren et al. 2010). The implication is that, in the case of phosphorus depletion (i.e., phosphate-limited waters) relative to other plant nutrients (e.g., nitrogen), arsenic can be an alternative candidate in the processes of synthetic metabolism of phytoplankton, with As (V) being transformed to As (III) in the surface waters of the East China Sea. In this case, a positive and linear relationship between As (III)/TDIAs and As (III)/DIP is found in the East China Sea (Fig. 5.9).

5.6 Biogeochemical Dynamics of Chemical Elements in the East China Sea

5.6.1 Anthropogenic Impacts on Coastal Ecosystems from Land-Based Activities

The input of plant nutrients (e.g., N and P) to coastal oceans has increased fivefold to 10-fold after 1850 as a result of expanding agriculture and the application of chemical fertilizers, raising concern internationally about eutrophication (Boesch 2002). In addition, engineering construction (e.g., dams and reservoirs) in large and small watersheds has dramatically modified the global land-source inputs into coastal ocean and has caused changes to the monthly and/or annual rhythm of freshwater discharge to estuaries, trapping of terrestrial sediments behind the dams, modification of the fluxes of chemical elements, and shifts in plant nutrient ratios (e.g., N/P and N/Si) (Zhang et al. 1999; Yang et al. 2007). For instance, previous studies have indicated that trapping the Huanghe (Yellow River) by dams and/or reservoirs in the upper and middle reaches caused a drop in the water supply to ca. 20–40 % of historical records, with no sediment being delivered to the Bohai in some cases, and a considerable reduction in offshore fish catches (Zhang et al. 2004; Wang et al. 2007). In the twentieth century, anthropogenic activity on the Changjiang Delta and the adjacent coastal environment had significant impacts (e.g., urbanization, hydraulic engineering, pollutant drainage and eutrophication, aquaculture, and overfishing), which had profound consequences for the structure and function of the marine ecosystem in the East China Sea (see Chap. 9).

It should be kept in mind that along with the construction of ca. 50×10^3 dams over the entire Changjiang drainage basin since the 1950s, the total reservoir water capacity had reached ca. 150×10^9 m^3 by 2001. This construction has a significant impact on the riverine water and sediment loads delivered to the ocean from the Changjiang (cf. Yang et al. 2007, 2010). The well-known TGD was enclosed in June 2003, and the water level of the reservoir was raised to 135 m. Water levels were raised again in October 2009 up to 170 m, and in October 2011 to 175 m. The potential impact of the TGD on the ecosystem in the East China Sea Shelf has caused extensive scientific debate and discussion in the public media, both nationally and internationally (see Chap. 4).

Using a box model approach, Chen (2000) showed that the nutrient cycle in the East China Sea responds to the dam construction in the Changjiang watershed (i.e., TGD) in a rather simple and linear way. Suppose that outflow from the Changjiang were reduced by 10 %, the cross-shelf water exchange would be reduced roughly by the same fraction

(Chen 2000). However, given the nonlinear nature of coastal circulation and the complex forcing–feedback interactions between the external fluxes and biogeochemical processes in the East China Sea, the linear perturbation–response model (i.e., the TGD-imposed change in the Changjiang watershed results in a direct change in the plant nutrient budget of the East China Sea Shelf) may well underestimate the effect of the TGD on the marine food web at various hierarchies. We consider that the change in the riverine flow regime and reduction in the sediment load due to trapping of terrestrial materials behind the TGD may have considerable impact on the stability of the delta region and the community structure and biodiversity of the ecosystem farther beyond, in the coastal and marine environment (see Chaps. 7 and 8).

The TGD may also cause retention of nutrients within the reservoir and hence a reduction in river budget to the East China Sea by 10 % for DIN and 50 % for DIP (Zhang et al. 1999). It has been estimated that 5–10 % of DIN and 20–40 % of DIP from upstream could be taken up by the growth of photosynthetic biomass within the TGR, without a remarkable change in annual water flow downstream of the dam; however, this would considerably increase the N/P ratio by 50–100 % in the lower reaches of the Changjiang (Zhang and Zhang 2003).

It should be noted, however, that all of the above-mentioned potential issues are rather speculative, and based on discrete observations with limited temporal and spatial scales. These hypotheses should be verified in the near future with process-oriented and modeling studies in combination.

Biological blooms are also characteristic of the coastal region off the river mouth of the Changjiang. Since the late 1990s, blooms of dinoflagellates (e.g., prorocentrum) have flourished in April–May each year. These overwhelm and inhabit the niche of diatoms that usually bloom in spring, which has unknown consequences for the marine food web. Blooms of diatoms are instead found in summer, mostly July–August each year. The number of harmful algal blooms (HABs) in the coastal waters of the East China Sea has increased sharply since the 1990s. In the first 10 years of this century, as many as 120–125 HABs were reported annually for the whole China, and up to 75 % of these "red-tide" events occurred in the coastal environment of the East China Sea (Fig. 5.10). Furthermore, it has been found that harmful algal blooms, composed of dinoflagellates, can cover an area as large as 5×10^3–10×10^3 km^2 in the coastal waters of the East China Sea (Zhou et al. 2001).

In the shelf region off the Changjiang Estuary, a seasonal depletion of DO is found in near-bottom waters in summer at depths of 20–50 m, which affects an area of up to 15×10^3–20×10^3 km^2 (Box 5.2). The low-oxygen water offshore from the Changjiang Estuary is comparable to the area affected by hypoxia in the Gulf of Mexico (Li et al. 2002; Rabalais et al. 2002). As shown in Fig. 5.11, the minimum level of DO in near-bottom waters offshore from the Changjiang Estuary in summer over the past 50 years is 1.0–2.0 mg l^{-1} or even lower. However, DO levels are higher in the KSSW across the shelf break with levels of 4–5 mg l^{-1}.

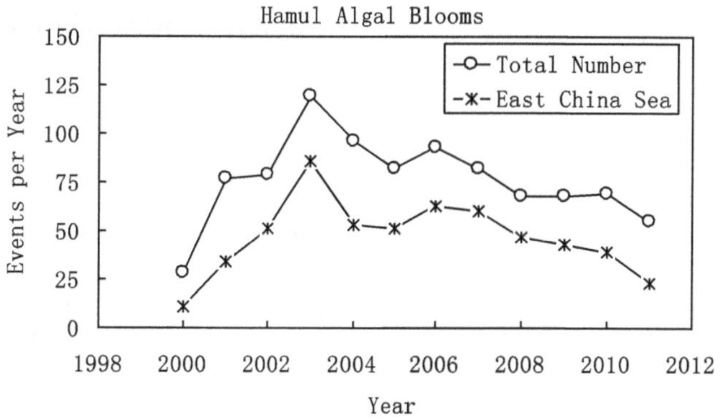

Fig. 5.10 Harmful algal blooms (i.e., "red tide") off the Changjiang Estuary and in the inner shelf of East China Sea, in comparison with total reported events for all sea areas off China. Data are from the State Oceanic Administration of China (web site: www.soa.gov.cn)

Box 5.2 Seasonal Hypoxia off the Changjiang Estuary and in the East China Sea

In the inner part of the East China Sea Shelf and off the Changjiang Estuary, depletion of dissolved oxygen (DO) and hypoxia (i.e., DO < 2 mg l^{-1}) occurs in summer seasons, usually between June and August of the year. This is also the time when rivers reach their maximum monthly water discharge in the subtropical climate zone, and the southerly monsoon prevails in East Asia and affects the surface circulation of the East China Sea.

Summer hypoxia in the East China Sea usually occurs in the near-bottom waters at depths of 20–50 m, but occasionally low DO values can be found in surface waters. The history of coastal hypoxia in this region can be traced back to the late 1950s, when China conducted its first national oceanography survey from the north to the south of the East China Sea.

An important issue in the evolution of this seasonal hypoxia in the coastal environment of the East China Sea is the rapid increase in the spatial extent of the low-DO waters. For instance, the area affected by hypoxic waters was 1×10^3 to 2×10^3 km^2 in the late 1950s and increased to 15×10^3 to 20×10^3 km^2 in 2006 (Zhu et al. 2011).

Seasonal hypoxia is related to the increase in the discharge of riverine organic matter and plant nutrients from the Changjiang; frequent eutrophication in adjacent coastal waters is fueled by an over enrichment of

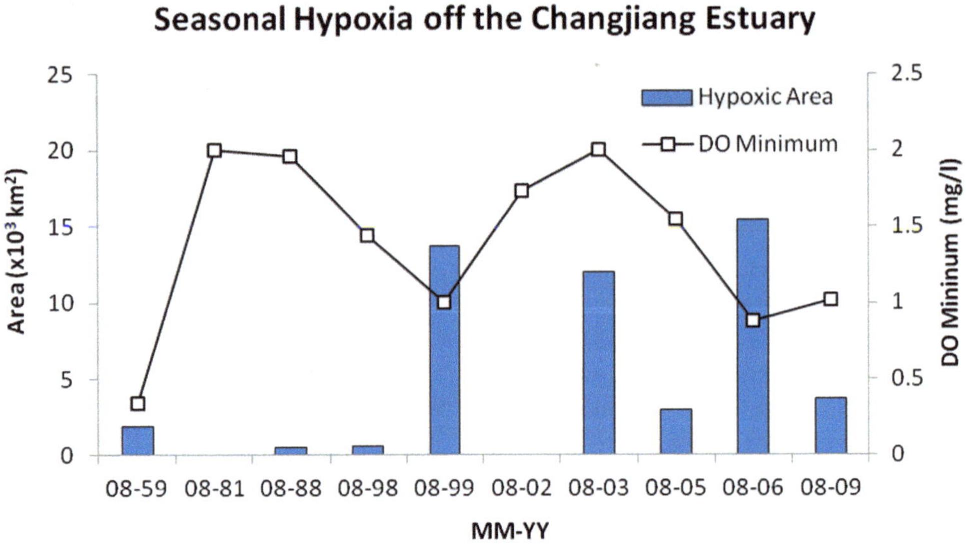

Fig. 5.11 Evolution of summer hypoxic events in the inner shelf of the East China Sea since the 1950s. This shows the area of DO depletion (i.e., DO < 2.0 mg l^{-1}) in near-bottom waters and the minimum DO values obtained from field observations. The figure is based on data from Zhu et al. (2011) and unpublished results. In order to examine the trend in seasonal hypoxia, only the observations in August are shown. Note that in some of the years, there were no seagoing observations in summer and/or DO data are not available

nutrients and is sustained by seasonal hydrographic conditions (e.g., stratification). The heterotrophic degradation of organic matter is related to eutrophication offshore from the Changjiang Estuary, which can consume a considerable amount of DO in near-bottom waters. Stratification in summer inhibits the exchange of DO across the isopycnal, which is associated with the change in temperature and salinity gradients in the water column. Moreover, the effect of Ekman suction induced by the southerly monsoon in the summer season can carry deep and low-DO waters to coastal areas. Consequently, the temporal and spatial distribution of seasonal hypoxia off the Changjiang Estuary and in the East China Sea has a very dynamic nature. Some observations have found a huge hypoxic water body off the Changjiang Estuary in summer, with DO in near-bottom waters as low as 0.5–1.0 mg l^{-1}, while in other studies the distribution of hypoxic waters has been observed to have a patchy character with two low-DO water areas along the coast, with different minimum DO values in near-bottom samples (Li et al. 2002).

The coastal hypoxia off the Changjiang Estuary is linked to the eutrophication with high rates of photosynthesis, and hence planktonic biomass in surface waters, followed by extensive degradation of organic matter in the water column. Eutrophication is regulated by a combination of high temperature in summer seasons and stratification due to the buoyancy effect of plume dispersal off the Changjiang (Li et al. 2002; Zhu et al. 2011). It has been found that the bacterial turnover rate increases linearly with higher primary production across the East China Sea Shelf, but is regulated by temperature, with elevated bacterial production offshore from the Changjiang Estuary (Shiah et al. 2003).

Given that the release of DOM from phytoplankton exudation is 10–15 % of bacterial production, it accounts for ca. 10 % of the bacterial carbon demand (i.e., 430–480 mg C m^{-2} d^{-1}) in the East China Sea (Ducklow and Carlson 1992; Shiah et al. 2003). This implies that allochthonous organic substrates (e.g., riverine input) could also be an important factor in sustaining bacterial demand and hence oxygen consumption in coastal waters of the East China Sea (see Chap. 6).

In summary, the continual intensification of eutrophication in the coastal environment of the East China Sea that occurs in response to the increase in terrestrial inputs (e.g., plant nutrients) accelerates the decomposition of terrestrial and marine organic matter. This decomposition occurs through microbial loops, such as a reduction in the autotrophic phytoplankton to heterotrophic bacteria biomass ratio, and hence results in the mineralization of plant nutrients and trace elements (Zhang et al. 2007a). Compared with other regional and global coastal ecosystems, the biomass composition from the near-shore to the open-shelf region of the East China Sea changes from dominantly heterotrophic to autotrophic, with strong feedbacks to the atmosphere, i.e., emission of greenhouse gases such as CO_2, CH_4, and N_2O (Zhang et al. 2008a; Naqvi et al. 2010), as well as the important effect of the "continental pump" to the NW Pacific Ocean (cf. Liu et al. 2010).

The coastal ecosystem of the East China Sea is shifting from a regime dominated by diatoms to one characterized by an increased concentration of dinoflagellates in spring when DIN/DIP and DIN/SiO_3^{2-} ratios in the water column are relatively low. This is followed by the dominance of diatoms when higher DIN/DIP and DIN/SiO_3^{2-} ratios occur in surface waters after the summer floods from the Changjiang. The impact of these seasonal cycles on the community structure of the East China Sea needs further investigation in the future.

The mechanism examined above can be used to explain the high concentrations of phosphate and silicate relative to nitrogen in near-bottom waters, together with the hypoxia in the inner shelf in summer relative to the winter. Another plausible explanation for elevated nutrient concentrations in near-bottom waters includes the exchange flux at the sediment–water interface, and remobilization of species along with frequent resuspension–deposition processes in coastal areas. These processes would induce relatively low DIN/DIP and DIN/SiO_3^{2-} ratios, as inorganic nitrogen can be lost through denitrification and/or annamox. For example, the difference in phosphate between near-bottom and surface samples, in the salinity range of 30–35, is ~0.5 μM, which corresponds to 5–10 μM of regenerated nitrogen in the water column, taking into account the Redfield ratio.

Data from in situ sediment core incubations show an exchange flux of 0.4–1.3 mmol m^{-2} d^{-1} for DIN, 0.006–0.078 mmol m^{-2} d^{-1} for PO_4^{3-}, and 0.7–3.2 mmol m^{-2} d^{-1} for dissolved silicate, with DIN/DIP ratios of 15–60 and DIN/Si ratios of 0.4–0.5. These values are induced by processes at the sediment–water interface in coastal environments off the Changjiang Estuary, indicating preferential regeneration and/or loss of nitrogen relative to other plant nutrients (Qi et al. 2003). The Chl-*a* in surface sediments averages 0.54 μg g^{-1} (dry weight) with phaeo-pigments (Pha-a) of 2.5 μg g^{-1}. This indicates the considerable uptake of nutrients by benthic plankton and algae, which also has an important influence on nutrient inventories in the water column (Li et al. 2003).

5.6.2 Chemical Element Budgets and Their Impacts on Ecosystems

The distribution of Ra isotopes shows that the mixing of different water masses over the broad East China Sea Shelf has a critical impact on the regulation of plant nutrient and trace element biogeochemical dynamics (Zhang and Su 2006; Zhang et al. 2007b). Briefly, the buoyant plumes from the Changjiang in summer cover an area up to 0.1×10^6 km^2 over the East China Sea, reaching as far as Cheju Island and even to the shelf break. In winter, the broad East China Sea Shelf is dominated by shelf-mixed waters, which consists of freshwater from the terrestrial influx (i.e., river and groundwater), rainfall, Taiwan Strait Warm Water, and incursion of Kuroshio waters (e.g., KSW and KSSW) across the shelf break in the preceding summer season.

Again, the northerly monsoon prevails in winter over East Asia, and riverine effluent from the Changjiang and other rivers of mainland China is restricted to the western side of the East China Sea. This discharge moves southward, forming a narrow but turbid belt with strong mesoscale front dynamics along the coast of China, and can reach locations south of Taiwan Strait. Similar to the riverine input, the incursion of Kuroshio waters to the East China Sea Shelf has dynamic spatial and temporal characteristics such as mesoscale surface curls and eddies. It has been shown that the incursion of Kuroshio waters into the East China Sea Shelf takes place year-round, but mainly in autumn and winter seasons (Guo et al. 2006).

The plant nutrient budget of the East China Sea Shelf has been described using box models based on water and solute conservation (e.g., hydrographic and chemical properties), with the assumption of steady-state conditions (cf. Zhang et al. 2007a). In terms of the summer (i.e., June to August) water budget, the East China Sea Shelf has a total inflow of ca. 3.4 Sv (1 Sv = 10^6 m^3/s), mainly through Taiwan Strait (70 %) and via the incursion of Kuroshio waters (26 %), with the rest of the input originating from riverine sources and atmospheric precipitation. The outflow is essentially through Tsushima/Korea Strait, with a limited amount of discharge into the Yellow Sea (Zhang et al. 2007a). In winter (i.e., December to February), the inflow to the East China Sea Shelf through Taiwan Strait is considerably reduced due to the strong northerly monsoon, while the Kuroshio incursion can be more than twice as high as the amount in summer, up to 60 %.

Our results show that water influx from Taiwan Strait and the incursion of Kuroshio waters, including KSW and KSSW, are of critical importance in adjusting the seasonal cycle of plant nutrients over the broad East China Sea Shelf, because of the regional geography and monsoonal climate. For instance, the collective contribution of Taiwan Strait and Kuroshio waters to the summer influx into the East China Sea Shelf accounts for 70–75 % of NO_3^-+ NH_4^+ and dissolved silicates, and up to 95 % of PO_4^{3-} (Table 5.5). The riverine input in this region as a whole is dominated by the Changjiang influx, which accounts for from ca. 2 % for phosphate to 15 % for SiO_3^{2-} and NO_3^-+NH_4^+, and hence for DIN (Table 5.5). The pathway from atmospheric wet and dry depositions is of secondary importance in terms of plant nutrient mass balance in the summer season. The influx into the water column from benthic and early-diagenetic processes through the sediment–water interface is 1–10 % (Table 5.5). The export of plant nutrients in summer essentially occurs through Tsushima/Korea Strait with a N/P

Table 5.5 Summary of nutrient budgets for the East China Sea Shelf in summer (i.e., June to August) and winter (i.e., December to February) (cf. Zhang et al. 2007a)

Pathways	NO_3^- +NH_4^+ (%)	PO_4^{3-} (%)	SiO_3^{2-} (%)
(a) **Summer season**			
Influx			
Kuroshio incursion	30.1	49.0	22.6
Taiwan Strait	43.2	45.8	53.7
Riverine influx	15.0	1.7	14.3
Bottom sediment	5.7	3.2	9.2
Atmospheric deposition	6.0	0.3	0.2
Outflow			
Shelf water export	99.8	99.3	99.5
Yellow Sea	0.2	0.7	0.5
(b) **Winter season**			
Influx			
Kuroshio incursion	56.3	64.8	52.0
Taiwan Strait	26.6	32.4	39.2
Riverine influx	4.0	0.4	3.0
Bottom sediment	4.6	1.8	5.5
Atmospheric deposition	8.2	0.2	0.05
Yellow Sea	0.3	0.4	0.2
Outflow			
Shelf water export	~100	~100	~100

To facilitate the comparison of data between different hydrographic regimes and/or observational studies, the nutrient fluxes (kmol s^{-1}) are transformed into percentage (%) of the total influx and/or outflow

ratio of 20–30. However, a limited amount of plant nutrients can be carried into the southern part of the Yellow Sea in summer because of the northeastward dispersal of Changjiang effluent plumes, driven by southerly monsoon-induced flow and tidal circulation (Liu et al. 2003b; Wu et al. 2011).

In winter, the Kuroshio incursion may alone contribute 50–65 % of the total influx of allochthonous plant nutrients to the East China Sea Shelf. This influx can be twice as high as that from Taiwan Strait because of the limited water flow and depleted concentration of TSWW relative to the incursion of Kuroshio Subsurface Waters (Table 5.5). The riverine flux is also considerably reduced in winter, to <5 % of the total inflow of plant nutrients, and is even smaller than the contribution through sediment–water exchange (Table 5.5). In winter, atmospheric depositions can also be an important source of DIN in the East China Sea Shelf relative to other terrestrial pathways, such as riverine inputs, but its contribution to the phosphate and dissolved silicate budgets is of minor significance.

Changing circulation patterns in the East China Sea Shelf during the northerly monsoon in the winter season can cause a certain amount (<0.5–1 %) of plant nutrients to be transported from the Yellow Sea into the East China Sea (Table 5.5). It has been found that the outflow of plant nutrients from the East China Sea Shelf (i.e., via Tsushima/Korea Strait) can be 50–100 % higher in winter than in summer. This is especially true for phosphate and dissolved silicate, as the major external sources of these nutrients are from oceanic waters and/or riverine input, with a limited contribution from atmospheric depositions (cf. Zhang et al. 2007a).

The seasonal nature of nutrient budgets of the East China Sea Shelf indicates that the ecosystem is highly dynamic. For instance, the dominance of the Kuroshio incursion on the plant nutrient influx to the East China Sea Shelf in autumn and winter underlines the fact that the replenishment of plant nutrient stocks by across-shelf processes can be of critical importance in regulating the spatial and temporal variability of phytoplankton spring blooms. Given the estimation of flushing time of ca. 1–2 years for the East China Sea Shelf, if the system is examined for chemical properties and even in the point of view of changes in food web structure and production, the strong seasonality of circulation and the budget of plant nutrients may have a "memory" effect. Hence, it can be expected that external forcings, such as the variability of waters on the eastern boundary (i.e., Kuroshio) of the study region, which is related to climate cycles such as the El Niño Southern Oscillation and the Pacific Decadal Oscillation, can control the interannual variability of primary productivity and the change in new versus recycled productions over the entire shelf, through the transmission and amplification of the "memory".

The existing knowledge of biogeochemical budgets of plant nutrients is mainly based on steady-state conceptual models, with outputs that include the net flux of chemical materials. Thus, the exchange processes and dynamic characteristics of fluxes are smoothed out when average values are used to characterize the stocks and specify the pathways between connected subsystems. In the case of the East China Sea, warm and saline waters have been observed to move farther north into the Yellow Sea each winter, known as the Yellow Sea Warm Water Mass, and this flux is compensated by the cold and less saline waters moving to the south along the coast of mainland China. The southward-moving waters have relatively high plant nutrient concentrations, which adds another uncertainty to the box model budget calculations. In this study, the exchange flux of plant nutrients at the bottom sediment–water interface remains fixed in the budget calculations, mainly due to the poor data coverage that limits the understanding of the temporal and spatial variability at seasonal resolutions.

Table 5.6 Summary of trace element budgets in the East China Sea Shelf (cf. Ren et al. 2006, 2010)

Pathways	Al (%)	As (%)
Input		
Kuroshio incursion	17.8	24.9
Taiwan Strait	31.9	73.2
Riverine influx	28.4	1.3
Atmospheric deposition	21.9	0.6
Output		
Sediment–water interface	0.2	–
Yellow Sea	–	0.3
Shelf water export	99.8	99.7

To facilitate the comparison of data between different hydrographic regimes and/or observational studies, the trace element fluxes are transformed into a percentage (%) of total influx and/or outflow

The same approach can be applied to understanding trace element (e.g., Al and As) budgets, but because these data are scarcer, only annual budgets can be examined (Table 5.6). With regard to dissolved aluminum, the contributions from the TSWW can be 50–100 % higher than the supply from the incursion of Kuroshio waters. However, this is comparable to the riverine influx, presumably due to the dominant water inflow through Taiwan Strait with moderate Al concentrations (Ren et al. 2006). In this region, atmospheric depositions can be an important pathway of Al into the surface waters, accounting for 21.9 % of the total inflow (Table 5.6), and wet and dry pathways may contribute equally to the total atmospheric flux (Ren et al. 2006). The export of dissolved Al out of the East China Sea Shelf is essentially through water exchange processes, e.g., via Tsushima/Korea Strait, with a minor sink from sediment–water interface processes (Table 5.6).

For arsenic, the overwhelming influx is via Taiwan Strait (73.2 %) due to the high flow of TSWW, which has a relatively high concentration of dissolved arsenic. The incursion of Kuroshio waters represents approximately a quarter of the total As influx to the East China Sea Shelf, owing to its low As levels compared with other pathways (Table 5.6). Although As levels in the Changjiang are higher than those of the TSWW and Kuroshio waters (i.e., KSW and KSSW), its impact on the budget of this area is rather limited (1.3 %) because the water influx from terrestrial sources as a whole is almost two orders of magnitude smaller than the combined contribution of the TSWW and the Kuroshio incursion into the East China Sea Shelf (Zhang and Su 2006). In contrast to Al, atmospheric wet and dry depositions are of minor importance (0.6 %) to the As budget, according to the most reliable estimates (Table 5.6). The export of As is mainly via water exchange through Tsushima/Korea Strait, with a limited amount entering the Yellow Sea, mostly in summer (Table 5.6).

5.7 Summary and Prospects for the Future

The present work examines biogeochemical dynamics of land–ocean interactions in the coastal zone of one of the world's largest river–estuary systems, the Changjiang, and the East China Sea. This area of high population density has experienced rapid economic growth over the past 30–40 years. The river empties into a broad continental shelf in the northwest Pacific Ocean (i.e., the East China Sea), and the climate is dominated by the East Asian monsoon. We compare our own work with data from previous studies and synthesize the knowledge about the biogeochemical dynamics from the Changjiang watersheds until the shelf-break region in the East China Sea, bounded by the western boundary current of the NW Pacific Ocean (i.e., the Kuroshio).

The watershed of the Changjiang has undergone dramatic changes due to anthropogenic forcings, such as hillside deforestation and reclamation of wetlands for agriculture in the upper and middle reaches of the river. Furthermore, the construction of dams and/or reservoirs in the drainage basin has modified water and sediment loads as well as the chemical composition of the river. Pollutants from farming (e.g., fertilizers and pesticides) and the rapid industrialization/urbanization of coastal areas over the last three to four decades have caused serious eutrophication and pollution in adjacent marine areas and altered the quality of the coastal environment. In coastal areas, overfishing, marine aquaculture (e.g., the conversion of salt marshes to shrimp ponds), and engineering activities (e.g., the construction of jetties and harbors, and dredging of water channels for navigation) have further damaged the already overstressed coastal and marine ecosystems.

In the middle and lower reaches of the Changjiang, the immediate consequences of these human activities include an increase in plant nutrient concentrations, a skewed ratio between nutrient species at the river mouth and in offshore waters, and a reduction in terrestrial sediment loads from the river, and hence a decrease in the capacity to buffer pollution delivery in coastal waters. These effects have a cumulative impact on the ecosystems in coastal and marine environments, including fluxes of chemical elements and plant nutrient ratios as well as the community structure of phytoplankton, which has a profound influence on the marine food web.

In the Changjiang Estuary, nitrate and dissolved silicate have a linear relationship with salinity gradients, illustrating the relatively short flushing time of the system due to strong tidal dynamics. However, other nutrient species (e.g., PO_4^{3-} and NH_4^+) have more complicated relationships with salinity, indicating different sources and/or pathways are involved.

A comparison of surface and near-bottom samples shows that regeneration of plant nutrients and trace elements can take place in the lower part of the Changjiang Estuary as well as the adjacent marine waters further offshore. This illustrates the heterotrophic nature of biogeochemical cycles in the adjacent coastal environment. Another plausible explanation of higher nutrient levels in the near-bottom samples in the offshore region is the incursion of KSSW, but this needs to be examined and quantified in future studies.

Although previous studies have shown evidence that the incursion of KSSW can be a significant component of plant nutrient turnover in the East China Sea, its spatial and temporal impacts have yet to be examined in full detail. It seems that the incursion of Kuroshio waters onto the East China Sea Shelf is most prevalent in autumn and winter, based on oceanic observations and numerical simulations. Riverine inflow, by contrast, is minimized in winter and/or has limited impact on the chemical properties of shelf waters due to the monsoonal climate. Given that the water residence time on the East China Sea Shelf is ca. 1–2 years, and the nutrient influx along with the Kuroshio shelf-incursion can play an important role in sustaining primary productivity (i.e., spring blooms), and hence the structure and function of ecosystems, as well as the production of food webs (e.g., fisheries) in this region. However, this also requires further investigation.

At seasonal timescales, budgetary studies have provided evidence that the East China Sea Shelf can accumulate a considerable amount of plant nutrients, and to lesser extent, trace elements (e.g., Al and As). Taking into account the fact that photosynthesis in the East China Sea is phosphorus limited in near-shore and coastal environment and becomes nitrogen limited and/or dissolved silicate limited in open-shelf waters, the accumulation of plant and trace nutrients suggests that the East China Sea as a whole can potentially function as a net carbon sink for atmospheric CO_2.

Acknowledgments This work is based on research started in the mid-1990s, with financial support from the Natural Science Foundation of China (Grant Numbers: 47021004 and 41021064), the Ministry of Education of China (Grant Number: ITR0472), the Municipalities of Shanghai, China (Grant Number: 088014192), and the Ministry of Science and Technology of China (Grant Numbers: 2006CB400600 and 2011CB409801). We are highly indebted to colleagues and students from the College of Chemistry and Chemical Engineering (OUC) and State Key Laboratory of Estuarine and Coastal Research (ECNU) for the support during field campaigns and cruises, sample collection, and laboratory analyses.

References

Abe K, Ishiki Y, Watanabe Y (2003) Dissolved copper in the Yellow Sea and the East China Sea as a tracer of the Changjiang discharge. Deep-Sea Res II 50:327–337

Balls PW (1989) The partition of trace metals between dissolved and particulate phases in European coastal waters: A compilation of field data and comparison with laboratory studies. Neth J Sea Res 23:7–14

Boesch DF (2002) Challenges and opportunities for science in reducing nutrients over-enrichment of coastal ecosystems. Estuaries 25 (4):886–900

Chen CTR (2000) The Three Gorges Dam: Reducing the upwelling and thus productivity in the East China Sea. Geophys Res Lett 27:381–383

Du JZ, Zhang J, Baskaran M (2011) Applications of short-lived radionuclides (^{7}Be, ^{210}Pb, ^{210}Po, ^{137}Cs and ^{234}Th) to trace sources, transport pathways and deposition of particles/sediments in rivers, estuaries and coasts. In: Baskaran M (ed) Handbook of environmental isotope geochemistry. Springer, Berlin, pp 305–329

Ducklow HW, Carlson CA (1992) Organic bacterial production. In: Marshall KC (ed) Advance in microbial ecology. Plenum, New York, pp 113–181

Edmond JM, Spivack A, Grant BC, Hu MH, Chen ZX, Chen S, Zeng XS (1985) Chemical dynamics of the Changjiang Estuary. Cont Shelf Res 4:17–36

Elbaz-Poulichet F, Huang WW, Martin JM, Seyler P, Zhong XM, Zhu JX (1990) Biogeochemical behavior of dissolved trace elements in the Changjiang Estuary. In: Yu GH, Martin JM, Zhou JY (eds) Biogeochemical study of the Changjiang Estuary. China Ocean Press, Beijing, pp 293–311

Grasshoff K, Kremling K, Ehrhardt M (eds) (1999) Methods of seawater analysis, 3rd edn. WILEY-VCH, Weinheim

Gu HK, Xiong XX, Liu MX, Li Y (1981) Marine geochemistry of nitrogen near estuary of Yangtze River I. Nitrate in seawater near estuary. J Shandong Coll Oceanol 11(4):37–46 (in Chinese)

Gu HK, Ma XN, Shen WR, Ren GF, Chen Z, Diao HX, Li GJ, Zhang LY (1982) Marine geochemistry of nitrogen near estuary of Changjiang II. Nitrite and ammonia in seawater near estuary. J Shandong Coll Oceanol 12(2):31–38 (in Chinese)

Guo XY, Miyazawa Y, Yamagata T (2006) The Kuroshio on shore intrusion along the shelf break of the East China Sea: The origin of the Tsushima Warm Current. J Phys Oceanogr 36:2205–2231

Han DS, Song JK, Batchelor B, Abdel-Wahab A (2013) Removal of arsenite [As (III)] and arsenate [As (V)] by synthetic pyrite (FeS2): Synthesis, effect of contact time, and sorption/desorption envelopes. J Colloid Interface Sci 392:311–318

Hatje V, Payne TE, Hill DM, McOrist G, Birch GF, Szymczak R (2003) Kinetics of trace element uptake and release by particles in estuarine waters: effects of pH, salinity, and particle loading. Environ Int 29:619–629

IMBER (2005) Integrated marine biogeochemistry and ecosystem research, science plan and implementation strategy. IGBP Report No. 52, IGBP Secretariat, Stockholm

Jiann KT, Wen LS (2012) Distribution and liability of dissolved iron in surface waters of marginal seas in southeastern Asia. Estuar Coast Shelf Sci 100:142–149

Jiann KT, Wen LS, Gong GC (2009) Distribution and behaviors of Cd, Cu, and Ni in the East China Sea surface water off the Changjiang Estuary. Terr Atmos Ocean Sci 20(2):433–443

Koshikawa MK, Takamatsu T, Takada J, Zhu MY, Xu BH, Chen ZY, Murakami S, Xu KQ, Watanabe M (2007) Distributions of dissolved and particulate elements in the Yangtze estuary in 1997–2002: background data before the closure of the Three Gorges Dam. Estuar Coast Shelf Sci 71:26–36

Li DJ, Zhang J, Huang DJ, Wu Y, Liang J (2002) Oxygen depletion off the Changjiang (Yangtze River) Estuary. Sci China 45D:1137–1146

Li XN, Zhou WH, Liu SM, Zhang J (2003) Sediment chlorophyll pigments in HABs (Harmful Algal Blooms) area of the East China Sea. Chin J Appl Ecol 14:1102–1106 (in Chinese)

Li MT, Xu KQ, Watanabe M, Chen ZY (2007) Long-term variations in dissolved silicate, nitrogen and phosphorus flux from the Yangtze River into the East China Sea and impacts on estuarine ecosystem. Estuar Coast Shelf Sci 71:3–12

Liu SM, Zhang J, Chen HT, Xiong H, Wu Y, Zhang ZF (2003a) Nutrients in the Changjiang and its tributaries. Biogeochemistry 62:1–18

Liu SM, Zhang J, Chen SZ, Chen HT, Hong GH, Wei H, Wu QM (2003b) Inventory of nutrient compounds in the Yellow Sea. Cont Shelf Res 23:1161–1174

Liu KK, Atkinson L, Quinones R, Talaue-McManus L (eds) (2010) Carbon and nutrient fluxes in continental margins: a global synthesis. Springer, Berlin

LOICZ (2005) Land-ocean interactions in the coastal zone, science plan and implementation strategy. IGBP Report No. 51/IHDP Report No. 18, Stockholm

Lopez P, Lluch XM, Vidal M, Morguí JA (1996) Adsorption of phosphorus on sediments of the Balearic Islands (Spain) related to their composition. Estuar Coast Shelf Sci 42:185–196

Müller B, Berg M, Yao ZP, Zhang XF, Wang D, Pfluger A (2008) How polluted is the Yangtze River? water quality downstream from the Three Gorges Dam. Sci Total Environ 402:232–247

Naqvi SWA, Bange HW, Farias L, Monteiro PMS, Scranton MI, Zhang J (2010) Marine hypoxia/anoxia as a source of CH_4 and N_2O. Biogeosciences 7:2159–2190

Qi XH, Liu SM, Zhang J, Chen HT (2003) Nutrient regeneration rate of sediments in HABs (Harmful Algal Blooms) area of the East China Sea. Chin J Appl Ecol 14:1112–1116 (in Chinese)

Rabalais NN, Turner RE, Dortch Q, Justic D, Bierman VJJr, Wiseman WJJr (2002) Nutrient-enhanced productivity in the northern Gulf of Mexico: past, present and future. Hydrobiologia 176:39–63

Ren JL, Zhang J, Li JB, Yu XY, Liu SM, Zhang ER (2006) Dissolved aluminum in the Yellow Sea and East China Sea—Al as a tracer of Changjiang (Yangtze River) discharge and Kuroshio incursion. Estuar Coast Shelf Sci 68:165–174

Ren JL, Zhang J, Li DD, Cheng Y, Liu SM (2010) Behavior of dissolved inorganic arsenic in the Yellow Sea and East China Sea. Deep-Sea Res II 57:1035–1046

Shen ZL (2006) Phosphorus and silicate fluxes in the Yangtze River. Acta Geograph Sinica 61(7):741–751 (in Chinese)

Shen ZL, Liu Q, Zhang SM (2003) Distribution, variation and removal patterns of total nitrogen and organic nitrogen in the Changjiang River. Oceanol Limnol Sinica 34(6):577–585 (in Chinese)

Shiah FK, Gong GC, Chen CC (2003) Seasonal and spatial variation of bacterial production in the continental shelf of the East China Sea: possible controlling mechanisms and potential roles in carbon cycling. Deep Sea Res II 50:1295–1309

Stumm W, Morgan JJ (1996) Aquatic chemistry: chemical equilibria and rates in natural waters. Wiley, New York

Wang ZL, Liu CQ (2003) Distribution and partition behavior of heavy metals between dissolved and acid-soluble fractions along a salinity gradient in the Changjiang Estuary, Eastern China. Chem Geol 202:383–396

Wang HJ, Yang ZS, Saito Y, Liu JP, Sun XX, Wang Y (2007) Stepwise decreases of the Huanghe (Yellow River) sediment load (1950–2005): impacts of climate changes and human activities. Global Planet Change 57:331–354

Wu H, Zhu JR, Shen J, Wang H (2011) Tidal modulation on the Changjiang river plume in summer. J Geophys Res 116:C08017. doi:10.1029/2011JC007209

Yan WJ, Zhang S, Sun P, Seitzinger SP (2003) How do nitrogen inputs to the Changjiang basin impact the Changjiang river nitrate: A temporal analysis for 1968–1997. Global Biogeochem Cycles 17:1091. doi:10.1029/2002GB002029

Yang SL, Zhang J, Xu XJ (2007) Influence of the Three Gorges Dam on downstream delivery of sediment and its environmental implications, Yangtze River. Geophys Res Lett 34:L10401. doi:10.1029/2007GL029472

Yang SL, Liu Z, Dai SB, Gao ZX, Zhang J, Wang HJ, Luo XX, Wu CS, Zhang Z (2010) Temporal variations in water resources in the Yangtze River (Changjiang) over the industrial period, based on reconstruction of missing monthly discharges. Water Resour Res 46:W10516. doi:10.1029/2009WR008589

Yao QZ (2005) Biogeochemical characteristics of arsenic and selenium in the Changjiang drainage area and its Estuary. PhD Dissertation, East China Normal University, Shanghai (in Chinese)

Yao QZ, Zhang J, Wu Y, Xiong H (2007) Hydrochemical processes controlling arsenic and selenium in the Changjiang (Yangtze River) system. Sci Total Environ 377:93–104

Zhang J (1995) Geochemistry of trace metals from Chinese river/estuary systems: an overview. Estuar Coast Shelf Sci 41:631–658

Zhang J (1996) Nutrient elements in large Chinese estuaries. Cont Shelf Res 16:1023–1045

Zhang J (2002) Biogeochemistry of Chinese estuarine and coastal waters: nutrients, trace metals and biomarkers. Reg Environ Change 3:65–76

Zhang ER (2003) Participation of trace elements in the Changjiang Estuary regulated by the adsorption – desorption with particulate matter. PhD Dissertation, East China Normal University, Shanghai (in Chinese)

Zhang YY (2007) Study on the control of nutrient phase transformation by adsorption – desorption in the Changjiang Estuary. PhD Dissertation, East China Normal University, Shanghai (in Chinese)

Zhang J, Su JL (2006) Nutrient dynamics of the Chinese Seas: The Bohai, Yellow Sea, East China Sea and South China Sea. In: Robinson AR, Brink KH (eds) The Sea, vol 14. Harvard University Press, Cambridge, pp 637–671

Zhang ER, Zhang J (2003) Analysis of the Three-Gorge Reservoir impacts on the retention of N and P in the Yangtze River. J Lake Sci 15:41–48 (in Chinese)

Zhang J, Huang WW, Wang JH (1994) Trace-metal chemistry of the Huanghe (Yellow River), China—examination of the data from in situ measurements and laboratory approach. Chem Geol 114:83–94

Zhang J, Zhang ZF, Liu SM, Wu Y, Xiong H, Chen HT (1999) Human impacts on the large world rivers: would the Changjiang (Yangtze River) be an illustration? Global Biogeochem Cycles 13:1099–1105

Zhang J, Xu H, Ren JL (2000) Fluorimetric determination of dissolved aluminum in natural waters after liquid-liquid extraction into *n*-hexanol. Anal Chim Acta 405:31–42

Zhang J, Ren JL, Liu SM, Zhang ZF, Wu Y, Xiong H, Chen HT (2003) Dissolved aluminum and silica in the Changjiang (Yangtze River): Impact of weathering in sub-continental scale. Global Biogeochem Cycles 17:1077. doi:10.1029/2001GB001600

Zhang J, Yu ZG, Raabe T, Liu SM, Starke A, Zou L, Gao HW, Brockmann U (2004) Dynamics of inorganic nutrient species in the Bohai Seawaters. J Mar Syst 44:189–212

Zhang J, Liu SM, Wu Y, Qi XH, Zhang GS, Li RX (2006) Dissolved silica in the Changjiang (Yangtze River) and adjacent coastal waters of the East China Sea. In: Ittekkot V, Unger D, Humborg C, Tac An N (eds) The silicon cycle: human perturbations and impacts on aquatic systems, SCOPE 66, Island Press, Washington DC, pp 71–80

Zhang J, Liu SM, Ren JL, Wu Y, Zhang GL (2007a) Nutrient gradients from the eutrophic Changjiang (Yangtze River) Estuary to the oligotrophic Kuroshio waters and re-evaluation of budgets for the East China Sea Shelf. Prog Oceanogr 74:449–478

Zhang L, Liu Z, Zhang J, Hong GH, Park Y, Zhang HF (2007b) Reevaluation of mixing among multiple water masses in the shelf: an example from the East China Sea. Cont Shelf Res 27:1969–1979

Zhang GL, Zhang J, Ren JL, Li JB, Liu SM (2008a) Distributions and sea-to-air fluxes of methane and nitrous oxide in the North East China Sea in summer. Mar Chem 110:42–55

Zhang YY, Zhang ER, Zhang J (2008b) Modeling on adsorption-desorption of trace metals to suspended particle matter in the Changjiang Estuary. Environ Geol 53:1751–1766

Zhang EF, Savenije HHG, Chen SL, Chen JY (2012) Water abstraction along the lower Yangtze River, China, and its impact on water discharge into the estuary. Phys Chem Earth 47–48:76–85

Zhou MJ, Zhu MY, Zhang J (2001) Status of harmful algal blooms and related research activities in China. Chinese Bull Life Sci 13:54–60 (in Chinese)

Zhu ZY, Zhang J, Wu Y, Lin J (2006) Bulk particulate organic carbon in the East China Sea: tidal influence and bottom transport. Prog Oceanogr 69:37–60

Zhu ZY, Zhang J, Wu Y, Zhang YY, Lin J, Liu SM (2011) Hypoxia off the Changjiang (Yangtze River) Estuary: oxygen depletion and organic matter decomposition. Mar Chem 125:108–116

Organic Matter and Biomarkers of the Changjiang Estuary and East China Sea Shelf

6

Ying Wu, Zhuoyi Zhu, Hongyan Bao, Shuchai Gan, and Jing Zhang

Abstract

As one of most populated river–estuary–shelf regions, the biogeochemistry of organic matter in the Changjiang Estuary and East China Sea (ECS) is critical for a quantitative understanding of global biogeochemical cycles. This chapter summarizes the spatial variation of organic matter and biomarkers from the watershed to the shelf. Monthly data collected from the downstream of the watershed help to elucidate the seasonal variation of biomarkers (e.g., pigments and amino acids) along with hydrological and biological processes. The geochemical characterization of biomarkers in the ECS is well studied, and this chapter provides an overview of the potential controls on biomarkers. Hydrological sorting, in situ primary production, and diagenetic processes contribute to the variable biomarker distribution, composition, and burial in the ECS. Anthropogenic disturbance is illustrated by an evaluation of the impact of the Three Gorges Dam (TGD) on the composition of terrigenous organic matter and pollutants in the study system. Climate change (e.g., extreme drought), impoundment by the TGD, and the effects of tributaries and lakes in the middle and lower streams of the river may play different roles in the delivery of organic carbon. Although there is high pressure from anthropogenic activities compared with other regions in the world, pollutant concentrations are relatively low and potential ecological risks are limited. With changing fluxes from rivers, and significant modification and burial of organic matter on the shelf due to global change, long-term observations are necessary to develop our knowledge of the biogeochemistry of organic matter in this highly dynamic shelf region.

Keywords

Organic carbon • Biomarker • Terrestrial organic matter • Changjiang • East China Sea

6.1 Introduction

Continental shelves are considered to be important repositories for organic carbon (OC) in marine sediments. Approximately 90 % of the OC buried globally in the ocean has been sequestered from shelf and slope sediments (Hedges and Keil 1995; Burdige 2005). Knowledge of the flux, compositions, and transformation of river-derived material, such as carbon and nitrogen, in the coastal ocean is critical for a quantitative understanding of global biogeochemical cycles (Hedges et al. 1997; Tesi et al. 2007).

Although the sources and fates of terrestrial inputs to the seabed have received attention in the marine organic

Y. Wu (✉) · Z. Zhu · H. Bao · S. Gan · J. Zhang
State Key Laboratory of Estuarine and Coastal Research,
East China Normal University, 3663 Zhongshan Road North,
Shanghai 200062, China
e-mail: wuying@sklec.ecnu.edu.cn

Z. Zhu
e-mail: zyzhu@sklec.ecnu.edu.cn

H. Bao
e-mail: shinybao1986@163.com

S. Gan
e-mail: gansuchai@163.com

J. Zhang
e-mail: jzhang@sklec.ecnu.edu.cn

J. Zhang (ed.), *Ecological Continuum from the Changjiang (Yangtze River) Watersheds to the East China Sea Continental Margin*, Estuaries of the World,
DOI 10.1007/978-3-319-16339-0_6,

geochemistry literature (Hedges and Keil 1995; Hedges et al. 1997; Tesi et al. 2007), information about organic matter in the Changjiang (Yangtze River) and adjacent East China Sea (ECS) is still very limited (Wu et al. 2003, 2007a; Zhu et al. 2008, 2011a, b; Li et al. 2012).

As the world's fourth-largest river in terms of sediment discharge, the Changjiang drains into a shallow (<130 m average depth) but broad (> 500 km wide) continental shelf in the ECS. The Changjiang–ECS shelf system provides a unique setting to examine the fate of terrigenous organic matter. Moreover, because of its proximity to human population centers, the Changjiang is easily impacted by anthropogenic activity (Box 6.1). For example, land use, agriculture, and hydrological engineering projects (e.g., dam construction and water diversion) have had a significant influence on the transport of river material to the sea (Yang et al. 2011). The potential impacts of anthropogenic activities on the composition and fluxes of riverine organic matter require more detailed study.

Box 6.1. Anthropogenic Disturbance

Anthropogenic disturbance means 'relating to or resulting from the influence of humans on the natural world.' Since large populations of people live in river and estuary regions, their activities unavoidably create disturbances to these environments. Human impacts on these systems include the following:

- Land use: land management practices, including impacts on water, soil, nutrients, plants, and animals.
- Dams: Changing river water flow from dams can restrict sediment deposits and nutrient availability downstream, and change fish migration patterns.
- Polluted discharge: Runoff from cities, farms, and factories can carry harmful pollutants to estuaries, such as excess nutrients, raw sewage and manure, and chemical waste.
- Dredging: Sand dredging in rivers changes water and sediment flow, and the dredging of deep navigation channels through estuaries and wetlands destroys and damages habitats.
- Reclamation: This is the process of creating new land from oceans, riverbeds, or lakes for agricultural activities, beach restoration, etc. Reclamation makes the land highly susceptible to soil liquefaction during earthquakes, and the land is less able to absorb storm surges and provide shelter or food for the animals and plants that exist in coastal regions.

Bulk, molecular, and isotopic techniques have been applied to the chemical characterization of organic matter from the watershed to the coast. Combined with previous literature, the goal of this work is to: (1) consider the spatial and temporal distribution of organic matter from the Changjiang to the ECS; (2) employ multiple biomarker tools to elucidate the complex processes that occur in the river and estuary system; and (3) evaluate the impact of the TGD on the delivery of terrestrial organic matter to the lower streams of the Changjiang.

6.2 Sampling and Methods

6.2.1 Field Observations and Sample Collection

This study synthesizes datasets from a series of field campaigns carried out within the Changjiang drainage basin and compares these data with other published results. These campaigns include monthly sampling at an anchor station (XLJ) at the seaward end of the river and cruises over the continental margin of the ECS (i.e., Okinawa Trough and Kuroshio) over past 15 years.

Detailed information about expeditions into the Changjiang watershed and cruises in the ECS can be found in Chap. 5 (see Fig. 5.1a, which shows the watershed sampling strategy, including the main channel and tributaries). Both wet and dry seasons and extreme events (e.g., droughts) were included in the sampling program.

Water samples were collected using either a pump system or a Niskin sampler for depth profiles. Bottom sediments, soils, and plants were also collected along the river. Samples were collected between 1999 and 2007 in the Changjiang Estuary and ECS shelf, from the river mouth to the Kuroshio waters with depth of ca. 2000 m (see Fig. 5.1b). On the cruises, water samples were collected with a CTD system, which provided in situ profiles of hydrographic parameters (e.g., temperature, salinity, epi-fluorescence, and pH). Bottom sediments for chemical and sedimentary analysis were collected using a box corer, multi-corer, and/or gravity corer. All filters, dissolved organic carbon (DOC) samples, and sediments were kept frozen after filtration or slicing.

6.2.2 Laboratory Analysis

DOC was determined by high-temperature catalytic oxidation with a total organic carbon analyzer (Shimadzu TOC V_{CPH}). The relative error for the concentration range was approximately ±2–3 % (Wu et al. 2003).

The analytical reproducibility of particulate organic carbon (POC) and particulate nitrogen (PN) measurements for particles, soil, and sediments samples were estimated to be ±2 and ±4 %, respectively. Stable isotopes of C-13 and N-15 from particular organic matter (e.g., total suspended matter

(TSM) and bottom sediments) were measured by gas isotope ratio mass spectrometer (Model: Finnigan Delta plus XP). For C-13 measurements, carbonates were removed using acid before analysis. The precision of $\delta^{13}C$ values is ±0.2‰. The precision of $\delta^{15}N$ for riverine TSM is ±0.5‰ due to the low N content in the samples, and the precision for plant and surface soils is ±0.2‰ (Wu et al. 2003).

Fluorescence measurements of dissolved organic matter (DOM) were carried out using a F-4500 fluorescence spectrophotometer (Hitachi) and were recorded in emission spectra spanning 300–550 nm (every 3 nm) and exciting wavelengths from 220 to 410 nm (by 3 nm). The fluorescence spectra measured from each sample were combined to create excitation emission matrices (EEMs). After the Milli-Q water spectra were subtracted, the EEMs were Raman-calibrated by normalizing to the integral of the Raman band (excitation 350 nm) from a Milli-Q water sample run the same day as the samples. The EEMs were modeled by parallel factor analysis (PARAFAC) in MATLAB software based on the least squares regression that splits the EEMs into a series of trilinear terms and a residual array. A total of 112 EEMs were collected and modeled in this study, and the appropriate number of components was determined by a split-half validation and residual analysis.

For the determination of particulate amino acids, the GF/F filters were hydrolyzed with HCl (16 %, 20 mL) in the presence of ascorbic acid at 110 °C (24 h) in precombusted sealed glass ampoules according to Fitznar et al. (1999). For D- and L-amino acids, the coefficients of variation between the duplicates were 1–8 %, and the relative standard deviation for the individual amino acids for each run was <3.5 % (Wu et al. 2007a).

Lignin phenols were analyzed using the CuO oxidation method (Hedges and Ertel 1982) as modified by Goñi and Montgomery (2000). When the concentration of an individual lignin phenol is >0.05 mg/100 mg OC, the precision is generally below 10 % (average, 5.2 %); when the concentration is <0.05 mg/100 mg OC, the precision is 15–24 % (Yu et al. 2011).

For pigment analysis, samples were first thawed and homogenized, and wet sediment was extracted with acetone. After probe sonication in an ice bath and centrifugation, the filtered extract was ready for injection. This analytical method follows Zapata et al. (2000) with some slight modifications (Zhu et al. 2009).

All sampling equipment for use in the field, on cruises, as well as in the laboratory for the analysis of bulk samples and biomarkers, were washed with hot detergent and rinsed with distilled water. Equipment was then soaked in HCl (v/v: 1:5) for 5–10 days and rinsed with Milli-Q water before use. Filters and glassware were precombusted at 450 °C for 5 h. Quality assurance and quality control (QA/QC) were managed by measuring national and international standards and reference materials (e.g., Washington Coast sediments for lignin phenols analysis). Care was taken during the entire analytical procedure to control background (blank) concentrations on a daily basis. All solvents were pre-extracted and were of high-performance liquid chromatographic or gas chromatographic (GC) grade. Glassware was kept in a sulfochromic bath for 24 h, and then rinsed with distilled water, Milli-Q water, and pure solvent before use.

6.3 Particulate and Dissolved Organic Carbon from the Watershed to the Shelf

6.3.1 Spatial and Temporal Variations in Particulate and Dissolved Organic Carbon in the Changjiang

The distribution of dissolved and particulate carbon along the main channel of the Changjiangin 1997, 2003, 2006, 2008, 2009, and 2010 is presented in Fig. 6.1. Average DOC concentrations in the main channel were 105 μM C in 1997, 108 μM C in 2003, and ~125 μM C in 2006 and 2008. The lowest average values (~90 μM C) occurred in 2009, and the highest values (~155 μM C) occurred in 2010 (Fig. 6.1). The lowest concentrations were observed in the upper pristine areas of the catchment (3000–3300 km from the estuary) in 1997 and 2003, and in the middle stream, in the dry season of 2008. In the middle and lower streams of the river, the DOC concentration remained almost constant in 1997 and 2003, but increased downstream in 2006 and 2008. Higher values were observed during the 2010 wet season, which may be due to efficient leaching from soil. Although the data were collected in various months, the seasonal variation is limited, especially in the upper stream.

POC, expressed as a percent of the TSM (POC (%) = 100 × POC/TSM), ranged from 0.5 to 2.5 % in 1997 and 2003 for the whole watershed. Higher POC levels were observed in the middle and lower streams of the river in 2006 and 2008, but POC dropped back to 2003 values in 2009 and 2010 (Fig. 6.1). The seasonal variation of POC is larger than the inter-annual variation. For the samples collected from 1997 and 2003, the POC distribution is similar to that of DOC (Fig. 6.2). Data from downstream stations, near the estuary, showed increasing POC from 1997 to 2010.

From the 1997 tributary OC data, higher DOC concentrations (>88 μM) were found in the upper stream, especially in the northern tributaries (>126 μM). In the middle stream, the lowest DOC value was observed in the southern most tributary, Yuanjiang (16 μM). The variation of DOC in various tributaries has a good correlation with population density in the watershed, which may shed light on the potential contribution of human activities to DOC levels

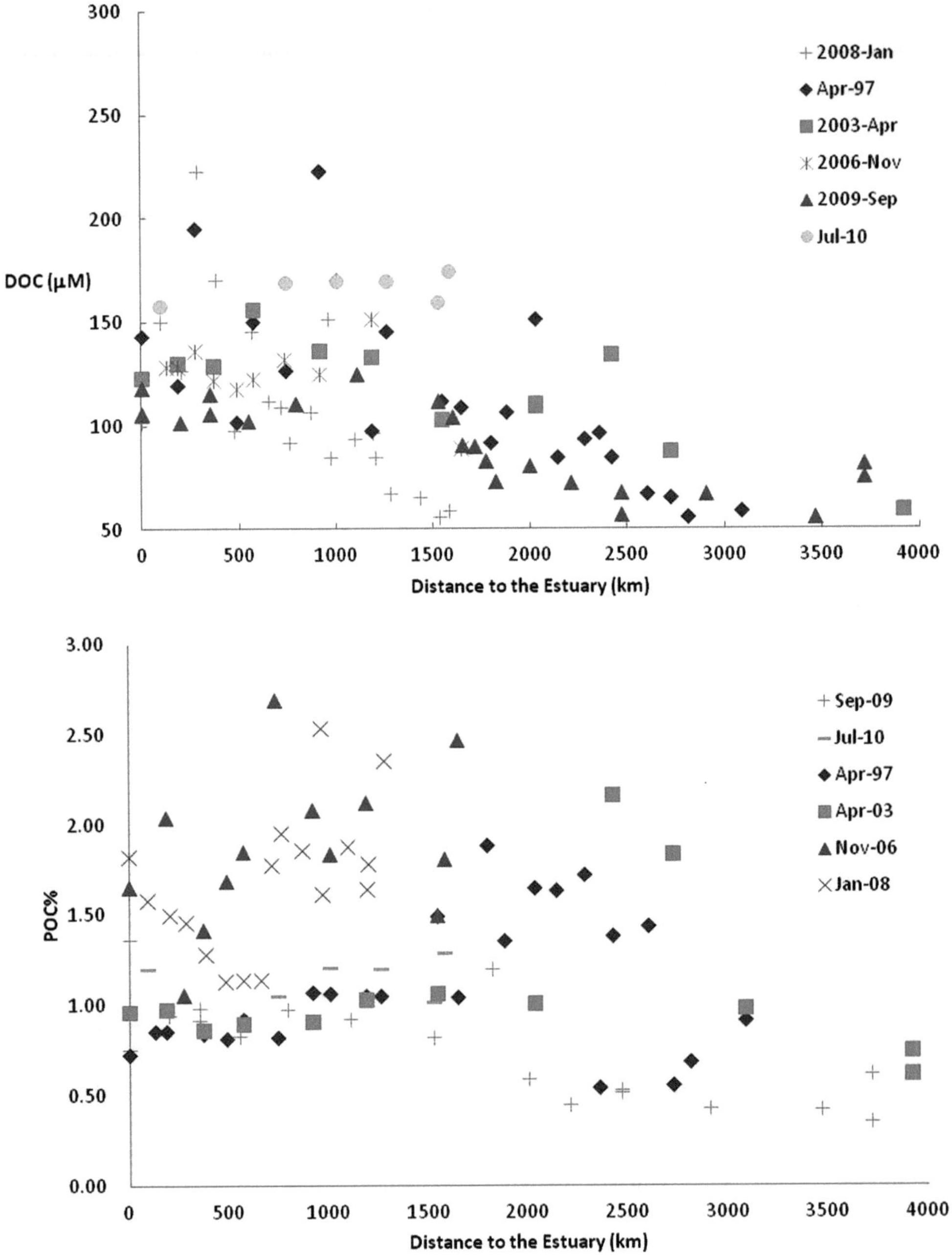

Fig. 6.1 Spatial variations in DOC and POC in the Changjiang during various years

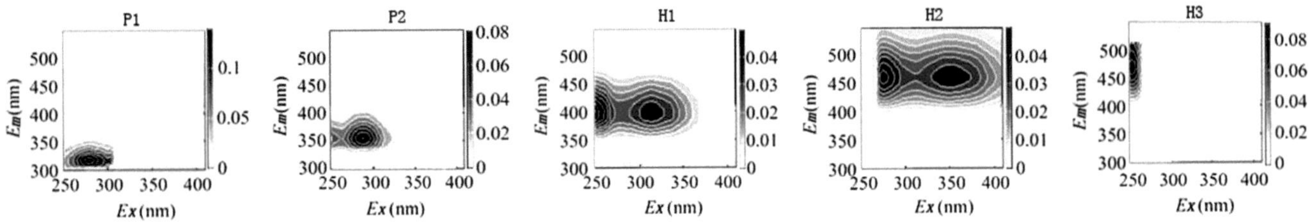

Fig. 6.2 Five fluorescent components (*P1*, *P2*, *H1*, *H2*, and *H3*) identified by the PARAFAC model

(Wu et al. 2007b). However, the correlation of POC and human populations reveals no obvious trend between them.

Average $\delta^{13}C$ and $\delta^{15}N$ values in plant material varied from −28.8 to −24.3‰ and −0.9 to 5.5‰, respectively. The surface-soil carbon isotope values are, in general, similar to those of C-3 plant leaves (Wu et al. 2007b). Autochthonous input from freshwater phytoplankton is minimal in the river and/or is masked by the input of terrestrial organic matter from the tributaries. The $\delta^{13}C$ and $\delta^{15}N$ values in tributaries and in the main channel varied from −26.8 to −25.1‰ and 2.8 to −6.0‰, respectively. As the isotopic signature of the particulate organic matter (POM) is very similar to that of surface soils along the riverbank, the predominant source of organic matter in the river is thought to be the soil (Wu et al. 2007b).

In the past four decades, POC fluxes decreased from $>5 \times 10^6$ t year^{-1} for 1960–1980 to about 2×10^6 t year^{-1} in 1997 (Wu et al. 2007b). Such fluctuation in POC fluxes may be explained by the decrease in sediment load in the Changjiang. Based on monthly observations at XLJ, DOC and POC fluxes in 2003 were estimated to be 1.32×10^6 t and 2.69×10^6 t, respectively. The DOC and POC fluxes in 2004 were only 1.20×10^6 and 1.63×10^6 t, respectively (Lin et al. 2007). During the extreme flood season of 1998, both the DOC and POC fluxes increased dramatically in a short period to 5.5×10^6 t and 9.0×10^6 t, respectively. Such huge DOC and POC fluxes transported to the coastal ECS in such a short period (about two months), with high levels of nutrients and pollutants, could be an important trigger for coastal environmental problems (Wu et al. 2007b).

6.3.2 Characterization of DOM with EEMs

Parallel factor analysis (PARAFAC) has revealed five individual fluorescent components in DOM samples collected from the 2009 field campaign. These include three humic-like (H1, H2, H3) and two protein-like components (P1, P2) (Fig. 6.2), which originate from a wide range of sources and are subjected to different removal and diagenetic processes. As these components suggest different sources and dynamics, EEMs are a powerful tool in understanding the biogeochemical cycles of DOM (Table 6.1).

The sum of all five components (ΣFluo) showed similar variability to DOC, which increased remarkably from the upper stream of the river to the TGD and then remained constant in the lower stream. ΣFluo was linearly correlated with DOC ($R^2 = 0.92$), indicating that EEMs are a powerful tool to trace DOM in the Changjiang. The protein-like component (ΣP) accounted for one quarter of ΣFluo, with weak correlations with DOC and chlorophyll-*a*. This implies that riverine DOM is not dominated by autochthonous

Table 6.1 Fluorescence groups and implications

Name	Type	Name from Coble	Property	Main source
P1	Protein-like	B	Labile	Autochthonous/ polluted water
P2	Protein-like	T	Labile	Autochthonous/ polluted water
H1	Humic-like	M	Less labile	Main humic-like peak in autochthonous/ polluted water
H2	Humic-like	C	Refractory	Allochthonous (soil-derived DOM)
H3	Humic-like	A	Refractory	Wide range of existence and resources

production, especially in the upper stream where there is higher TSM than in other regions.

Humic-like (H1, H2, H3) components have different correlations with DOC, especially in the lower stream where H3 (Ex/Em: 250/450–485 nm) was accumulated. The other components (H1, H2) and a_{350} seemed to degrade faster than H3. This indicates that DOM might be subjected to selective biological and photochemical degradation processes in the Changjiang. Significant differences in molecule weight and aromaticity, as indicated by the spatial distributions of Sr and $S_{290-350}$, suggest that there is potential for stronger photo degradation in the lower stream and for newly produced DOM to be retained by the TGD.

6.3.3 Dynamics of POC and DOC in the East China Sea

Dissolved organic matter and particulate organic matter have a profound role in the marine biogeochemical cycle of carbon (Druffel et al. 1992). The continental margins occupy less than 20 % of the surface area of the world's oceans, but play an equally important role to that of the deep sea in terms of the carbon biogeochemical cycle (Walsh 1991).

Limited DOC data are available for the ECS (Hung et al. 2000; Ogawa et al. 2003; Lin 2007), but it is known that DOC concentrations are controlled by the hydrographic setting and have a strong correlation with salinity. Highest DOC values are observed in the Changjiang Estuary, and the lowest values occur in Kuroshio samples (Hung et al. 2000).

The spatial distribution of DOC is relatively invariable on a seasonal basis in the ECS region. DOC depth profiles show that concentrations are moderately high in the surface layer and generally decrease with depth. The DOC concentration in shelf waters decreases with increasing salinity, which indicates the mixing of the DOC-rich Chinese coastal water in the inner shelf and the DOC-poor Kuroshio water from the shelf break (Hung et al. 2000).

POC has a higher spatial variability than DOC in the ECS. The POC maximum occurs near the Changjiang Estuary, where turbidity is high (50–60 μg/L), but on the inner shelf, the concentration is homogenous in the surface layer. POC abundance generally decreases seaward from the Changjiang Estuary to the Kuroshio waters (Wu et al. 2003). The concentration is generally the greatest in the surface layers and decreases with depth, but there are exceptions to this general trend.

The anomalously high concentrations of POC in the bottom water in the mid and inner shelves resulted from resuspension of bottom sediments, particularly during rough sea conditions (Zhu et al. 2006). Besides wind mixing, strong tidal currents can also induce resuspension in the inner shelf (Hung et al. 2000; Zhu et al. 2006). The significant correlations between POC and TSM, as well as between POC and chlorophyll-*a* in the euphotic zone, suggest that resuspension and in situ production are two dominant processes controlling the POC load in the bottom and euphotic layers of water column, respectively.

The biogenic nature of POC is also supported by a highly significant correlation between POC and PN in the euphotic zone of the study area (Hung et al. 2000; Zhu et al. 2006). A recent study based on the observation of typhoon Fengwong in the southern ECS found that the in situ biological responses, including phytoplankton blooms and POC fluxes, were associated with the storm. The passage of the typhoon was followed by a dramatic increase in chlorophyll-*a* concentrations, indicating that the POC export flux was enhanced due to a phytoplankton bloom (Hung et al. 2010).

The residence time of POC in the ECS water column is on the order of weeks and decreases seaward from the coast (e.g., off the Changjiang Estuary) to the shelf break. The residence time is inversely related to the organic carbon concentration in surface sediment (%). Such short POC residence times indicate that the POC of the ECS undergoes rapid removal (i.e., vertical export or decomposition) from euphotic waters (Zhu et al. 2006).

6.4 Characterization of Organic Matter with Various Biomarkers

6.4.1 Temporal Variation in Riverine Organic Matter from Pigments and Amino Acids

Long-term monitoring studies of rivers are valuable for assessing the effects of climatic change *versus* anthropogenic impacts. However, previous research on riverine organic matter is still very limited. We established the monitoring station at XLJ to collect monthly samples (from 2003 to the present) of organic matter. The detailed characterization of POM from this station is presented here, with a focus on pigments and amino acids.

Overall, the chlorophyll-*a* concentrations at XLJ increased during our sampling period (from 2004 to 2007). The mean chlorophyll-*a* concentrations for 2005 and 2006 were 0.9 and 1.1 μg/L, respectively. Chlorophyll-*a* concentrations during the wet (May to October) and dry (November to April) seasons did not show any clear differences, but other diagnostic pigments (e.g., chlorophyll-*b* and fucoxanthin) were higher in the wet season relative to the dry season (Table 6.2).

Nutrients are considered an important factor for phytoplankton growth, and hence pigment concentrations, but at XLJ, phytoplankton pigments are more likely influenced by other factors such as light illumination and water stability (i.e., the flushing effect). Surface chlorophyll-*a* at XLJ is limited to TSM, and the relationship can be expressed as follows:

$$\text{Chlorophyll-}a = -0.2917 \times \text{Ln(TSM)} + 2.0367 \qquad (6.1)$$

It is interesting that when TSM was lower than 100 mg/L, chlorophyll-*a* concentrations also had high variability, suggesting that when TSM < 100 mg/L, other factors limit chlorophyll-*a* concentrations (e.g., temperature, nutrients, water stability, and the flushing effect). When TSM was higher than 100 mg/L, chlorophyll-*a* concentrations were

Table 6.2 Phytoplankton pigments from XLJ from August 2004 to February 2007

	Per	Fuco	Neo	Viola	Diadnio	Allo	Zea	Lut	Chlb	Chla	Pras	Hex-fuco
Average	0.1	0.33	0.02	0.03	0.04	0.04	0.02	0.05	0.09	0.88	bdl	bdl
Max	0.34	0.98	0.19	0.08	0.45	0.21	0.09	0.13	0.8	3.46	bdl	bdl
Min	bdl	0.03	bdl	bdl	bdl	0.01	bdl	bdl	0.02	0.01	bdl	bdl

Notes (1) *Unit* μg/L; *bdl* below detection limit. (2) Abbreviation: *Per* peridinin, *Fuco* fucoxanthin, *Neo* neoxanthin, *Viola* violaxanthin, *Diadino* diadinoxanthin, *Allo* alloxanthin, *Zea* zeaxanthin, *Lut* lutein, *Chlb* chlorophyll-*b*, *Chla* chlorophyll-*a*, *Pras* prasinoxanthin, *Hex* fuco-19′-hexanoyloxyfucoxanthin

Table 6.3 Viola/Chla ratios for different end members

	Taihu Lake	Higher plant	XLJ (summer)	XLJ (winter)
Viola/Chla	0.014	0.072	0.024	0.034

Note Viola-violaxanthin, Chla-chlorophyll-*a*

negatively related to TSM concentrations, indicating that when TSM > 100 mg/L, light becomes the limiting factor for phytoplankton growth.

Unlike marine systems, riverine systems are exposed to strong terrestrial inputs, and plant pigment contributions to riverine pigments can be significant. Table 6.3 gives violaxanthin/chlorophyll-*a* ratios, which can partly reflect the plant contributions to riverine systems. For severe-bloom samples (i.e., Taihu samples), the violaxanthin/chlorophyll-*a* ratio was as low as 0.014, whereas the violaxanthin/chlorophyll-*a* ratio for higher plants can be up to 0.072 (Table 6.2). Thus, higher plants have violaxanthin/chlorophyll-*a* ratios five times greater than that for algae. It is interesting that the violaxanthin/chlorophyll-*a* ratios of XLJ samples are within the range expected for algae blooms and higher plants (Table 6.3). Pigments in XLJ samples are, therefore, derived from in situ production (algae growth or advection flux) and higher-plant detritus.

Furthermore, a quantitative estimate of both source contributions to XLJ pigments can be made by taking the bloom ratio (0.014) and the higher plant ratio (0.072) as two end members. The result is that, in summer, higher plants contribute 17 % of the total XLJ pigments, and 34 % in winter. In other words, phytoplankton contributes much more to the total observed pigments in summer (83 %), whereas in winter, their contribution is reduced to 66 %.

CHEMTAX (Box 6.2) (Mackey et al. 1996) is a program that is widely used in studies of marine and riverine environments to estimate the phytoplankton community structure based on phytoplankton pigments (e.g., Mackey et al. 1996; Valdes-Weaver et al. 2006; Gameiro et al. 2007). This program is not applied here, primarily because it is mainly designed to estimate the phytoplankton community structure, rather than applying pigments as biomarkers to characterize the organic matter, which is the main focus of this chapter.

Box 6.2. CHEMTAX and Pigments

In addition to classic microscopic methods, phytoplankton pigments are powerful tools for studying phytoplankton. Chlorophyll-*a* has long been used as a unique molecular marker of phytoplankton biomass, whereas diagnostic pigments are useful markers in studies of phytoplankton community structure. High-performance liquid chromatography (HPLC) is widely applied to analyze pigments in investigations of phytoplankton community structure, especially since the introduction of CHEMTAX. CHEMTAX uses a steepest-descent algorithm to identify the best fit to the data matrix based on the suggested pigment/chlorophyll-*a* ratio. This method does not require the presence of unambiguous marker pigments and can yield reasonable results when compared with classic taxonomic methods. CHEMTAX was first introduced by Mackey et al. (1996). In addition to its use in marine studies, CHEMTAX has been successfully applied in rivers, lakes, and estuaries. To yield reasonable results via CHEMTAX, the following points should be considered:

1. Sample size: There should be enough samples to run the calculation (≥15).
2. Because the program assumes that the pigment ratios within a class are the same, the samples should be from the same phytoplankton community and live under similar physiological conditions.
3. Pigments that are present in nearly all algal classes are unlikely to yield useful information.

For a detailed description of CHEMTAX, see the CHEMTAX manual (Mackey et al. 1997).

Total particulate amino acids (TPAA) at XLJ have strong relationships with TSM ($r = 0.78, p < 0.005$), suggesting that riverine suspended particles are key contributors to amino acids (Shao et al. 2011). This relationship also suggests that soil organic matter within the river basin (including bacterial degradation) is one of the main sources of organic matter (Wu et al. 2007a). Amino acid enantiomers also varied with TSM (Shao et al. 2011), which means that in situ production must only play a minor role in sources of organic matter.

This is further proved by the weak relationship between the diagenetic index (DI) and the POC/chlorophyll-*a* ratio. In the case of amino acids driven by in situ production, low POC/chlorophyll-*a* values should correspond to high DI values. At XLJ in April, when primary production can be high while river discharge is still low, the opposite relationship occurs, i.e., low POC/chlorophyll-*a* ratios correspond with high DI values. In July, when river discharge is high, this relationship does not exist (Shao et al. 2011).

This trend suggests that hydrological and weathering processes are the main factors in determining the nature of riverine organic matter, rather than riverine in situ production processes (Shao et al. 2011). This interpretation is supported by the DI value of the whole Changjiang, which is >−1.0, indicating that organic matter is highly degraded (Wu et al. 2007a).

Monthly observations reveal that the D form of amino acids enantiomers makes up 6 % of the TPAA (Shao et al. 2011), which is similar to the results of Wu et al. (2007a) for a basin-wide investigation. As for respective amino acids enantiomers, D-Asx is lower in the Changjiang (6 %) relative to Russian Arctic rivers (11 %). D-Glx, by contrast, is 12 % for the Changjiang and 7.5 % for Russian Arctic rivers (Dittmar et al. 2001; Wu et al. 2007a).

6.4.2 Delivery of Particulate Terrigenous Organic Matter from the River to the Shelf

In the early 1990s, bulk geochemical indicators (e.g., carbon isotopes and the C/N ratio) and biomarkers began to be used to elucidate the behavior and composition of organic matter in the Changjiang Estuary (Cauwet et al. 1993; Sicre et al. 1993; Bouloubassi et al. 2001). Those earlier data showed that the large range of carbon isotopic compositions observed during the period of high water-discharge (June) in the Changjiang Estuary could be attributed to the mixing of terrestrial and marine organic carbon. A dominant contribution of terrestrial organic carbon was observed during November, when low water-discharge occurs.

The limited sedimentary organic carbon in the ECS exhibits a relatively narrow isotopic range, suggesting a dominant marine source (Lin et al. 2002). Particulate sterols demonstrated a good correlation with chlorophyll-*a* in the Changjiang Estuary. The composition of sterols indicates that diatoms are the main phytoplankton species in the saline water environment (Sicre et al. 1993). Changes in the concentrations and compositions of sterols between flood and ebb periods reveal the migratory nature of the turbidity maximum zone in the estuary (Sicre et al. 1993).

Limited information about POM behavior in the watershed is available in the literature (Wu et al. 2007b). From our recent research of particulate lignin phenols in the basin, we found high spatial and temporal variability in the composition and diagenetic status of lignin phenols in the middle and lower streams of the river (Yu et al. 2011). The abundance of vanillyl, syringyl, and cinnamyl (V + S + C) lignin-derived phenols (Λ8) (mg/100 mg OC) in 2006 were much lower than in 2003, with average values of 0.90 and 1.91 mg/100 mg OC, respectively (Table 6.4). Most Ó8 values (the abundance of (V + S + C) normalized to suspended matter (mg/10 g) of SPM) from 2006 were also lower than those from 2003, except for samples collected nearest to the estuary. The S/V and C/V ratios of SPM from both years were much greater than zero, indicating that non-woody angiosperm tissue was the source of the SPM (Yu et al. 2011). Moreover, the ratios were similar in both years, indicating a consistent source of terrestrial OC.

Table 6.4 Comparison of variables between April–May 2003, September 2006, and January 2008 in the middle stream of the Changjiang

	April–May 2003	September 2006	January 2008
	Pre-impoundment	Post-impoundment	Post-impoundment
TSM (mg/L)	93.8 ± 23.9[a]	14.59 ± 7.87[a]	48.5 ± 51.8
Grain size (μm)	15.4 ± 3.8[a]	6.85 ± 1.95[a]	7.2 ± 1.6
OC%	1.00 ± 0.08[a]	2.14 ± 0.54[a]	2.41 ± 1.14
C/N	n.a.	n.a.	7.8 ± 1.0
$\delta^{13}C$	−24.8 ± 0.1[a]	−24.8 ± 0.4[a]	−25.4 ± 0.5
Σ8	1.98 ± 0.27[a]	0.96 ± 0.62[a]	2.20 ± 0.76
Λ8	2.00 ± 0.44[a]	0.78 ± 0.066[a]	1.08 ± 0.61
(Ad/Al)v	0.25 ± 0.04[a]	0.49 ± 0.34[a]	0.43 ± 0.17

[a]Data from Yu et al. (2011)
n.a. not available

Most SPM (Ad/Al)v ratios (the ratio of the acid to aldehyde of the vanillyl series) were higher in 2006 than in 2003, with values >0.4. This shows that most of the SPM from 2006 was degraded. However, although most SPM samples from 2006 were defined as degraded, and those from 2003 were considered to be fresher, the difference in (Ad/Al)v between years was not statistically significant. Based on average values, the difference between the two years was higher in the middle stream than in the lower stream of the river (Table 6.4) (Yu et al. 2011).

To better understand the carbon cycle and biogeochemical processes in the East China Sea, it is important to study the contribution of terrigenous POM from the Changjiang to the shelf region. Figure 6.3 shows the percentage of terrigenous POM, estimated by isotope composition end-member calculations (Wu et al. 2003). Within a distance of <100 km from the river mouth, the biogeochemical cycling of POM is controlled by water/sediment processes, and the terrigenous POM contributes 70–80 % of the total POM. The percentage of terrestrial POM decreases quickly at a distance around 150 km from the mouth, and only half of the continental POM could be traced to this region. This area is also a productive fishing ground, and the terrigenous POM is important in nourishing fish stocks. Farther offshore (>250 km), the contribution of terrigenous POM is limited, particularly where the Kuroshio water intrudes from the shelf edge and brings organic matter from the marine pool (Wu et al. 2003).

The highest concentrations of particulate amino acid (PAA) were found in the Changjiang Estuary (Wu et al. 2007a). From the upper to the middle stream of the Changjiang, the concentrations of PAA steadily decreased from 2.3 to 1.3 μM, but increased from the middle to the lower stream to reach maximum concentrations in the estuary (4.4 μM on average). Across the shelf, the PAA

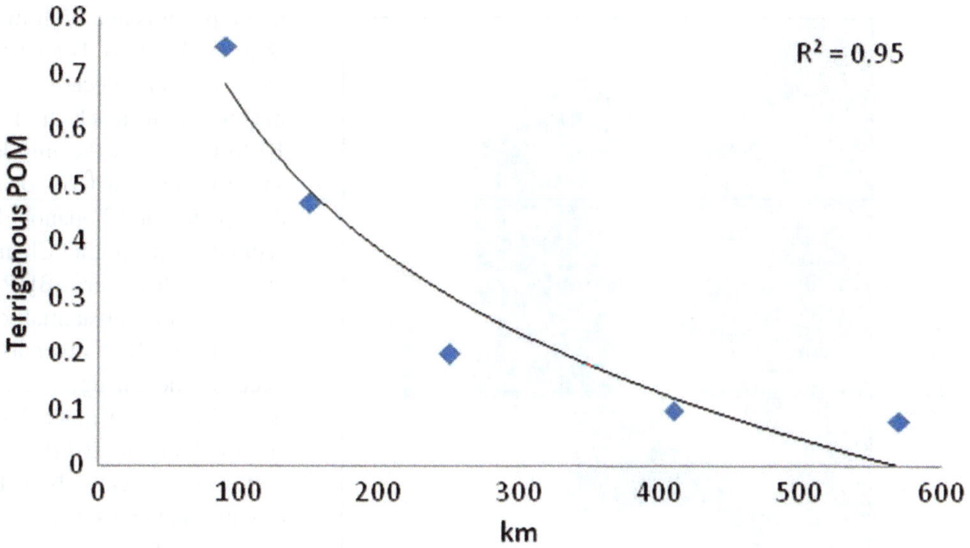

Fig. 6.3 Distribution of the percentage of terrigenous POM from the Changjiang Estuary to the ECS shelf based on the isotopic composition (after Wu et al. 2003)

concentration steadily decreased to 0.23 μM at the shelf edge (Wu et al. 2007a). The molecular composition of PAA also differed significantly between riverine and oceanic samples.

The amino acid composition of the POM in the Changjiang indicates DI values ranging from −0.8 to −1.3, suggesting that the POM is in an advanced stage of degradation (Wu et al. 2007a). As suggested by Wu et al. (2007a), highly degraded organic matter from soils is probably a main source of POM in the Changjiang, but the relatively high $\delta^{13}C$ values and low C/N ratios (7.7 ± 1.6) also indicate a contribution from anthropogenic sources.

It is interesting that the DI values of the whole river have a relatively low range (around −1), suggesting that the organic matter has a rather homogenous degradation status. When the organic matter enters the estuary, the DI value quickly increases to over −0.5 and then increases steadily on the shelf with distance offshore to 0.8 ± 0.1 at the shelf edge due to in situ production (Wu et al. 2007a). This complex pattern of DI from the lower stream of the river to the estuary suggests that the organic matter is affected by dynamic processes in this region. Possible explanations for the observed trend include flocculation and sorption/desorption due to changes in saline waters. More estuarine studies are needed to further understand organic matter processes in this environment.

6.4.3 Distribution of Organic Matter in the Sediments Over the East China Sea

Due to the wide and shallow shelf, with large river inputs, the ECS provides an ideal setting to evaluate the burial of terrigenous organic matter (Deng et al. 2006). The Changjiang Delta is an important geographic feature of the inner continental shelf of the ECS (up to ~60 m in depth). It receives large amounts of material from the Changjiang during the flood season, and much of this is eroded during winter storms and transported south to the Taiwan Strait (Liu et al. 2007).

The outer continental shelf of the ECS (water depths of 100–200 m) is supplied with only a little terrestrial material, and most of the area, known as the 'relict deposit,' retains the morphological features and sedimentary characteristics of the late Pleistocene, when sea levels were lower (Emery 1968). Since the 1980s, research has focused on the fate of Changjiang material once it reaches the ECS shelf (DeMaster et al. 1985; Milliman et al. 1985; Hori et al. 2002; Liu et al. 2006, 2007). This work was mainly conducted using either bulk parameter analysis or sedimentologic tools. Detailed source determination, using special biomarkers (e.g., lignin phenols) or radiocarbon ages of organic materials, is still limited (Zhu et al. 2011a; Li et al. 2012). These studies can be challenging and sometimes inconclusive because organic proxies may be modified by shelf processes, such as hydrological sorting, selective degradation, and lateral transport (Goñi et al. 1998; Ohkouchi et al. 2002; Zonneveld et al. 2010). In addition, sources are highly heterogeneous, and processes in marginal seas are highly dynamic.

The distribution of organic carbon in the ECS continental shelf sediments shows a banded pattern (Lin et al. 2002; Zhu et al. 2008). In general, the organic carbon concentrations decrease seawards, reflecting a corresponding increase in the sediment grain size and proportion of marine organic material. Lower concentrations of organic carbon, observed in the middle and outer continental shelf sediments, are probably a combined result of limited fine-grained sediment deposition and the rapid oxidation of organic carbon in the sediments. Organic carbon in the delta and inner shelf

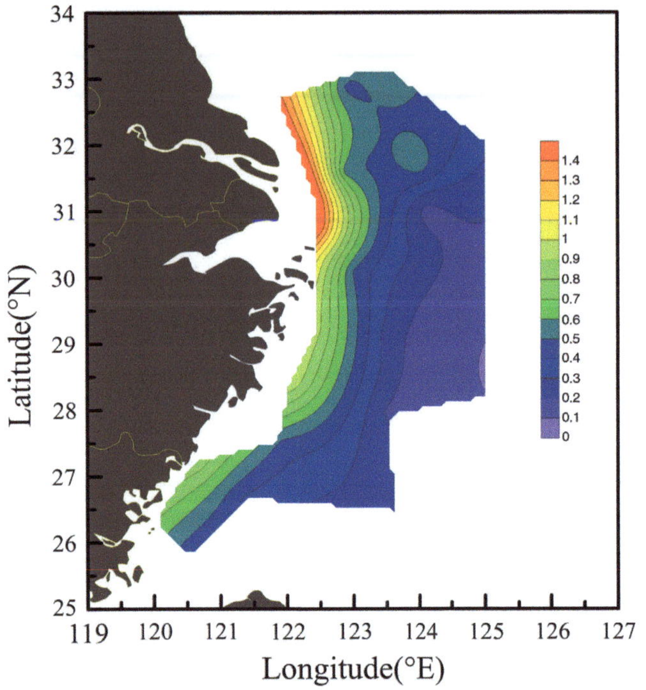

Fig. 6.4 Spatial distribution of Λ8 (mg/100 mg OC) of lignin phenols in the ECS shelf (modified from Yang et al. 2008)

sediments is composed primarily of terrigenous material, and $\delta^{13}C$ values are as low as −25‰ (Lin et al. 2002).

The abundance of lignin-derived phenols (V + S + C) (Λ8) (mg/100 mg OC) decreased seaward from the river mouth to the shelf, but there was no discernible decrease southward along the Zhejiang–Fujian Coast (Yang et al. 2008) (Fig. 6.4). The highest value was detected at the river mouth station, with a value of 2.50 mg/100 mg OC, which is also confirmed by Li et al. (2012). The average Λ8 value differed between inner and outer shelf sediments, with values of 0.95 and 0.29 mg/100 mg OC, respectively, which are comparable to values from the Gulf of Mexico and the Peru margin (Sampere et al. 2008; Bergamaschi et al. 1997).

Average values of syringyl to vanillin (S/V) and cinnamyl to vanillin (C/V) ratios were 0.70 and 0.10, respectively, for the inner shelf, which is close to the values observed in the lower stream of the Changjiang (Yu 2007). S/V and C/V values of 0.95 and 0.16, respectively, for the outer shelf indicate that samples are mainly composed of lignin compounds from angiosperm/gymnosperm woods with a mixture of non-woody angiosperm tissue. The seaward increase in S/V and C/V ratios indicates increasing proportions of angiosperm leaves- and grasses-derived lignin in the outer shelf region. The average (Ad/Al)v values are much higher on the outer shelf (1.21) than on the inner shelf (0.59).

The abundances of glycerol dialkyl glycerol tetraethers (GDGT) also show a clear land–sea trend, consistent with the distributions of other terrestrial biomarkers, such as lignin phenols and high-molecular weight (HMW) n-alkanes (Zhu et al. 2008). Branched GDGT abundance is highest in the river and decreases substantially, by two orders of magnitude, to reach the lowest values (decreasing from ca. 10 to 0.1 %) in the mid-lower estuary (Zhu et al. 2011a). Twenty-four bacteriohopanepolyols (BHPs), diplopterol, diploptene, and hopanols have been identified in soils and sediments from the Changjiang–ECS system (Zhu et al. 2011b). However, BHP concentrations and structural diversity are substantially lower in shelf samples than in soils. High BHP diversity and enhanced BHP production occur in the Changjiang Estuary, whereas BHP distribution is uniform, with much less structural diversity in the ECS shelf (Zhu et al. 2011b).

Limited black carbon (BC) data and radiocarbon ages of organic matter from the ECS are available in the literature (Wang and Li 2007; Li et al. 2012). However, a significantly higher (>30 %) proportion of BC/OC was observed in a recent study (Li et al. 2012), which is higher than in most other coastal regions. Combined with isotope values, this indicates that the sources of BC in the ECS may be derived from combustion/pyrolysis by-products of fossil fuels, biomass burning, soil-derived polymers, and ancient rock weathering (Li et al. 2012). The radiocarbon ages of surface sediments of the ECS shelf show temporal and spatial variability in the ECS. The age data collected during the floods of 2010 are older due to the inputs of older material from the Changjiang (Li et al. 2012). Planktonic sources of OC may also cause the slightly younger ages at some stations, when higher pigment values are detected.

Chlorophyll-*a* in surface sediments averaged 65 mg/g OC, and elevated values were found at the edge of the study area (Fig. 6.5). Elevated chlorophyll-*b*, alloxanthin, and neoxanthin were found close to the northeastern region, while lutein was elevated in the southeastern region. Other diagnostic pigments (i.e., fucoxanthin, peridinin, violaxanthin, and zeaxanthin) were elevated in the vicinity of the Zhejiang Province coast, and in the northern and/or the eastern region (Fig. 6.5). Fucoxanthin/chlorophyll-*a* and the Peridinin/chlorophyll-*a* ratio (mass ratio) had average values of 0.56 and 0.25 in surface sediments, respectively, which is consistent with values in the water column (Fig. 6.6). At XLJ station, located in the lower stream of the Changjiang, these two ratios were 0.41 and 0.06, respectively, indicating a terrestrial influence (Fig. 6.6).

Fucoxanthin was better correlated with chlorophyll-*a* in surface sediments (R^2 = 0.88, p < 0.0001, n = 31) than peridinin (R^2 = 0.26, p = 0.008, n = 31), because it is influenced by benthic diatoms and biological processes in the water column. Lutein is one of the pigments contained in higher plants (Jeffrey et al. 1997) and, among the phytoplankton commonly observed in the Changjiang Estuary, lutein is mainly related to chlorophytes, which also contain

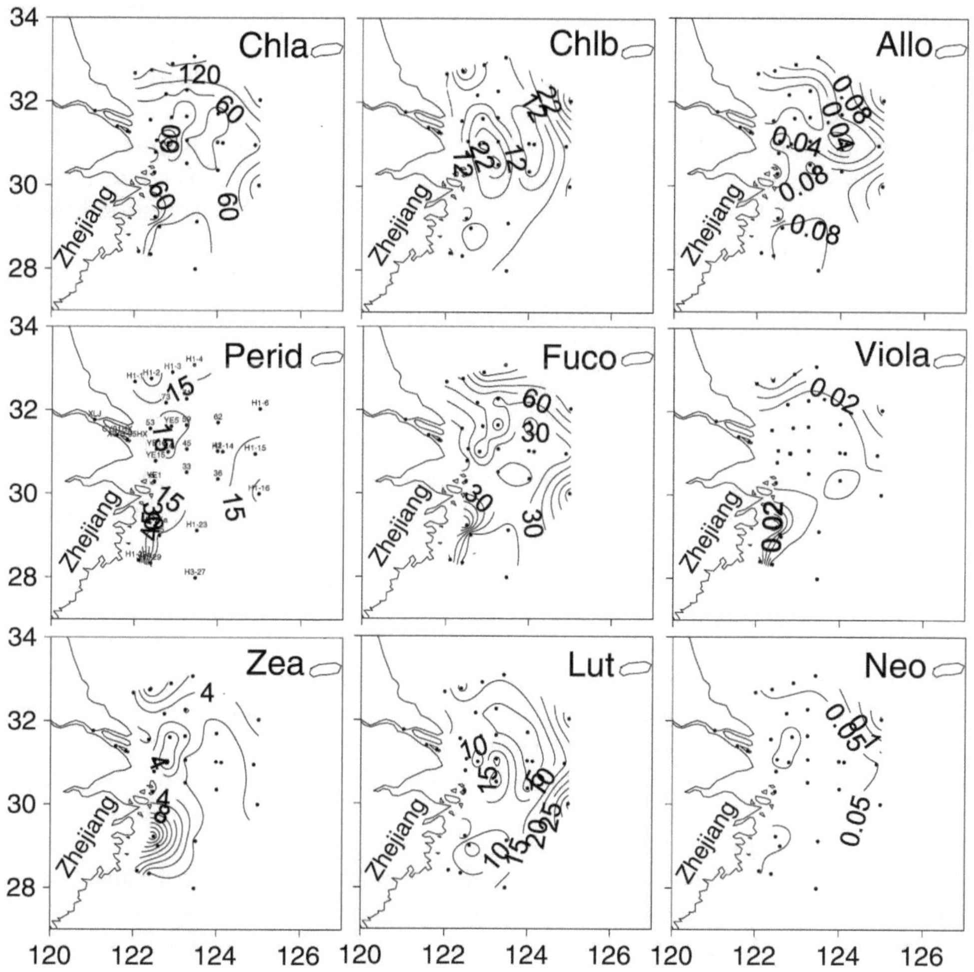

Fig. 6.5 Pigment distributions (mg/g OC) in surface sediment of the ECS shelf. Abbreviations are listed in Table 6.2

chlorophyll-*b* (Mackey et al. 1996; Jeffrey et al. 1997). Low abundances of lutein in the open estuary resulted in much higher chlorophyll-*b*/lutein ratios in the Changjiang Estuary than those found in the Changjiang and Hudson rivers, where lutein could partly originate from terrestrial higher plants (Bianchi and Findlay 1990). This elevated ratio consequently resulted in higher chlorophyll-*b*/lutein ratios in the Changjiang Estuary sediment (1.6) than in the Hudson River's sediment (0.4–1.0) (Bianchi et al. 1993). The ratio of chlorophyll-*b*/lutein was greatly reduced in the sediment (with an average of 1.6) (Fig. 6.6), probably because chlorophyll-*b* decays much faster than lutein (Bianchi and Findlay 1990).

High sedimentation rates (Henrichs and Reeburgh 1987), elevated primary production (Calvert and Peterson 1991), and anoxic conditions (Ding and Sun 2005) are favorable for the preservation of phytoplankton pigments in sediment. The majority of organic matter is derived from in situ primary production (e.g., pigments) and is easily degraded within the water column before being deposited in sediment (Hurley and Armstrong 1990; Bianchi et al. 2002), due to oxidation and herbivory and/or enzymatic metabolism during senescence (Chen et al. 2001). As reported previously, however,

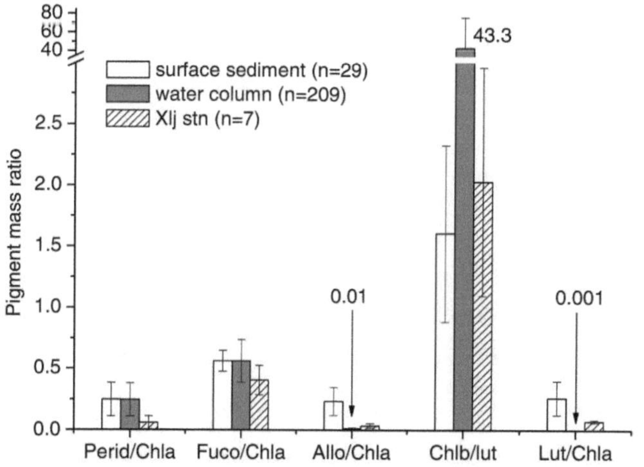

Fig. 6.6 Average summer pigment mass ratios in surface sediments of the ECS, corresponding upper water columns, and at XLJ. Numbers above the columns indicate the corresponding average value. Pigment data from the water column and XLJ are from Zhu (2007). Only summer data were used, and the error bars are the standard deviations of the data. Abbreviations are listed in Table 6.2

pigment concentration in surface sediments usually fails to have strong relationships with water depth in estuarine areas such as the Mississippi River Estuary (Wysocki et al. 2006), the Rhone River Estuary (Alliot et al. 2003), and the Palmones River Estuary (Moreno and Niell 2004). The water depth at all studied stations ranged from 10 to 88 m, with average values of 39 m.

There was no clear relationship between pigment concentration and water depth ($R^2 < 0.2$ in all cases) in the Changjiang Estuary. Hence, in areas with strong hydrodynamic conditions and sharp biogeochemical gradients, such as estuaries, pigment distribution in the surface sediment is more regulated by the extension and variation of the riverine plume (i.e., terrestrial/riverine influence), and corresponding autochthonous biological processes, than by water depth alone (van Iperen et al. 1987; Alliot et al. 2003; Wysocki et al. 2006).

Available organic proxy data in the ECS have elucidated the difference between bulk TOC and variable biomarkers (e.g., HMW alkanes, lignin phenols, and pigments) over the shelf (Yang et al. 2008; Zhu et al. 2008, 2011a, b). This may reflect hydrodynamic sorting and the reworking processes of variable biomarkers. Although the dominant source of terrestrial OC is derived from the Changjiang, in some small regions, the contribution from other rivers along the Zhejiang and Fujian coasts cannot be ignored. Lateral transport may be quite important due to the special character of sediment transport mechanisms in the Changjiang Estuary (sediment is buried in summer and redeposited in winter) (Liu et al. 2007). Flood events from the watershed and storms over the shelf may modify the distribution and flux of organic matter, and further impact the regional diagenetic and burial processes over the ECS.

6.5 Anthropogenic Impacts on the River, Estuary, and Adjacent Regions

6.5.1 Impacts of Dams on the Transportation of Terrestrial Organic Matter to the Lower Stream

To identify the impact of dams on the transportation of terrestrial organic matter (OM) in the lower Changjiang, POC, total nitrogen (TN%), stable carbon isotopes ($\delta^{13}C$), and lignin phenols were determined for samples collected before and after the closing of the TGD. The POM in the Changjiang may consist of C-3 plant material, soil OM, and phytoplankton. Soil OM is the predominant fraction, while the contribution from C-4 plant material was insignificant, which is in agreement with the distribution of C-3 and C-4 plants in the Changjiang drainage basin (Still et al. 2003).

Figure 6.7 clearly shows that the post-TGD period (October 2006 and January 2008) had lower Λ8 values compared with the pre-TGD period (April 2003), in both the middle and lower stream. This indicates a decrease in the terrestrial OM and/or an increase in in situ production in these streams. The POC sources in the river often relate to total suspended particle concentrations (Ludwig et al. 1996; Sempr et al. 2000; Contrim et al. 2001; Gao et al. 2002). The significant positive linear correlation between TSM and Λ8 ($R^2 = 0.64$, $p = 0.06$) indicates that in Changjiang, the factors that cause the variation in TSM are also responsible for the changes in POM compositions.

Before the construction of the TGD, seasonal variation was the major factor that controlled TSM in the Changjiang, e.g., the highest TSM in Datong occurred in July and was

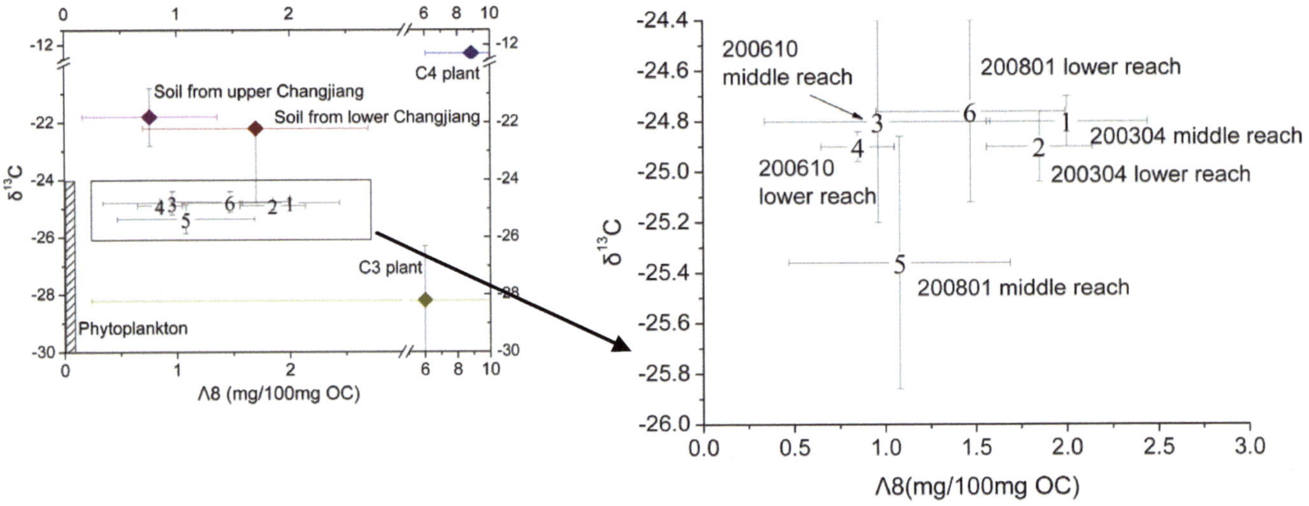

Fig. 6.7 Distribution of $\delta^{13}C$ and lignin phenols of suspended particles from the middle and lower Changjiang, collected in different sampling campaigns: April 2003, October 2006, and January 2008 represent April–May 2003, October 2006, and January 2008, respectively

approximately seven times greater than the lowest TSM (Chen et al. 2001). In the Changjiang, sediments are sourced mainly from the upper basin and the northern region of the watershed (Xu et al. 2007). Thus, after the construction of the TGD, most sediment was transported from the upper stream and was trapped within the reservoir (Yang et al. 2007), which caused a significant reduction in the Changjiang sediment load. Water discharge did not exhibit any significant changes (Yang et al. 2011), but the TSM in the middle and lower Changjiang decreased drastically.

Other dams on the Changjiang have also had an important influence on the decrease in sediment load. Yang et al. (2011) pointed out that the decrease of sediment load in the Changjiang started in the 1960s, when many dams were built on the tributaries of the Changjiang. Danjiangkou Dam on the Hanjiang River, for example, caused the sediment load from the Hanjiang River to the Changjiang to decrease from 100 Mt/year (1951–1972) to 18 Mt/year (1974–2004) (Changjiang Sediment Bulletin 2005).

The TGD also affected the suspended particle size, which further influenced the composition of POM. Studies of the Amazon, Congo, and Piracicaba rivers suggested that the OM sources varied significantly in particle size, with finer particles tending to have lower $\Lambda8$ and higher (Ad/Al)v (Hedge et al. 1986; Krusche et al. 2002; Spencer et al. 2012). In the Changjiang, the smaller grain size in January 2008 partly explains the lower $\Lambda8$ and higher (Ad/Al)v compared with the data from April to May 2003 (Table 6.4). The mean grain size at Yichang during the pre-TGD period was ~20 mm in the January and 28–37 mm in April–May, respectively (Wu 2004). The observed difference between January 2008 and April–May 2003 was greater than in the pre-TGD period (Yu et al. 2011), which suggests that (except for the seasonal variation) the trapping of coarse particles within the TGD was another factor contributing to the smaller grain size in January 2008.

Sediment trapping within the Three Gorgens Reservoir (TGR) caused not only a decrease in TSM and a change in particle size, but also a change in the relative contribution of suspended particles from southern and northern tributaries. Yu (2007) measured the lignin phenol concentration of soils in the Changjiang drainage basin and showed that soils in the southern tributaries tend to have higher $\Sigma8$ and slightly lower $\Lambda8$ values. Since most of the northern tributaries are in the upper stream, this fraction of sediment could not be transported to the ECS after the TGD was built; thus, the southern tributaries might have played a more important role in the post-TGD period.

Table 6.4 shows that $\Sigma8$ in the middle stream of the Changjiang was higher in 2008 than in 2003, which is partly because of the higher proportion of terrestrial OM from the southern tributaries in 2008. Dongting Lake and Poyang Lake are the two most important tributaries in the southern Changjiang. Moreover, because of the trapping of suspended particles within the TGR, erosion of the middle and lower channel has been observed (Yang et al. 2007), which would result in the retransport of organic matter that settled before the closing of the TGD.

Gauging data from Datong station show significant decreases in the Changjiang sediment load and associated terrestrial OM, and this is interpreted to have been induced by the building of dams (Yang et al. 2007). As mentioned, after the building of the TGD, terrestrial OM exported from Dongting Lake and Poyang Lake, as well as OM eroded from the middle and lower riverbed and/or bank, substantially increased.

Table 6.5 shows the average sediment and lignin flux at Yichang and Datong stations from 2004 to 2008. The lignin phenol concentration at the stations is the average value of different sampling campaigns during the post-TGD period, in which lignin data of Xuliujing were used to represent those of Datong, and lignin data for Cuntan were the average value of the upper Changjiang sampled in September 2009 (unpublished data). Approximately 50 % of lignin phenols were removed between Cuntan and Yichang, and most of them were deposited within the TGR. Moreover, approximately 60 % of the lignin phenols that were transported to the coastal region were derived from the middle and lower streams, which further suggests that after the closing of the TGD, the middle and lower stream played a more important role in the terrestrial OM transportation of the Changjiang.

The factors influencing the transport of organic matter in the Changjiang have become more complicated over time. Climate change (e.g., extreme drought), impoundment by the TGD, and the effects of tributaries and lakes in the middle and lower streams of the river may play different roles in the delivery of OC. In the middle stream, the TGD and extreme drought were the primary factors influencing both terrestrial and autochthonous OC, whereas in the lower stream, the impact of the TGD was weaker, and the contributions of tributaries and lakes were enhanced (Yu et al. 2011). Moreover, different tributaries and lakes make

Table 6.5 Average sediment and lignin fluxes at different gauging stations in the Changjiang

Average (2004–2008)	Cuntan	Yichang	Datong	Cuntan-Yichang	Datong-Yichang
Sediment load (Mt/year)	196.8	55.0	144.7	141.8	89.7
Lignin flux (10^3 t/year)	14.6	7.5	18.4	7.1	10.9

different contributions to terrestrial and autochthonous OC. Cursory calculations reveal that tributaries and lakes in the middle and lower streams of the river supplied a great deal of sediment and terrestrial OC to the main channel, and this material was more important in the extreme drought year of 2006 (Yu et al. 2011).

6.5.2 Distribution of Pollutants and Their Impacts

Traditionally, the Changjiang is thought of as a lifeline for the most populated basin in China. However, the quantity of pollution released into the Changjiang by industries, households, and agriculture is among the highest in the world. Surprisingly, the concentrations of many anthropogenic compounds in the river are comparable with those in other major rivers in the world because of dilution effects (Muller et al. 2008). Sediments serve as the sink of most pollutants, and studies have detected pollutants in riverine sediments (Chen et al., 2002).

As China was the world's second-largest producer of organochlorine pesticides (OCPs) from 1950 to 1980, DDTs and HCHs have been extensively used as insecticides in Chinese agricultural activities. Almost half of all OCPs produced ($\sim 40 \times 10^4$ t) were used in the Changjiang basin during 1976–1980 (Chen et al. 2002). The OCP records were used to trace the historical usage in the basin (Chen et al. 2002). OCP concentrations were at their lowest values before 1952, but rapidly increased and reached maximum values in the 1970s, before decreasing again, especially after 1983.

Although OCPs were banned in 1983, they were often detected in sediment due to their low solubility and high affinity to riverine particles. The residues of HCHs in Shanghai soils were reported to be as high as 15–337 ng/g in 1987 (Yang et al. 2002). However, the HCHs in the tidal flat of the Changjiang Estuary were lower, with an average value of 0.073 ng/g. The accumulation of OCPs in fish and mussels shows limited variability between the 1980s and the twenty-first century (Kong et al. 2007). Due to bioaccumulation and slow metabolite rates, however, OCPs may have a long-term impact on high trophic-level species.

The coastal environment adjacent to the ECS has been under huge stress from anthropogenic activities and population growth in the Changjiang drainage basin and coastal areas (Li and Daler 2004). The major pollutants are PCBs, PAHs, and OCPs. The concentration of PCBs in sediments and biotas is in the range of 11.8–28.6 and 43.8–334.5 ng/g (wet weight), respectively, which is considered to be mid-contamination levels (Wu et al. 1999; Liu et al. 2005). The main compounds of PAHs in the sediments were phenanthrene, 2-methylanthracene, 4,6-dimethyldibenzothiophene, fluoranthene, pyrene, and chrysene + triphenylene. Total concentrations of PAHs (including 50 compounds) in the coastal sediments of the ECS ranged from 22.1 to 262 ng/g, with the highest levels (181–262 ng/g) in the coastal areas and lower levels (112–118 ng/g) in the outer shelf region.

Distribution patterns of POC and BC in the sediments were similar to those of PAHs. Molecular ratios, such as anthracene/178 and Fl/Fl + pyrene, suggest that terrestrial sources are important for PAH contamination of the ECS and that long-range atmospheric transport is not very important for contributing PAHs to the study area (Bouloubassi et al. 2001; Hung et al. 2010). Levels of organochlorine pesticides (OCPs) were determined for surface sediments in the region adjacent to the ECS. The concentrations of DDTs and HCHs were in the range of 0.03–3.43 and 0.01–1.97 ng/g, respectively (Hu et al. 2011). OCPs may be derived from earlier residues in the sediment and atmospheric transport. The mud deposits on the shelf could be the main sink of OCPs in the ECS.

Compared with other worldwide regions, pollutant (PAHs, OCPs, etc.) concentrations are in the low range in the Changjiang basin and offshore seas, and the potential ecologic risk is limited. However, contaminated agricultural soil can still act as a reservoir of those chemicals, which will drain into the marine environment during extreme flood events. Moreover, a mixture of inorganic nitrogen, phosphorus, oil hydrocarbons, organic matter, and heavy metals is expected to fuel algae blooms (Li and Daler 2004). Trace elements and persistent organic chemicals, especially those related to suspended particles, may also accumulate in the food chain of the ECS (Muller et al. 2008).

6.6 Summary

This chapter summarizes the spatial variations in organic matter and biomarkers from the Changjiang watershed to the shelf and shows the spatial and temporal variations in POC and DOC. Climate change (e.g., extreme drought), impoundment by the TGD, and the effects of tributaries and lakes in the middle and lower streams of the river may play different roles in the delivery of organic carbon. Long-term monitoring data help to elucidate the seasonal variations in biomarkers (e.g., pigments and amino acids) in relation to hydrological and biological processes in the lower stream of the river. Geochemical characterization of biomarkers in the ECS is well studied, and potential control factors are considered in this study. Hydrological sorting, in situ primary production, and diagenetic processes contributed to the variable biomarker distributions and to their composition and burial in the ECS. Although there is high pressure from anthropogenic activities, compared with other regions worldwide, pollutant concentrations are relatively low and the potential ecological risk is limited.

As one of most populated river basins, with increasing anthropogenic pressure, the biogeochemistry of the Changjiang and estuary system is being increasingly altered from its natural state. These trends need long-term monitoring to evaluate the changes due to natural and human factors. Increasing global temperatures, changes in watershed hydrology, land-use change, and other factors may modify the composition and fluxes of organic matter from the river to the coast.

The processes in the Changjiang Estuary are highly dynamic and also sensitive to human impacts on the geomorphology of the estuary–ocean interface (e.g., dredging, diking, and loss of marshes). Those processes may have a significant effect on sediment transport, carbon burial, and water/particle residence time. The evaluation of submarine groundwater should be considered in future studies because of its important influence on the fluxes of terrestrial DOC and nitrogen in the estuarine region. With changing fluxes from rivers, and significant modification and burial of organic matter in the shelf, a comprehensive standardized monitoring network, combining remote sensing, modeling, and long-term observations, is necessary to understand the biogeochemistry of organic matter in the highly dynamic ECS shelf.

Acknowledgments This work was financially supported by the Ministry of Science and Technology of China (grant 2011CB409801/2) and the Natural Science Foundation of China (grant 41021064,41276081).

References

Alliot E, Younes WAN, Romano JC et al (2003) Biogeochemical impact of a dilution plume (Rhone River) on coastal sediments: comparison between a surface water survey (1996–2000) and sediment composition. Estua Coast Shelf Sci 57:357–367

Bergamaschi B, Tsamakis E, Keil R et al (1997) The effect of grain size and surface area on organic matter, lignin and carbohydrate concentration and molecular compositions in Peru Margin sediments. Geochim Cosmochim Acta 71:1247–1260

Bianchi TS, Findlay S (1990) Plant pigments as tracers of emergent and submergent macrophytes from the Hudson River. Can J Fisher Aqua Sci 47:492–494

Bianchi TS, Findlay S, Dawson R (1993) Organic matter sources in the water column and sediments of the Hudson River estuary: the use of plant pigments as tracers. Estua Coast Shelf Sci 36:359–376

Bianchi TS, Rolff C, Wildbom B et al (2002) Phytoplankton pigments in Baltic Sea seston and sediments: seasonal variability, fluxes, and transformations. Estua Coast Shelf Sci 55:369–383

Bouloubassi I, Fillaux J, Saliot A (2001) Hydrocarbons in surface sediments from the Changjiang (Yangtze River) estuary, East China Sea. Mari Pollut Bull 42:1335–1346

Burdige DJ (2005) Burial of terrestrial organic matter in marine sediments: a re-assessment. Global Biogeochem Cycl 19. doi:10.1029/2004GB002368

Calvert SE, Peterson TF (1991) Organic carbon accumulation and preservation in marine sediments: how important is anoxia? In: Whelan JK, Farrington JW (eds) Organic matter: productivity, accumulation and preservation in recent and ancient sediments. Column University Press, New York, pp 231–263

Cauwet G, Mackenzie FT (1993) Carbon inputs and distribution in estuaries of turbid rivers: the Yangtze and Yellow rivers (China). Mar Chem 43:235–246

Changjiang Sediment Bulletin (2005) Press of Ministry of Water Resources of the People's Republic of China. Website http://www.cjh.com.cn/

Chen JF, Xia XM, Ye XR et al (2002) Marine organic pollution history in the Changjiang Estuary and Zhejiang coastal area-HCHs and DDTs stratigraphical records. Mar Pollu Bull 45:391–396

Chen Z, Li J, Shen HT et al (2001) Yangtze River of China: historical analysis of discharge variability and sediment flux. Geomorph 41:77–91

Contrim L, Gadel Ë, Blazi JL (2001) Lignin-derived phenolic compounds in the particulate organic matter of a French Mediterranean river: seasonal and spatial variations. Org Geochem 32:305–320

Deng B, Zhang J, Wu Y (2006) Recent sediment accumulation and carbon burial in the East China Sea. Global Biogeochem Cycl 20. doi:10.1029/2005GB002559

De Master D, McKee B, Nittrouer C et al (1985) Rates of sediment accumulation and particle reworking based on radiochemical measurements from continental shelf deposits in the East China Sea. Cont Shelf Res 4:143–158. doi:10.1016/0278-4343(85)90026-3

Descy JP, Sarmento H, Higgins HW (2009) Variability of phytoplankton pigment ratios across aquatic environments. Eur J Phycol 44 (3):319–330

Ding HB, Sun MY (2005) Effects of intracellular structural associations on degradation of algal chloropigments in natural oxic and anoxic seawaters. Geochim Cosmochim Acta 69:4237–4252

Dittmar T, Fitznar HP, Kattner G (2001) Origin and biogeochemical cycling of organic nitrogen in the eastern Arctic Ocean as evident from D- and L-amino acids. Geochim Cosmochim Acta 65:4103–4114

Druffel ERM, Williams PM, Bauer JE et al (1992) Cycling of dissovled and particulate organic matter in the open ocean. J Geophy Res 97:15639–15659

Duan SW, Bianchi TS (2006) Seasonal changes in the abundance and composition of plant pigments in particulate organic carbon in the lower Mississippi and Pearl Rivers (USA). Estuaries Coasts 29 (3):427–442

Emery K (1968) Relict sediments on continental shelves of the world. Americ Assoc Petro Geol Bull 52:45–464

Fitznar HP, Lobbes JM, Kattner G (1999) Determination of enantiomeric amino acids with high-performance liquid chromatography and pre-column derivatisation with o-phthaldialdehyde and N-isobutyrylcysteine in seawater and fossil samples (mollusks). J Chromatogr A 832:123–132

Gameiro C, Cartaxana P, Brotas V (2007) Environmental drivers of phytoplankton distribution and composition in Tagus Estuary, Portugal. Estuar Coast Shelf Sci 75(1–2):21–34

Gao Q, Tao Z, Shen C et al (2002) Riverine organic carbon in the Xijiang River (South China): seasonal variation in content and flux budget. Environ Geol 41:826–832

Goñi M, Ruttenberg K, Eglinton T (1998) A reassessment of the sources and importance of land-derived organic matter in surface sediments from the Gulf of Mexico. Geochim Cosmochim Acta 62:3055–3075. doi:10.1016/S0016-7037(98)00217-8

Goñi MA, Montgomery S (2000) Alkaline CuO oxidation with a microwave digestion system: lignin analyses of geochemical samples. Analy Chem 72:3116–3121

Hedges JI, Ertel JR (1982) Characterization of lignin by gas capillary chromatography of cupric oxide oxidation products. Analy Chem 54:174–178

Hedges JI, Keil RG (1995) Sedimentary organic matter preservation: an assessment and speculative synthesis. Mar Chem 49:81–115

Hedges JI, Clark WA, Quay PD et al (1986) Compositions and fluxes of particulate organic material in the Amazon River. Limnol Ocean 31:717–738

Hedges JI, Keil RG, Benner R (1997) What happens to terrestrial organic matter in the ocean? Org Geochem 27:195–212

Henrichs SM, Reeburgh WS (1987) Anaerobic mineralization of marine sediment organic matter: rates and role of anaerobic processes in the oceanic carbon economy. J Geomicrobio 5:191–237

Hori K, Saito Y, Zhao Q, et al (2002) Architecture and evolution of the tide-dominated Changjiang (Yangtze) River delta, China. Sediment Geol 146:249–264. doi:10.1016/S0037-0738(01)00122-1

Hu LM, Lin T, Shi XF et al (2011) The role of shelf mud depositional process and large river inputs on the fate of organochlorine pesticides in sediments of the Yellow and East China Seas. Geophy Res Lett 38. doi:10.1029/2010GL045723

Hung JJ, Lin PL, Liu KK (2000) Dissolved and particulate organic carbon in the Southern East China Sea. Contin Shelf Res 20:545–569

Hung CC, Gong GC, Chou WC et al (2010) The effect of typhoon on particulate organic carbon flux in the southern East China Sea. Biogeosci 7:3007–3018

Hurley JP, Armstrong DE (1990) Fluxes and transformations of aquatic pigments in lake Mendota, Wisconsin. Limnol Oceanogr 35:384–398

Jeffrey SW, Mantoura RFC, Wright SW (eds) (1997) Phytoplankton pigments in oceanography: guidelines to modern methods. UNESCO Publishing, Paris 638 p

Kong DJ, Li DJ, Wu Y (2007) Evoluation of organic pollution in the Changjiang estuary in the past 50 years. Trans Oceano Limnol 2:96–103 (in Chinese with English abstract)

Krusche AV, Martinelli LA, Victoria RL et al (2002) Composition of particulate and dissolved organic matter in a disturbed watershed of southeast Brazil (Piracicaba River basin). Water Res 36:2743–2752

Li D, Daler D (2004) Ocean pollution from land-based sources East China Sea, China. Ambio 33:107–113

Li X, Bianchi T, Allison M, et al (2012) Composition, abundance and age of total organic carbon in surface sediments from the inner shelf of the East Chins Sea. Mar Chem 145–147:37–52

Lin J (2007) Distributions of dissolved organic carbon and particulate organic carbon in the Changjiang Estuary and its adjacent area. MS thesis, East China Normal University

Lin J, Wu Y, Zhang J et al (2007) Seasonal variation of organic carbon fluxes in the Yangtze River and influence of Three-Gorges engineering. China Environ Sci 27:246–249 (in Chinese with English abstract)

Lin S, Hsieh I, Huang K et al (2002) Influence of the Yangtze River and grain size on the spatial variations of heavy metals and organic carbon in the East China Sea continental shelf sediments. Chem Geol 182:377–394

Liu HL, Liu M, Yang Y (2005) Distribution of PCBs and OCPs in biotas from sediments of Changjiang tidal flat. Environ Sci 25:69–73 (in Chinese with English abstract)

Liu J, Li A, Xu K et al (2006) Sedimentary features of the Yangtze River-derived along-shelf clinoform deposit in the East China Sea. Cont Shelf Res 26:2141–2156. doi:10.1016/j.csr.2006.07.013

Liu J, Xu K, Li A et al (2007) Flux and fate of Yangtze River sediment delivered to the East China Sea. Geomorphology 85:208–224. doi:10.1016/j.geomorph.2006.03.023

Ludwig W, Probst JL, Kempe S (1996) Predicting the oceanic input of organic carbon by continental erosion. Global Biogeo Cycles 10:23–41

Mackey MD, Mackey DJ, Higgins HW et al (1996) CHEMTAX—a program for estimating class abundances from chemical markers: application to HPLC measurements of phytoplankton. Mar Ecol Progr Series 144:265–283

Mackey MD, Higgins HW, Mackey DJ, Wright SW (1997) CHEMTAX user's manual: a program for estimating class abundances from chemical markers—application to HPLC measurements of phytoplankton pigments. CSIRO Marine Laboratories, Australia

Milliman J, Beardsley R, Yang Z, et al (1985) Modern Huanghe-derived muds on the outer shelf of the East China Sea: identification and potential transport mechanisms. Cont Shelf Res 4:175–188. doi:10.1016/0278-4343(85)90028-7

Moreno S, Niell FX (2004) Scales of variability in the sediment chlorophyll content of the shallow Palmones River Estuary, Spain. Estua Coastal Shelf Sci 60:49–57

Muller B, Berg M, Yao ZP et al (2008) How polluted is the Yangtze river? Water quality downstream from the Three Gorges Dam. Sci Total Environ 402:232–247

Ogawa H, Usui T, Koike I (2003) Distribution of dissolved organic carbon in the East China Sea. Deep Sea Res Part II 50:353–366

Ohkouchi N, Eglinton T, Keigwin L, et al (2002) Spatial and temporal offsets between proxy records in a sediment drift. Science 298:1224–1227. doi:10.1126/science.1075287

Sampere T, Bianchi T, Wakeham S et al (2008) Sources of organic matter in surface sediments of the Louisiana Continental margin: effects of major depositional/transport pathways and Hurricane Ivan. Cont Shelf Res 28:2472–2487. doi:10.1016/j.csr.2008.06.009

Sempr R, Charribre B, Wambeke FV et al (2000) Carbon inputs of the Rhone River to the Mediterranean Sea: biogeochemical implications. Global Biogeo Cycles 14:669–681

Shao L, Wu Y, Zhu ZY et al (2011) Monthly changes of particulate organic nitrogen in Xuliujing (Changjiang estuary): example of the D&L enantiomeric amino acids. Mar Environ Sci 30:554–558 (in Chinese with English abstract)

Sicre MA, Broyelle I, Lorre A et al (1993) Sources and transport of particulate hydrocarbons in the meso-tidal Changjiang Estuary. Estua Coastal Shelf Sci 37:557–573

Spencer RGM, Hernes PJ, Aufdenkampe AK (2012) An initial investigation into the organic matter biogeochemistry of the Congo River. Geochim Cosmochim Acta 84:614–627

Still CJ, Berry JA, Collatz GJ (2003) Global distribution of C 3 and C 4 vegetation: carbon cycle implications. Global Biogeochem Cycles 17. doi:10.1029/2001GB001807

Tesi T, Miserocchi S, Goñi MA, Langone L, Boldrin A, Turchetto M (2007) Organic matter origin and distribution in suspended particulate materials and surficial sediments from the western Adriatic Sea (Italy). Estuar Coast Shelf Sci 73:431–446

Valdes-Weaver LM, Piehler MF, Pinckney JL, Howe KE, Rossignol K, Paerl HW (2006) Long-term temporal and spatial trends in phytoplankton biomass and class-level taxonomic composition in the hydrologically variable Neuse-Pamlico estuarine continuum, North Carolina, USA. Limnol Oceanogr 51(3):1410–1420

van Iperen JM, van Weering TCE, Jansen JHF et al (1987) Diatoms in surface sediments of the Zaire deep-sea fan (Se atlantic ocean) and their relation to overlying water masses. Netherlands J Sea Res 21:203–217

Walsh JJ (1991) Importance of continental margins in the marine biogeochemical cycling of carbon and nitrogen. Lett Nat 350:53–44

Wang XC, Li AC (2007) Preservation of black carbon in the shelf sediments of the East China Sea. Chin Sci Bull 52:3155–3161

Wu YY (2004) Study on the changes of suspended sediment grain-size distribution in the middle and lower Yangtze mainstream. Master thesis. East China Normal University, Shanghai, China (in Chinese with English abstract)

Wu Y, Zhang J, Zhou Q (1999) Persisiten organochlorine residues in sediments from Chiense river/estuary system. Environ Pollu 105:143–150

Wu Y, Zhang J, Li DJ et al (2003) Isotope variability of particulate organic matter at the PN section in the East China Sea. Biogeochemistry 65:31–49

Wu Y, Dittmar T, Ludwichowski KU et al (2007a) Tracing suspended organic nitrogen from the Yangtze River catchment into the East China Sea. Mar Chem 107:367–377

Wu Y, Zhang J, Liu SM et al (2007b) Sources and distribution of carbon within the Yangtze River system. Estua Coast Shelf Sci 71:13–25

Wysocki LA, Bianchi TS, Powell RT et al (2006) Spatial variability in the coupling of organic carbon, nutrients, and phytoplankton pigments in surface waters and sediments of the Mississippi River plume. Estua Coast Shelf Sci 69:47–63

Xu KH, Milliman JD, Yang ZS (2007) Climate and anthropogenic impacts on water and sediments discharges from the Yangtze River, 1950–2005. In: Gupta A (ed) Large rivers: geomorphology and management. Wiley, West Sussex, pp 609–626

Yang Y, Liu M, Hou LJ (2002) Distribution of OCPs in sediments of Changjiang tidal flat. Shanghai Environ Sci 21:530–532

Yang SL, Zhang J, Dai SB et al (2007) Effect of deposition and erosion within the main river channel and large lakes on sediment delivery to the estuary of the Yangtze River. J Geophy Res 112. doi:10.1029/2006JF000484

Yang L, Wu Y, Zhang J et al (2008) Distribution of lignin and sources of organic matter in surface sediments from the adjacent area of the Changjiang Estuary in China. Acta Oceanol Sinica 30:35–42 (in Chinese with English Abstract)

Yang S, Milliman J, Li J et al (2011) 50,000 dams later: erosion of the Yangtze River and its delta. Glob Planet Change 75:14–20

Yu H (2007) Transport and burial of particulate terrigenous organic matter in Changjiang (Yangtze River) and East China Sea shelf—illustrated by lignin. Dissertation, East China Normal University

Yu H, Wu Y, Zhang J, et al (2011) Impact of extreme drought and the Three Gorges Dam on transport of particulate terrestrial organic carbon in the Changjiang (Yangtze) River. J Geophy Res 116. doi:10.1029/2011JF002012

Zapata M, Rodríguez F, Garrido JL (2000) Separation of chlorophylls and carotenoids from marine phytoplankton: a new HPLC method using a reversed phase C8 column and pyridine-containing mobile phases. Mar Ecol Progr Series 195:29–45

Zhu ZY (2007) Hypoxia in the Changjiang Estuary and its adjacent area—started with phytoplankton pigments. Dissertation, East China Normal University

Zhu ZY, Zhang J, Wu Y et al (2006) Bulk particulate organic carbon in the East China Sea: tidal influence and bottom transport. Progr in Oceanogr 69:37–60

Zhu C, Xue B, Pan J et al (2008) The dispersal of sedimentary terrestrial organic matter in the East China Sea (ECS) as revealed by biomarkers and hydro-chemical characteristics. Org Geochem 39:952–957

Zhu ZY, Ng WM, Liu SM et al (2009) Estuarine phytoplankton dynamics and shift of limiting factors: a study in the Changjiang (Yangtze River) Estuary and adjacent area. Estuar Coast Shelf Sci 84:393–401

Zhu C, Wagner T, Pan J et al (2011a) Multiple sources and extensive degradation of terrestrial sedimentary organic matter across the energetic, wide continental shelf. Geochem Geophys Geosyst 12. doi:10.1029/2011GC003506

Zhu C, Wang Z, Xue B et al (2011b) Characterizing the depositional setting for sedimentary organic matter distributions in the lower Yangtze River-East China Sea shelf system. Estua Coast Shelf Sci 93:182–191

Zonneveld K, Versteegh G, Kasten S et al (2010) Selective preservation of organic matter in marine environments; processes and impact on the sedimentary record. Biogeosci 7:483–511. doi:10.5194/bg-7-483-2010

Coastal Wetlands in the Changjiang Estuary

7

Liquan Zhang, Lin Yuan, and Huamei Huang

Abstract

The enormous quantities of sediment produced by the Changjiang (Yangtze River) have created extensive coastal wetlands in the estuarine region, which have been colonized by various types of salt marsh vegetation. As the Changjiang Estuary is located adjacent to the megacity of Shanghai, human activities exert substantial pressures on the coastal wetlands and cause coastal ecosystem degeneration and consequent decreases or loss in ecosystem services. The major stresses from human activities on the coastal wetlands in the Changjiang Estuary include urbanization, reclamation, the introduction of invasive species, overfishing, and environmental pollution. In this chapter, the succession of plant communities in the coastal wetlands of the Changjiang Estuary is described. The spatial and temporal variation of plant communities on salt marshes are analyzed on the basis of remote sensing (RS) and mapping, combined with field inventories of sampling plots. The population expansion of an exotic species, *Spartina alterniflora*, after its introduction to the coastal wetlands in the Changjiang Estuary in the 1990s, is described and analyzed. The impacts of the invasion of *S. alterniflara* on benthic communities are examined along intertidal salt marsh gradients, and the invasion history is described. A cellular automaton (CA) model was developed to simulate the expansion of *S. alterniflora* and to study the interactions between the spatial pattern of expansion and salt marsh vegetation ecosystems. Field experiments and a demonstration project were carried out to control the invasion of *S. alterniflora.* Finally, this chapter discusses the future development, management, and conservation of coastal wetlands and biodiversity in the Changjiang Estuary.

Keywords

Changjiang estuary • Coastal wetlands • Salt marsh dynamics • Biological invasion • Conservation

7.1 General Characteristics of Coastal Wetlands in the Changjiang Estuary

The Changjiang (Yangtze River) Estuary is located on the eastern coast of China (Fig. 7.1). It is neighbored by Hangzhou Bay to the south, Jiangsu Province to the north, and opens to the East China Sea (30°50′–31°50′N; 121°0′–122°05′E). Due to intensive interactions between runoff and tidal currents, the hydrodynamic processes of the estuary have produced unique sedimentological features (Yun 2004; Chap. 4). The river mouth area and the submerged delta are

L. Zhang (✉) · L. Yuan
State Key Laboratory of Estuarine and Coastal Research, East China Normal University, 3663 Zhongshan Road North, Shanghai 200062, China
e-mail: lqzhang@sklec.ecnu.edu.cn

L. Yuan
e-mail: lyuan@sklec.ecnu.edu.cn

H. Huang
South China Sea Marine Engineering and Environment Institute, State Oceanic Administration, Xingangzhong Road 353, Guangzhou 510300, China
e-mail: hmhuang2007@gmail.com

J. Zhang (ed.), *Ecological Continuum from the Changjiang (Yangtze River) Watersheds to the East China Sea Continental Margin*, Estuaries of the World,
DOI 10.1007/978-3-319-16339-0_7,

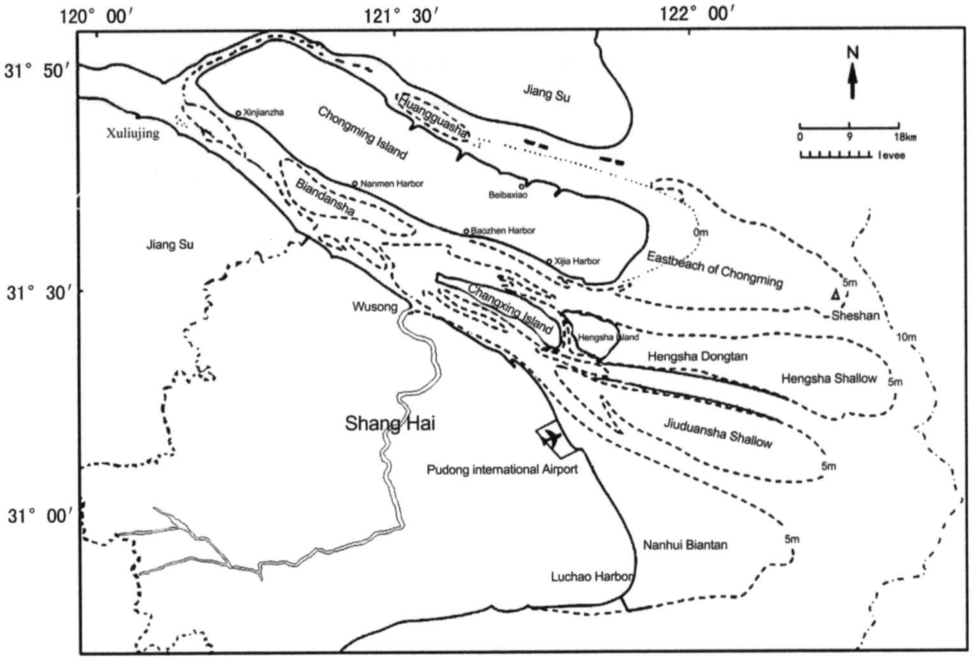

Fig. 7.1 Location of the Changjiang Estuary and the distribution of coastal wetlands in the Shanghai region

the depocenter and receive tremendous quantities of fine-grained sediments from the Changjiang River. This sediment forms extensive shoals and tidal flats in the estuarine region (Fig. 7.1), which have been colonized by various types of salt marsh vegetation (Gao and Zhang 2006).

The Changjiang Estuary region has a northern subtropical monsoon climate, with an average annual temperature of 16 °C, average summer temperatures of 28 °C, and average winter temperatures of 4 °C. Average annual precipitation is approximately 1200 mm, with 60 % of rainfall occurring between May and September, with additional contributions from typhoons during summer and autumn.

The dynamic and geomorphologic processes of the coastal wetlands are profoundly influenced by the interactions between runoff and tidal (semidiurnal) currents. The tidal range exhibits significant monthly, semimonthly, and semidiurnal signals (see Chap. 2). Spring tides generally have amplitudes of more than 450 cm, which is approximately twice that of neap tides, which typically have amplitudes less than 250 cm. The maximum and average tidal range is 4.62 and 2.67 m, respectively. The strong tidal range introduces a significant tidal current, which can oscillate at the estuarine front and mix dissolved materials (see Chap. 2).

According to the 1971 Ramsar International Wetland Conservation treaty, wetlands are defined as: "…areas of marsh, fen, peatland or water, whether natural or artificial, permanent or temporary, with water that is static or flowing, fresh, brackish or salt, including areas of marine water the depth of which at low tide does not exceed 6 m…", and "…wetlands may incorporate riparian and coastal zones adjacent to the wetlands, and islands or bodies of marine water deeper than six meters at low tide lying within the wetlands." (Mitsch and Gosselink 2007).

There are two different types of coastal wetlands in the Changjiang Estuary: mainland and island wetlands. The wetlands along the mainland coast, such as Nanhui Biantan, are primarily located on the southern coast of the Changjiang Estuary, from Xuliujing in the west to Luchao Harbor in the southeast. The wetlands along the island coast mainly include the eastern and northern coasts of Chongming Island, the western coast of Hengsha Island, the northern coast of Changxing Island, and the Jiuduansha Shoals (Fig. 7.1). According to the Shanghai Water Authority (2010), the total area of coastal wetlands in 1990–2008, stretching from the inland dyke seaward to the −5 m isobath, was 2,318.5–2,892.5 km^2 (Table 7.1).

By their very nature, coastal wetlands in the Changjiang Estuary are dynamic and provide various ecosystem services, such as improving water quality, mitigating climate change, and conserving biodiversity. Two national nature reserves have been established in the Changjiang Estuary: Chongming Dongtan National Birds Nature Reserve, which

Table 7.1 Total area (km^2) of coastal wetlands in the Changjiang Estuary during 1990–2008 (from Huang 2009)

Year	>0 m	−2 to 0 m	−5 to −2 m	Total
1990	828.6	713.2	1,350.7	2,892.5
2000	841.4	723.3	1,221.8	2,786.4
2004	753.0	691.9	1,152.8	2,597.8
2008	559.0	666.5	1,093.0	2,318.5

Box 7.1 Coastal wetlands and salt marsh succession series along the intertidal zones in the Changjiang Estuary (modified from Zhang and Yong 1992).

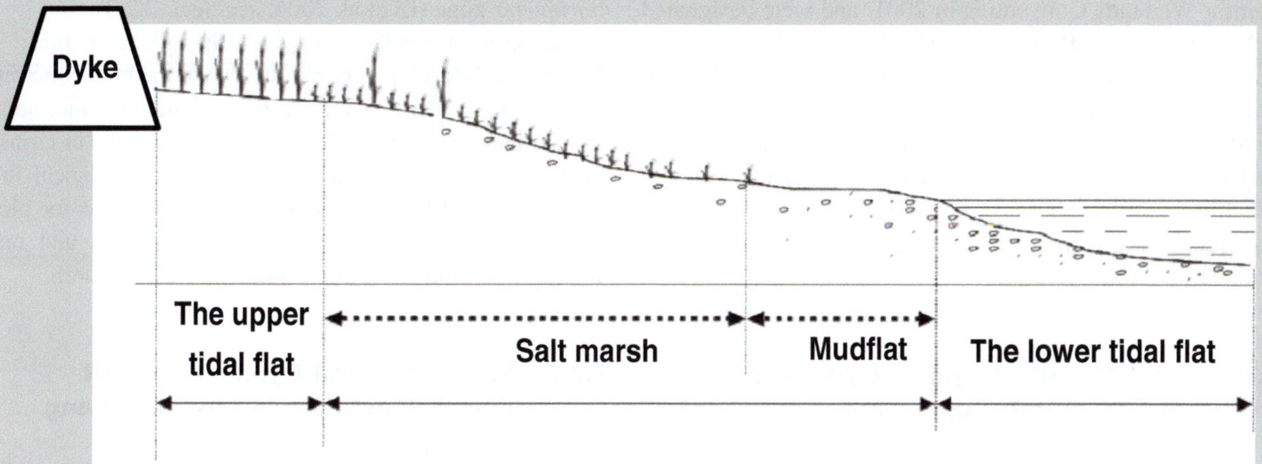

Phragmites australis *Scirpus mariquete* Mudflat The lower tidal flat

Spartina alterniflora

The enormous quantities of sediment produced by the Changjiang River have created extensive coastal wetland areas in the estuarine region, which have been colonized by various types of salt marsh vegetation. Zonation of vegetation is widespread in the coastal wetlands. Within the intertidal zone, three distinct salt marsh vegetation zones have been identified at different elevations, stretching from the inland dyke toward the low tidemark (Zhang and Yong 1992). The tidal flats closest to the low watermark, with elevations less than 2 m asl (using the local Wusong bathymetric benchmark), are characterized by mudflats which are devoid of any higher plants (angiosperms). The tidal flats between 2.0 and 2.9 m in elevation are dominated by *S. mariqueter* communities. Above 2.9 m elevation, plant communities are dominated by *P. australis*. An additional species in this zone is an exotic species, *S. alterniflora* (smooth cordgrass). This species is a perennial salt marsh grass that is native to the Atlantic and Gulf coasts of North America. It has been widely introduced to many coastal and estuarine regions of the world for ecological engineering, due to its great capacity to reduce tidal wave energy, mitigate erosion, trap sediments, and protect the seacoast (Chung 2006). *S. alterniflora* was introduced to the coastal wetlands in the Changjiang Estuary during the 1990s (Huang 2009). Over the past 20 years, this species has gradually invaded large areas formerly covered by *P. australis* and has also started to invade the upper parts of the *S. mariqueter* zone (Li et al. 2006; Sect. 7.2.3).

covers 242 km^2 (Fig. 7.6), and Jiuduansha Wetland Nature Reserve, which covers 402 km^2 (Fig. 7.7). These two wetlands were listed as *Chinese Protected Wetlands* in 1992, were designated as internationally important under the Ramsar Wetlands Convention in 2001, and were designated as National Nature Reserves in 2005 (Gao and Zhang 2006). The area of nature reserves amounts to ca. 25 % of the total area of coastal wetlands in the Changjiang Estuary.

Studies of coastal wetlands can provide valuable insights into the ecology of wetlands in the Changjiang Estuary, which is important for wetland biodiversity conservation and resource management, both in this region and in other areas affected by human impacts.

7.2 Spatial and Temporal Dynamics of Salt marsh Vegetation

7.2.1 Succession of Salt marsh Vegetation in the Changjiang Estuary

A salt marsh is essentially a muddy seashore colonized by vegetation. These marshy areas form a transitional zone between land and salty (or brackish) water. Salt marshes develop on depositional coasts, bays, and estuaries where tidal movements are gentle and erosion is light (Mitsch and Gosselink 2007) and where the climate, substrate and geomorphologic processes are suitable for the development of salt marsh vegetation.

Zonation of salt marsh vegetation on the coastal wetlands in the Changjiang Estuary has been described in detail. Within the tidal flats of the intertidal zone, three distinct zones of salt marsh vegetation at different elevations have been identified (Zhang and Yong 1992; Box 7.1). The tidal flats closest to the low watermark, with elevations of less than 2 m above sea level (asl) (using the local Wusong bathymetric benchmark), are characterized by mudflats that are devoid of any higher plants (angiosperms). The tidal flats between 2.0 and 2.9 m in elevation are dominated by *Scirpus mariqueter* communities, with some rarer *Scirpus triqueter* communities. Above 2.9 m elevation, plant communities are dominated by *Phragmites australis*, with small areas of *Imperata cylindrica, Suaeda glauca, Juncus setchuensis*, and *Carex scabrifolia*, with the latter tending to border the various creeks crossing the salt marshes. An additional species in this zone is an exotic species, *Spartina alterniflora*, or smooth cordgrass,which was introduced to the Changjiang Estuary during the 1990s. This species is a perennial salt marsh grass that is native to the Atlantic and Gulf coasts of North America. It has been widely introduced to many coastal and estuarine regions of the world as a species for ecological engineering, due to its great capacity to reduce tidal wave energy, mitigate erosion, trap sediments, and protect the seacoast (Chung 2006). Over the past 20 years, this species has gradually invaded large areas formerly covered by *P. australis* in the coastal wetlands in the Changjiang Estuary and has also started to invade the upper parts of the *S. mariqueter* zone (Li et al. 2006; see Sect. 7.2.3).

The coastal wetlands in the Changjiang Estuary are dynamic and develop very rapidly due to the huge amount of silt brought by the Changjiang River. The dynamic, neonatal, and natural characteristics of these wetlands, in terms of silt deposition, vegetation succession, and biological invasion, are linked to their importance as a wetland for biodiversity conservation and resource management and make this an ideal study area for coastal wetland research.

7.2.2 Spatiotemporal Dynamics of Salt marsh Vegetation in the Changjiang Estuary

Salt marsh vegetation is one of the most important components of coastal wetlands, providing biodiversity and valuable ecosystem services. By their very nature, salt marshes and intertidal zones are dynamic, and their management relies on up-to-date spatial information. Remote sensing (RS), in conjunction with geographical information systems (GIS) and global positioning systems (GPS), is becoming the main technique used in land-cover inventory data collection and has been widely used as a tool for mapping vegetation and monitoring vegetation dynamics and biodiversity conservation at large scales (Huang et al. 2007). The relatively flat topography of intertidal zones, the limited number of species, and generally the homogeneous nature of the vegetation mosaic make salt marshes ideal candidates for mapping and monitoring using RS techniques.

The spatial and temporal dynamics of salt marsh vegetation in the Changjiang Estuary have been studied from 1990 to 2008 (Huang 2009). A general framework for the study is presented in Fig. 7.2. A set of multi-temporal Landsat thematic mapper (TM) images was used, which basically covered the state of the low tide at the time the images were taken and the salt marshes of different stages under the impacts of human activities. These satellite images were geometrically corrected using a series of nautical charts and ERDAS Imagine software. Two spectral enhancement methods, Tasseled Cap (KT) Transform and Normal Difference Vegetation Index (NDVI), were used to interpret the satellite images more efficiently (Huang 2009). Moreover, several in situ field surveys were carried out to validate and revise the image classifications and to conduct a classification accuracy assessment.

The main results of our study are summarized in Table 7.2 and Fig. 7.3. In 1990, the total salt marsh vegetation area in the coastal wetlands of the Shanghai region was 17,882 ha. At this time, *P. australis* was the dominant

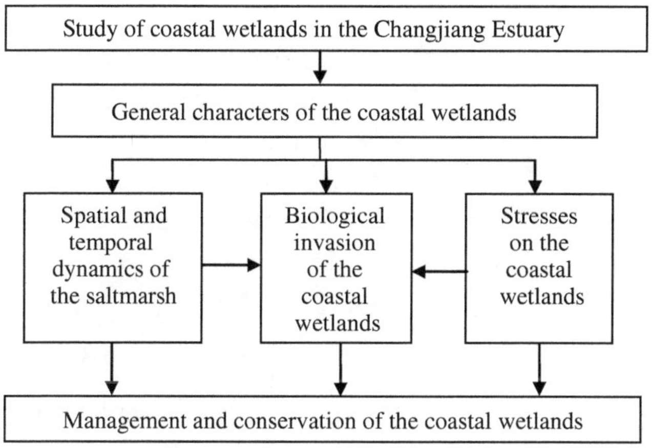

Fig. 7.2 General framework for study of the coastal wetlands in the Changjiang Estuary

Table 7.2 Dynamics of salt marsh vegetation (in hectares) in the Changjiang Estuary during 1990–2008 (from Huang 2009)

Year	1990	2000	2003	2008
Scirpus mariqueter	3,782	4,834	7,602	4,235
Phragmites australis	14,100	7,086	10,075	5,618
Spartina alterniflora	0	1,903	4,276	5,698
Others	0	0	0	2,765
Total	17,882	13,822	21,954	18,315

community and accounted for 14,100 ha, or 79 %, of the total salt marsh area (Table 7.2; Fig. 7.3). However, almost half of the *P. australis* salt marsh was lost during 1990 and 2000 as a result of large-scale reclamation of coastal wetlands. At the same time, an exotic species, *S. alterniflora*, was introduced and began to invade the wetlands (Table 7.2; Fig. 7.4).

By 2003, the total area of salt marsh vegetation had reached 21,954 ha, and the area of *S. alterniflora* was 4276 ha, or 20 %, of the total salt marsh area. After 2003, the total salt marsh area decreased, mainly due to continued and intensive reclamation. Both *S. mariqueter* and *P. australis* communities declined almost by half, while the *S. alterniflora* community continued to expand. By 2008, the *S. alterniflora* community had colonized 5698 ha and accounted for more than 25 % of the total salt marsh vegetation area (Table 7.2; Fig. 7.5).

This study of the spatial and temporal dynamics of salt marsh vegetation in the Changjiang Estuary has provided up-to-date spatial and temporal data, which is an important first step in providing a scientific basis for coastal wetland management and biodiversity conservation.

7.2.3 Spatiotemporal Salt marsh Dynamics in the Chongming Dontan National Nature Reserve

The Chongming Dongtan Nature Reserve is situated at the mouth of the Changjiang Estuary in eastern China (31°25′–31°38′N, 121°50′–122°50′E) (Fig. 7.1). The total area of the nature reserve is 242 km^2, including 100 and 140 km^2 of tidal flats above and below 0 m asl, respectively. The Chongming Dongtan Nature Reserve is one of the largest nature reserves for migratory birds in East Asia and is in the midsection of the eastern migratory routes of the Asia–Pacific region (Gao and Zhang 2006).

The mouth bar region in the Changjiang Estuary and the submerged delta offshore are the major locations for the sedimentation of very large amounts of silt brought by the Changjiang. For this reason, the eastern fringe of the nature reserve has been advancing at a rate of between 100 and 300 m per year in the past century. However, the progradation of intertidal wetlands has slowed down drastically in recent decades in response to the decline of riverine sediment discharge (see Chap. 4).

Based on analyses and interpretations of Landsat TM remotely sensed images, integrated with historical data and field surveys, the spatial and temporal dynamics of salt marshes in the Chongming Dongtan National Nature Reserve were investigated for the period 1990–2008 (Huang 2009). The results show that the total area of salt marsh vegetation in 1990 was 7566 ha, of which the *P. australis* and *S. mariqueter* communities accounted for 6762 ha and 804 ha, respectively.

During the 1990s, the *P. australis* community experienced a drastic decrease due to the reclamation that occurred in 1992, 1998, and 2001. While the total decrease of the *P. australis* community was 969 ha in 1998, the *S. mariqueter* community increased to 1,509 ha. After establishing the nature reserve, the native salt marsh vegetation began to recover slowly, along with accretion of the tidal flats. The *P. australis* community increased from 969 ha in 1998 to 1,060 ha in 2008, and the *S. mariqueter* community increased from 1,509 ha in 1998 to 2,091 ha in 2008 (Table 7.3; Fig. 7.6).

Our study of the spatial and temporal dynamics of salt marsh vegetation in the Chongming Dongtan Nature Reserve indicated that, after the introduction of *S. alterniflora* in 1995, this species has rapidly invaded large areas of the nature reserve, outcompeting the native species, *S. mariqueter* and *P. australis*, due to its stronger competitive capacity, wider ecological niche, and the rapid accretion of intertidal flats. After its initial introduction, the area of the mono-dominant *S. alterniflora* community increased rapidly, reaching 1378 ha by 2008 (Table 7.3; Fig. 7.6), or almost one-third of the total intertidal salt marsh in the Chongming Dongtan Nature

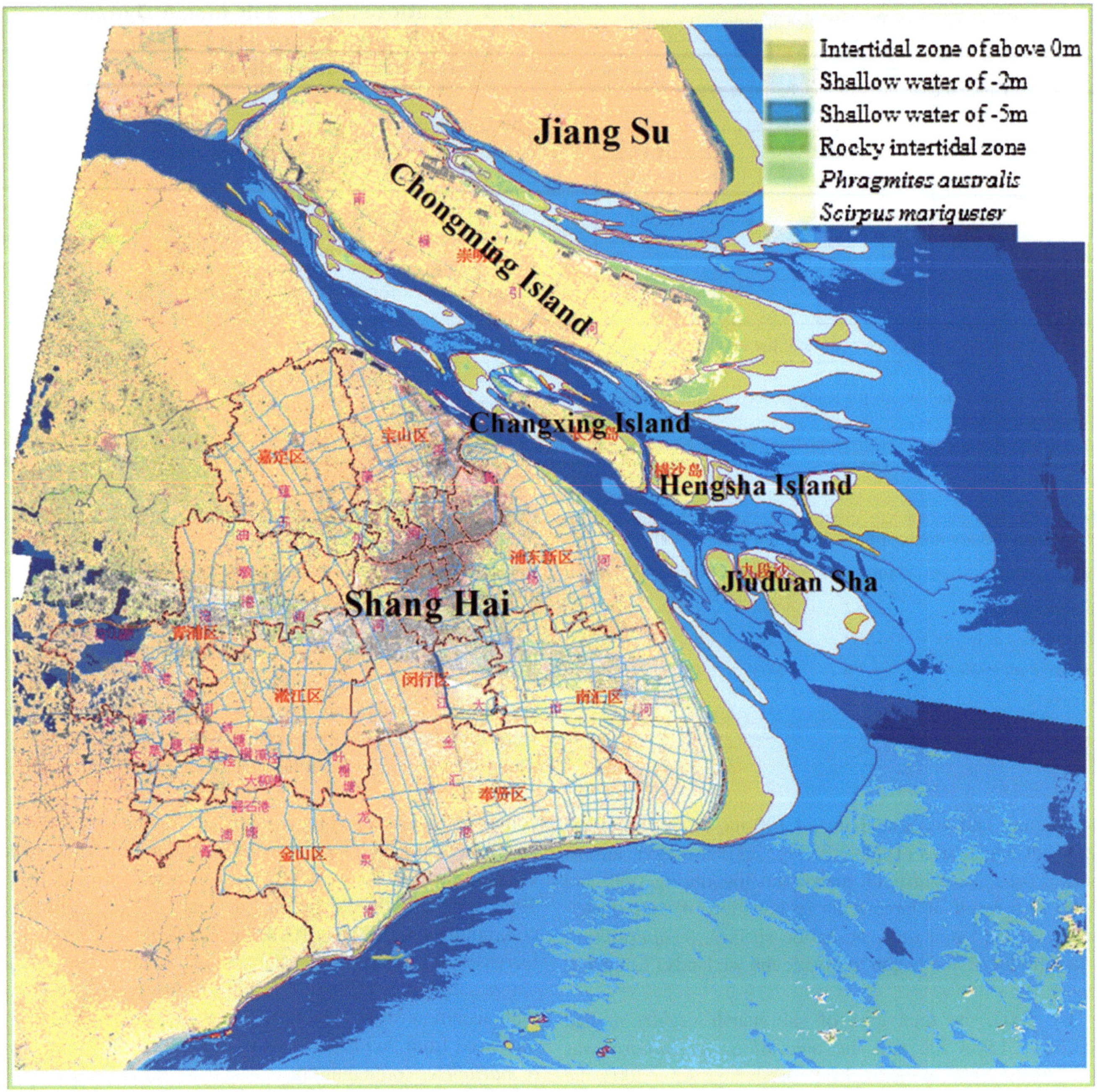

Fig. 7.3 Distribution of coastal wetlands and salt marsh vegetation (i.e., the *Phragmites australis* community and the *Scirpus mariqueter* community) in the Changjiang Estuary in 1990 (from Huang 2009)

Reserve. We recommend that measures be taken to manage and control this invasive species at the nature reserve.

7.2.4 Spatiotemporal Dynamics of Salt marsh in the Jiuduansha National Nature Reserve

The Jiuduansha National Wetland Nature Reserve is situated at the mouth of the Changjiang Estuary in eastern China (31°03′–31°17′N, 121°46′–122°15′E). The nature reserve covers 423.2 km^2, with an east–west distance of 25.9 km and north–south length of 46.3 km. The reserve consists of Upper Shoal, Middle Shoal, and Lower Shoal (Fig. 7.7a) according to the time that each area emerged and their location from west to east (Huang and Zhang 2007). The neonatal shoals at Jiuduansha initially emerged above the water surface in the 1920s and have been growing rapidly since then. According to measurements in 2004, the shoal areas above the 0 and −5 m isobaths have areas of 127.3 and 348 km^2,

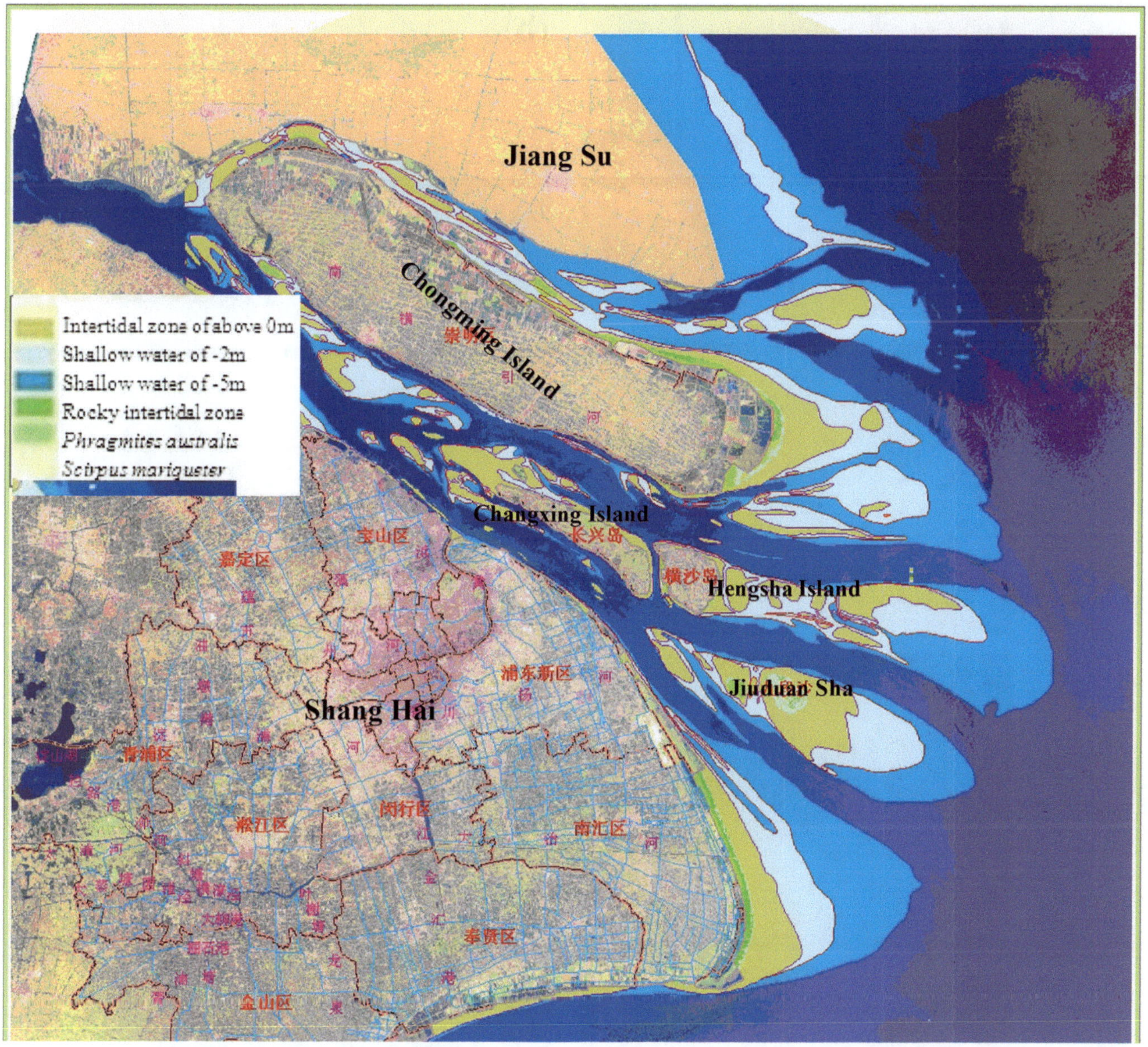

Fig. 7.4 Distribution of coastal wetlands and salt marsh vegetation (i.e., the *Phragmites australis* community, *Spartina alterniflora* community, and *Scirpus mariqueter* community) in the Changjiang Estuary in 2000 (from Huang 2009)

respectively, with a maximum elevation of 4 m (Huang and Zhang 2007). These shoals have never been colonized by humans and have been in their natural state since they were formed. The Shanghai Municipality designated the shoals as a wetland nature reserve in 2003, and they became the Jiuduansha National Wetland Nature Reserve in 2005. The dynamic, neonatal, and natural characteristics of the Jiuduansha Shoals, in terms of silt deposition, salt marsh development and biological invasion, are linked to their importance for wetland biodiversity conservation and resource management and make these an ideal study area.

The pioneer *S. mariqueter* salt marsh community can be seen in satellite images taken in 1987. These images show the community first colonizing the tidal flats above 2 m asl. In April and May of 1997, 40 ha of *P. australis* and 50 ha of *S. alterniflora* were transplanted on the Middle Shoal as an ecological engineering project for Pudong International Airport (Huang and Zhang 2007). Clumps of these two species were planted along 100-m-wide belts, in a regular arrangement with 1 m spacing. In addition, ca 5 ha of *S. alterniflora* was transplanted above 2.5 m asl on the Lower Shoal (Huang and Zhang 2007).

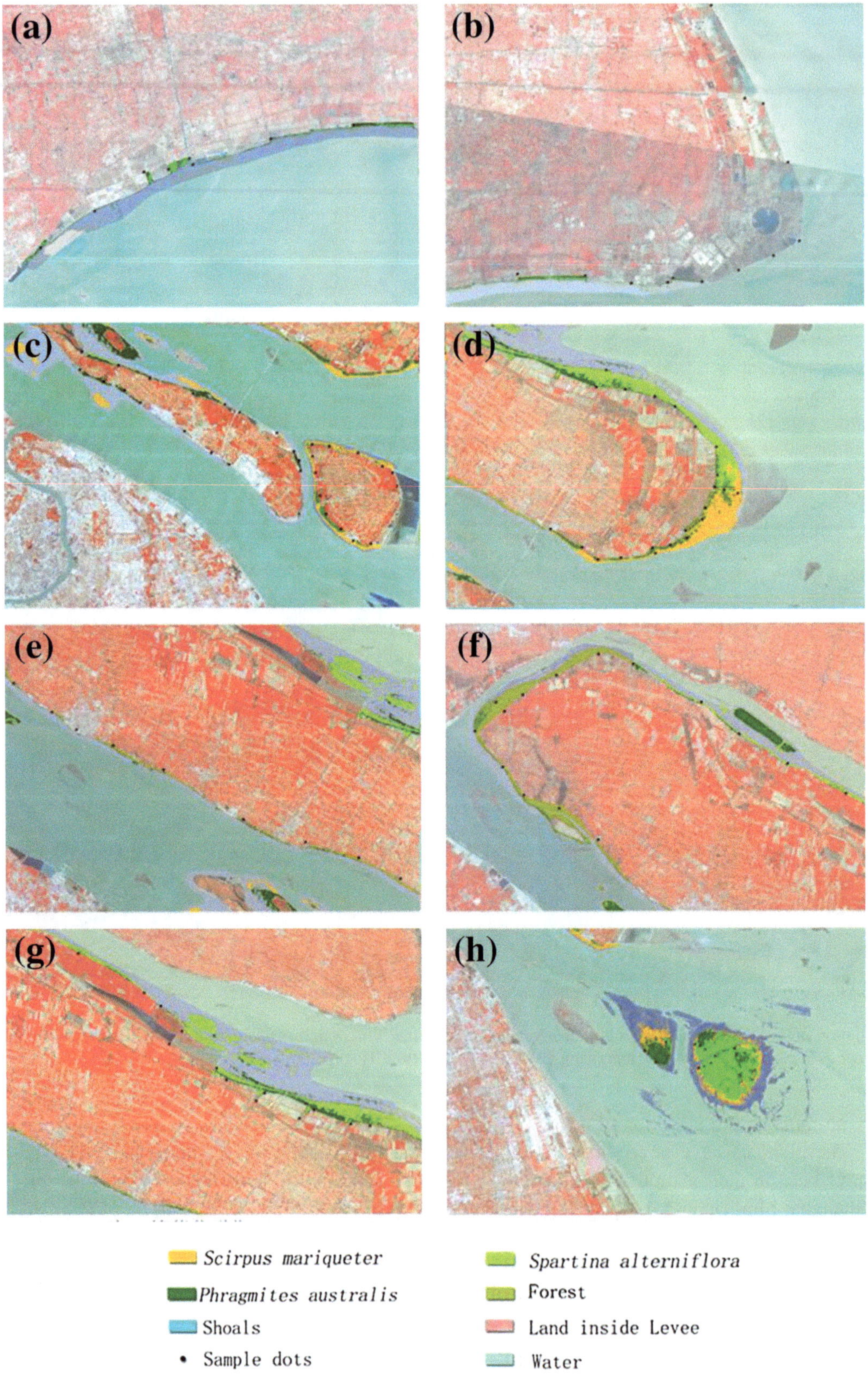

Fig. 7.5 Distribution of salt marsh vegetation (i.e., the *Phragmites australis* community, *Spartina alterniflora* community, and *Scirpus mariqueter* community) in the Changjiang Estuary in 2008 (from Huang 2009)

Table 7.3 Dynamics of salt marshes during 1990–2008 at the Chongming Dongtan Nature Reserve (from Huang 2009)

Year	Salt marsh vegetation (hectares)			
	Scirpus mariqueter	*Phragmites australis*	*Spartina alterniflora*	Total area
1990	804	6,762	0	7,566
1998	1,509	969	0	2,478
2000	2,322	1,034	187.5	3,543
2003	2,713	542	932	4,187
2005	2,953	452	1,283	4,688
2008	2,091	1,060	1,378	45,289

The spatial and temporal dynamics of salt marshes in the Jiuduansha Shoals are presented in Fig. 7.7. The distribution of each type of salt marsh vegetation for the 1997–2008 period is summarized in Table 7.4. The total area of salt marsh vegetation increased from 1,095 ha in 1997 to 3,601 ha in 2008, along with accretion of the shoals. The *P. australis* and *S. mariqueter* communities continued to spread, but at a much lower rate than for *S. alterniflora*. The *P. australis* community increased from 168 ha in 1997 to 924 ha in 2008. The *S. mariqueter* community increased from 967 ha in 1997 to 1,850 ha in 2003 and thereafter decreased to 968 ha in 2008. The *S. alterniflora* showed a strong competitive capacity in the Jiuduansha Shoals and increased to 1,709 ha by 2008, accounting for 47 % of the total salt marsh vegetation (Table 7.4).

One common feature of biological invasions is a lag time between initial colonization and the onset of rapid population growth and range expansion (Sakai et al. 2001). This is evident in the expansion pattern of *S. alterniflora* on the Jiuduansha Shoals. Due to the small size of the clumps, the newly planted *S. alterniflora* and *P. australis* were in the initial colonization stage and could not be discerned in 1997 satellite images (with a resolution of 30 m × 30 m). For this reason, the plantation locations (from GPS data) were imposed on the salt marsh vegetation map (Fig. 7.7a).

The *S. alterniflora* community was not detectable on satellite imagery until 2000, when the plantation belt on the Middle Shoal and many small patches on the Lower Shoal were clearly visible (Fig. 7.7b). Thereafter, *S. alterniflora* clumps began to invade and successfully colonize their surroundings, which indicates that the species was expanding in conjunction with sedimentation and accretion. The lag time in the colonization of *S. alterniflora* was observed in the population introduced to the Jiuduansha Shoals between 1998 and 2000 and is mainly due to the adaptation of the species to the new habitat.

The 2000–2001 period marked the onset of rapid population growth and range expansion for *S. alterniflora* (Fig. 7.7c–h). However, as most of the suitable niche areas on the Middle Shoal had been occupied by 2005, the expansion rates of both the *S. alterniflora* and *P. australis* communities began to slow down. At the same time, an enormous range expansion of *S. alterniflora* occurred on the Lower Shoal, which is mainly attributed to rapid sedimentation and accretion in this area. It is anticipated that the rapid range expansion of *S. alterniflora* will last for a considerable time in the future on the Jiuduansha Shoals.

Factors influencing the number of dispersal modes and vital rates (e.g., births and deaths) are critical in regulating the spread of an invasive species. Compared with other native species, *S. alterniflora* has a strong reproductive capacity and wide ecological niche. The native *P. australis* also showed a similar pattern of range expansion to *S. alterniflora*, although at a slower rate. Once either of these two species occupied a new habitat, they could not out-compete and replace each other, which is clearly compatible with the classical space-preemption model for plant community dynamics (Watt 1947).

The Jiuduansha Shoals provide an ideal setting for the analysis of the sequence of events associated with the initial colonization, lag time, rate of geographic spread, and pattern of expansion of the exotic *S. alterniflora* and the native *P. australis*. The spatial community and population pattern are a product of the interactions among processes such as colonization, establishment, competition, renewal, and succession, as well as their living environment (Zhu et al. 2012). Invasive species provide opportunities to better understand aspects of community dynamics, are crucial for applications of community theory in restoration biology, and are important for wetland biodiversity conservation and resource management (Zhu et al. 2012).

7.3 Biological Invasion in the Coastal Wetlands

Biological invasion has received considerable attention recently because of its increasing impacts on local ecosystems (Mooney et al. 2005). Expansion of *S. alterniflora*, a non-native species, on the coastal wetlands in the Changjiang Estuary, is a prime example of such biological invasion (Huang and Zhang 2007). It has been reported that this exotic species may out-compete native plants, threaten native ecosystems and coastal aquaculture, and cause a decline in the diversity of native species in the coastal wetlands of the Changjiang Estuary (Callaway and Josselyn 1992; Chen et al. 2005; Li et al. 2006).

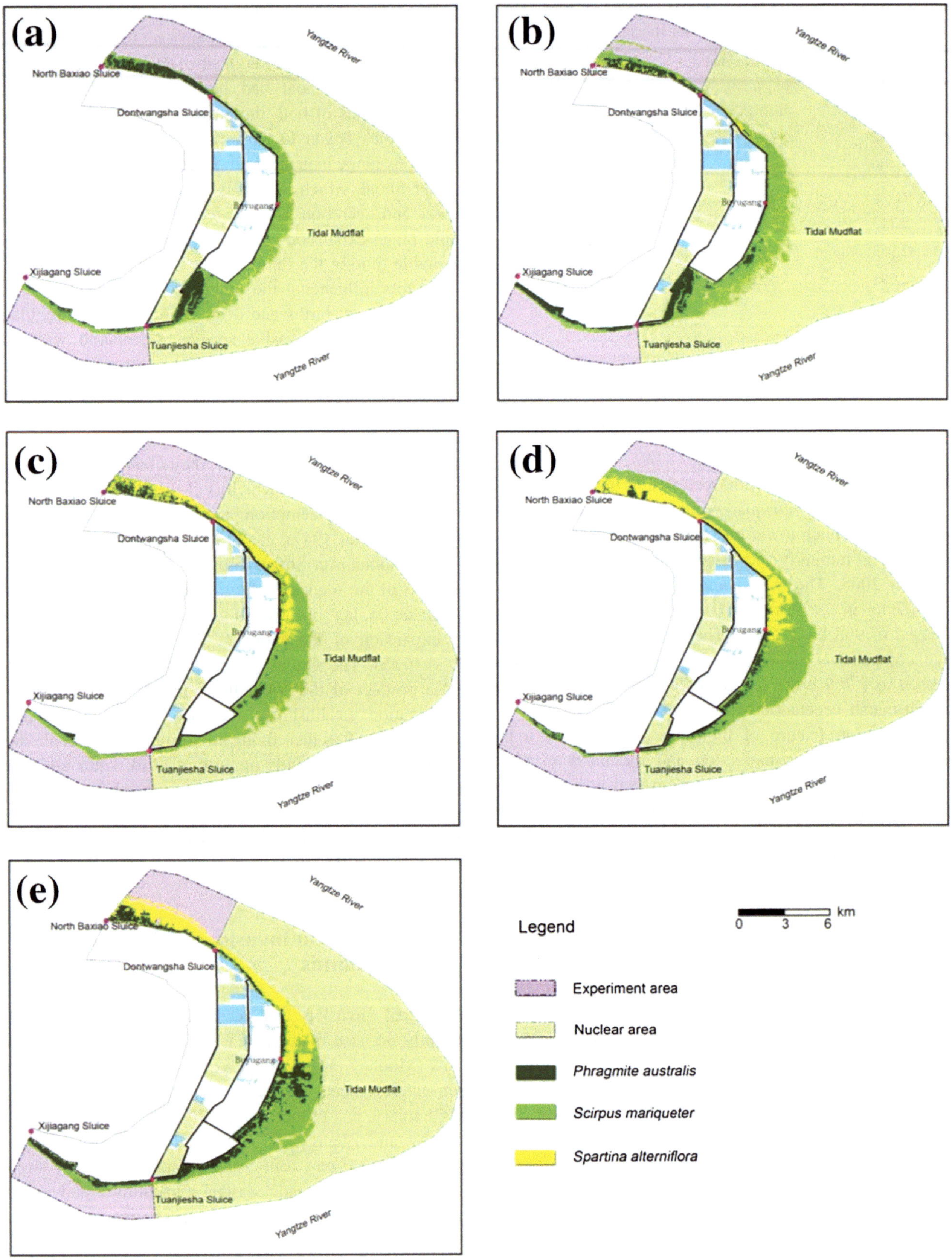

Fig. 7.6 Spatiotemporal dynamics of salt marsh vegetation (i.e., the *Phragmites australis* community, *Spartina alterniflora* community, and *Scirpus mariqueter* community) during 1998–2008 in the Chongming Dongtan Nature Reserve (from Huang 2009)

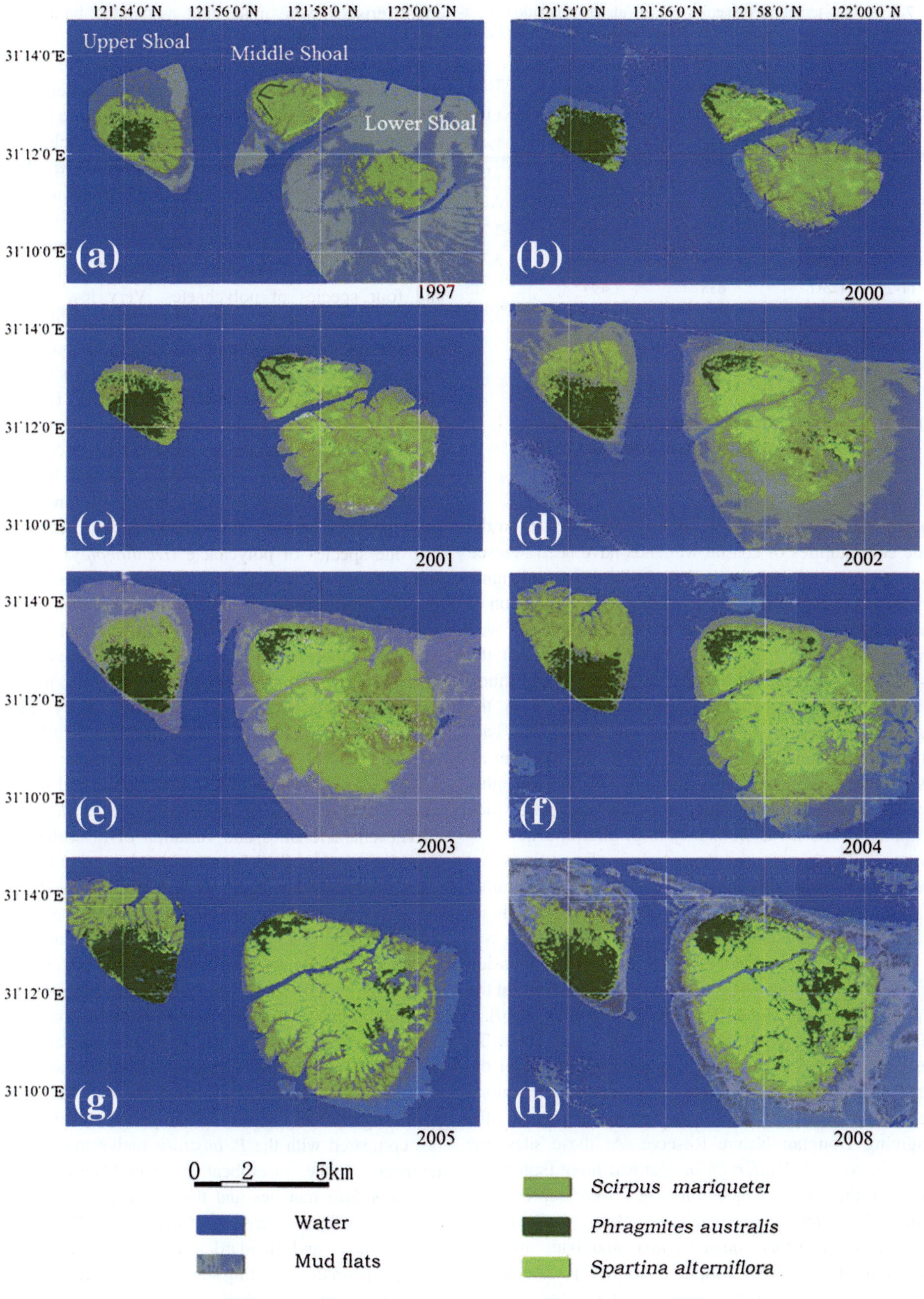

Fig. 7.7 Spatiotemporal dynamics of salt marshes (i.e., *Phragmites australis* community, *Spartina alterniflora* community, and *Scirpus mariqueter* community) during 1997–2008 in the Jiuduansha Shoals; i.e., the Upper Shoal, Middle Shoal, and Lower Shoal (modified from Huang 2009)

Table 7.4 Spatial–temporal dynamics of salt marsh vegetation (hectares) at the Jiuduansha Shoals (Upper Shoal, Middle Shoal, and Lower Shoal) (from Huang 2009)

TM date	*S. alterniflora*	*P. australis*	*S. mariqueter*	Total
1997-10-20	55	168	967	1,094
2000-05-24	102	354	1,017	1,473
2001-07-26	284	369	1,383	2,036
2002-11-11	377	402	1,608	2,387
2003-08-02	470	463	1,850	2,783
2004-07-19	1,014	564	1,789	3,367
2005-11-27	1,281	638	1,494	3,413
2008-04-25	1,709	924	968	3,601

Planted in 1997 (see text for details)

7.3.1 Impacts of the *S. alterniflora* Invasion on Benthic Communities in the Wetlands

Studies of the impacts of the *S. alterniflora* invasion on the benthic communities of coastal wetlands have been focused on the density, composition, and diversity of macrobenthic communities; accessibility to consumers; the modification of trophic function; and changes to the remaining food web (Wang et al. 2010). Chen et al. (2007) reported that the invasion of *S. alterniflora* changed the species composition of benthic communities in the coastal wetlands in the Changjiang Estuary, in particular through increases in bacterial-feeding nematodes. Xie et al. (2008) and Zhao et al. (2009) found significant differences in the macrobenthic composition and abundance along intertidal gradients, which were linked to the invasion history of *S. alterniflora*.

Body size is an important characteristic of organisms and is linked with many physiological and ecological parameters. The normalized biomass size spectrum (NBSS) is an indicator of habitat change (Quiroga et al. 2005) and was used to evaluate the impacts of the *S. alterniflora* invasion on benthic communities along intertidal zone gradients at the Chongming Dongtan Nature Reserve (Wang et al. 2010).

Fieldwork was carried out in late 2007 and 2008. To represent the range of benthic communities present in the coastal wetlands, seven sampling sites were selected along an elevation gradient within the northern part of the Chongming Dongtan Nature Reserve. At these sites, the invasion history of *S. alterniflora* spanned more than seven years, and the major vegetation types were represented, including *P. australis* (sites PA1 and PA2), *S. alterniflora* (SA1 and SA2), *S. mariqueter* (SM), and bare mudflats (BM1 and BM2). Another sampling site (SA3) was selected at the eastern end of the nature reserve, where the invasion history of *S. alterniflora* spanned less than two years. At each sampling site, paired measurements of habitat characteristics were made and associated benthic samples were collected (Wang et al. 2010).

A total of 29 macrobenthic species were recorded at the eight sampling sites, along an elevation gradient that represented the invasion history of *S. alterniflora* (Table 7.5). Gastropods were the dominant group and included 11 different species. Within this group, *Stenothyra glabra*, *Elachisina* sp., *Assiminea* sp., and *Assiminea latericea* accounted for more than 85 % of the total macrobenthic density. The more minor macrobenthic communities included five species of bivalves, eight species of crustaceans, and four species of polychaetes. Very low-concentration species (<0.2 % of the total density) included three types of crustaceans (*Deiratonotus cristatum*, *Exopalaemon annandalei*, and *Metaplax longipes*) and a species of nemertean (Wang et al. 2010). On closer inspection of Table 7.5, the characters of macrobenthic communities in relation to the elevation gradient and the invasion history of *S. alterniflora* could be recognized. The mudflats at the low-elevation (BM1 and BM2) sites were dominated by three species of gastropods (*S. glabra*, *Elachisina* sp., and *Assiminea* sp.) and one species of polychaete (*Dentinephtys glabra*). The same dominant species of *S. glabra* and *Elachisina* sp. occurred at the *S. mariqueter* marshes (SM). The *P. australis* marshes (PA1 and PA2) were characterized by gastropod species, *A. latericea* and *Assiminea* sp. The *S. alterniflora* marshes with a relatively long invasion history (SA1 and SA2) were characterized by gastropods, *A. latericea*, *Assiminea* sp., and *Cerithidea largillierti*, along with a polychaete, *Notomastus latericeus*. The newer *S. alterniflora* marshes (SA3) were dominated by a smaller species of *Assiminea* sp. and *Heteromastus filiforms*.

The characteristics and variation of macrobenthic communities are described by a multivariate technique, Canonical Correspondence Analysis (CCA). This clearly revealed the elevation gradients and invasion history of *S. alterniflora* (Fig. 7.8). The first axis of the CCA is significantly related to variations in elevation, and the second axis is related to variations in salt marsh vegetation. The macrobenthic density and biomass decreased from *P. australis* marshes to bare mudflats, which could be attributed to a decrease in the size of the dominant benthic species along the elevation gradient.

The results showed clear evidence that the invasion of *S. alterniflora* had a remarkable impact on benthic communities compared with the *P. australis* native marshes. The differences in the macrobenthic assemblages between the *S. alterniflara* marshes and the native *P. australis* marshes decreased with an increase in *S. alterniflara* (Wang et al. 2010).

The mean and standard deviation of the macrobenthic biomass from the sampling sites are presented in Fig. 7.9. The macrobenthic biomass had a decreasing trend along the elevation gradient, from 24.39 ± 2.66 g dwt/m^{-2} in *P. australis* marshes to 1.84 ± 0.57 g dwt/m^{-2} in bare mudflats (Fig. 7.9a).

Table 7.5 Composition and abundance of macrobenthos in the sampling sites

Taxon	PA1	PA2	SA1	SA2	SA3	SM	BM1	BM2
Polychaeta								
Harmothoe imbricata	–	–	–	–	–	–	–	+
Dentinephtys glabra	–	–	+	–	–	–	–	++
Notomastus latericeus	–	+	–	++	+	–	–	+
Heteromastus filiforms	+	+	–	++	++	+	–	+
Gastropoda								
Neritidae sp.	–	–	–	–	–	+	–	–
Assiminea violacea	–	–	–	+	–	–	–	–
Assiminea latericea	+++	+	+++	++	+	–	–	–
Assiminea sp.	–	+++	++	++	+++	+	–	++
Elachisina sp.	–	+	–	+	–	+++	++	+
Cerithidea sinensis	+	+	+	+	–	–	–	–
Cerithidea largillierti	++	+	+	++	+	–	–	–
Stenothyra glabra	–	–	–	–	–	++	+++	++
Rissonia sp.	–	–	–	+	–	+	–	–
Pseudoringicula sinensis	–	+	–	+	–	–	–	–
Onchidium verruculatum	–	–	–	+	–	–	–	–
Bivalvia								
Glauconome chinensi	–	–	–	–	+	+	+	+
Moerella iridescens	–	–	–	–	–	–	–	+
Sinonovacula constricta	–	–	–	–	+	+	–	+
Siliqua minima	–	–	+	–	–	+	–	+
Potamocorbula amurensis	–	–	–	–	–	+	+	+
Crustacean								
Gnorimosphaerosphaeroma rayi	–	–	–	–	–	+	+	–
Exopalaemon annandalei	–	–	+	–	–	–	–	–
Deiratonotus cristatum	–	–	–	+	–	–	–	–
Ilyoplax deschampsi	–	–	+	+	+	+	–	–
Helice tientsinensis	–	–	–	+	+	–	–	–
Chinomantes dehaani	+	+	–	+	+	–	–	–
Helicana wuana	–	–	–	–	+	–	–	–
Metaplax longipes	–	+	+	–	–	–	–	–
Nemertean	–	–	–	–	–	+	–	+

Symbols –, +, ++, and +++ indicate the abundance of 0, <10, 10–50, and >50 % of the total, respectively
SA1 and SA2 are the *Spartina alterniflara* marshes with a long invasion history; SA3 is the recently invaded *S. alterniflara* marshes; SM is the *Scirpus mariqueter* marshes; and BM1 and BM2 are the bare mudflats (from Wang et al. 2010)

The macrobenthic biomass was 20.17 ± 1.61 g dwt/m^{-2} at the *S. alterniflora* marshes with a longer invasion history. This was higher than in the recently invaded *S. alterniflora* marshes, where the macrobenthic biomass was 13.15 ± 1.82 g dwt/m^{-2}. In contrast, the meiobenthic biomass showed a reverse trend along the elevation gradient, from 0.40 ± 0.04 g dwt/m^{-2} in *P. australis* marshes to 0.66 ± 0.05 g dwt/m^{-2} in bare mudflats (Fig. 7.9b). The meiobenthic biomass at the *S. alterniflora* marshes with a longer invasion history (0.23 ± 0.02 g dwt/m^{-2}) was significantly lower ($p < 0.05$) than in the recently invaded *S. alterniflora* marshes (0.43 ± 0.09 g dwt/m^{-2}) (Wang et al. 2010).

The NBSS values of the benthic communities from the sampling sites are presented in Fig. 7.10. The regressions of the $\log_2$ (normalized biomass) against $\log_2$ (size class) were statistically significant in all cases ($p < 0.05$), and the R^2 values were mostly greater than 0.8. The slopes of NBSS for the sampling sites ranged from −0.982 to −1.270, and

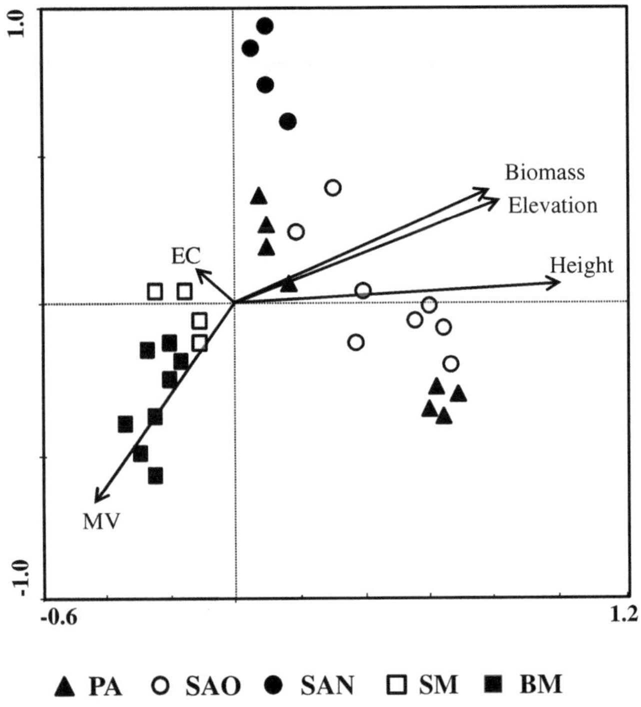

Fig. 7.8 Canonical correspondence analysis (CCA) ordination for the macrobenthic assemblages and environmental variables from the sampling sites at the Chongming Dongtan Nature Reserve. *PA* is the *Phragmites australis* marshes; *SAO* is the *Spartina alterniflora* marshes with a long invasion history; *SAN* is the recently invasion marshes; *SM* is the *Scirpus mariqueter* marshes; and *BM* is the bare mudflats. *EC* and *MV* indicate electronic conductivity and redox potential of the soil, respectively, and "height" is the height of the vegetation (from Wang et al. 2010)

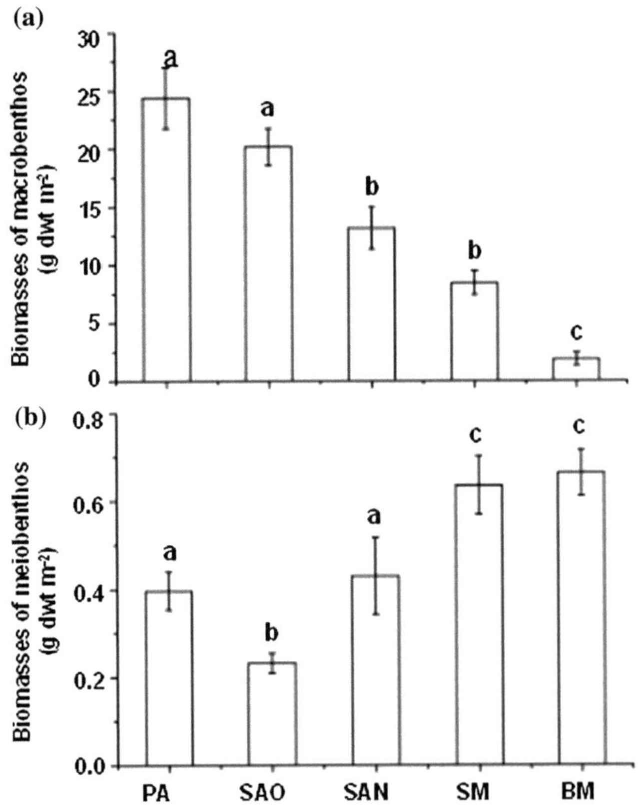

Fig. 7.9 Biomass of **a** macrobenthos and **b** meiobenthos from the sampling sites at the Chongming Dongtan Nature Reserve. The abbreviations for the sites are the same as in Table 7.5 (from Wang et al. 2010)

steeper slopes occurred with decreasing elevation, i.e., from *P. australis* marshes to mudflats (Fig. 7.10). This trend indicates a higher contribution of smaller benthos at the lower elevation sites. The NBSS slopes at recently invaded *S. alterniflora* marshes (Fig. 7.10c) were significantly steeper ($p < 0.05$) than for marshes with a longer invasion history (Fig. 7.10b). However, the difference between *P. australis* marshes (Fig. 7.10a) and *S. alterniflora* marshes with a longer invasion history (Fig. 7.10b) was not significant ($p > 0.05$).

The present results show clear evidence that the invasion of *S. alterniflora* had a remarkable impact on benthic communities compared with the native marshes. However, the differences tended to be smaller between native marshes and *S. alterniflora* marshes with a long invasion history. The lower redox potential in the recently invaded *S. alterniflora* marshes could indicate a positive response by the small-sized benthos, which have a higher ratio of surface area to body volume and are better adapted to the hypoxic environment (Wang et al. 2010). As the *S. alterniflora* marshes develop, the benthic assemblages appear to adapt gradually to the changing environment, as reflected by the increase in species number and the body size of the benthic assemblages in *S. alterniflora* marshes with a longer invasion history. However, further work using a longer sequence of invasion stages is needed to confirm this conclusion.

7.3.2 CA Model for Simulation of the Expansion of *S. alterniflora*

Like other complex biological processes, the expansion of an invasive species can be numerically intractable. One potential method to simplify and understand the process is to use cellular automaton (CA) models. The CA model introduced by Ulam and von Neumann in the 1950s has provided a formal framework for investigating the behavior of complex, extended systems (Sipper 2002). Due to their ease of implementation, ability to replicate spatial forms, and ease of use in GIS, CA models have been widely applied to different ecological problems. Invasion of the coastal wetlands in the Changjiang Estuary by *S. alterniflora* is a prime example of a spatially structured invasion, in a relatively simple habitat, for which strategic control efforts can be modeled and applied (Huang and Zhang 2007).

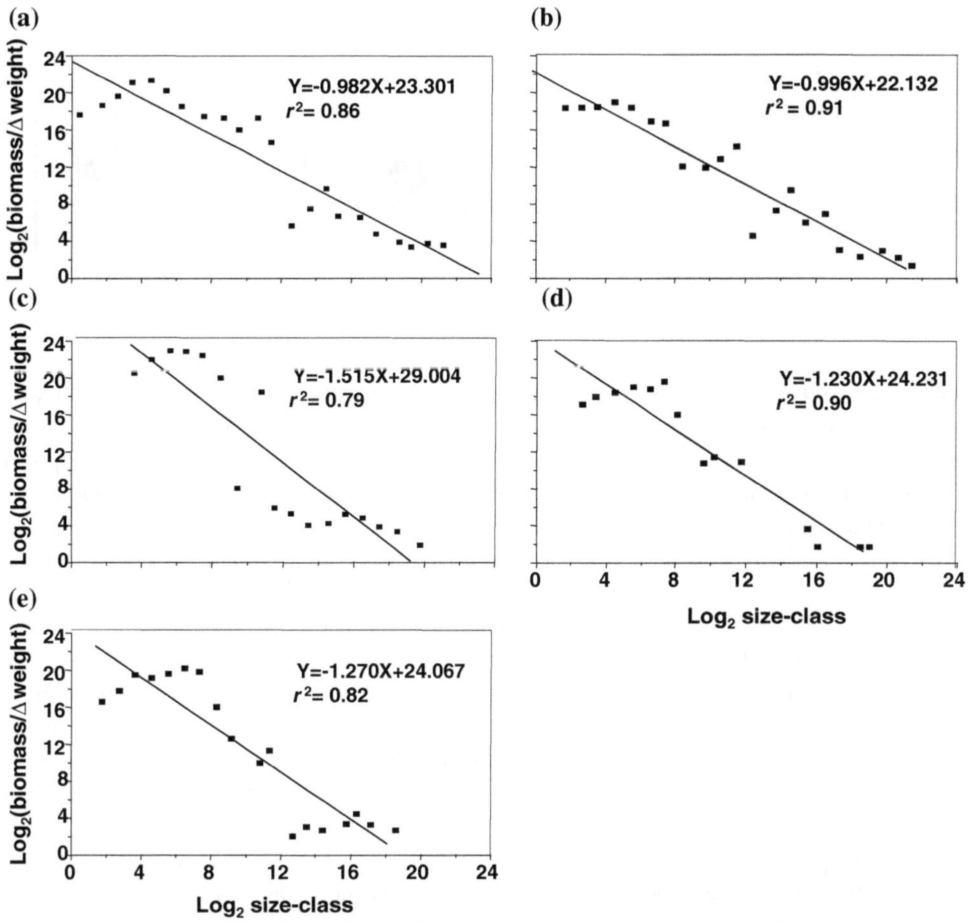

Fig. 7.10 Normalized biomass size spectra of benthic assemblages at the sampling sites. **a** *Phragmites australis* marshes; **b** *Spartina alterniflora* marshes with a long invasion history; **c** recently invaded *Spartina alterniflora* marshes; **d** *Scirpus mariqueter* marshes; and **e** bare mudflats (from Wang et al. 2010)

A CA model to simulate the expansion of *S. alterniflora* was developed to study the interactions between the spatial pattern and ecosystem processes of the salt marsh vegetation on Jiuduansha Shoals. It was intended to yield insight into the population and community ecology of invasive species in the Changjiang Estuary, which could be important for wetland biodiversity conservation and resource management.

A CA is defined by (*S, N, and f*) and with set language as follows:

$$S_{t+1} = f(S_t, N), \quad (7.1)$$

where *S* is a finite set and represents the cell state, *t* denotes the transition step, *f* stands for the transition rule or function, and *N* is the cell neighbors (Wolfram 1984).

A standard CA consists of an array of identical cells, a set of discrete cell states, a cell neighborhood of a defined size and shape, a set of transition rules, and discrete time steps. Each cell can be in one of a finite number of possible states, updated synchronously in discrete time steps according to a local and identical interaction rule. The state of a cell is determined by the previous states of a surrounding neighborhood of cells (Wolfram 1984).

A Euclidean space was divided into square grids, and the tessellation of these spatiotemporal grids was depicted as a lattice of cells, each with eight neighbors. The cells are equivalent to the pixels in the satellite imagery used in this study, which are not only discrete spaces but also have geographical significance after the pixels are integrated. The cells were used to define the spatiotemporal distribution of salt marsh vegetation on Jiuduansha Shoals, derived from the vegetation classification raster images with a resolution of 30 m × 30 m. There were four possible states for each cell: *P. australis* community, *S. alterniflora* community, *S. mariqueter* community, and mudflat with elevations of ≤2 m. These states were identified by their gray values and denoted as $S_{\mathrm{p.a}}$, $S_{\mathrm{s.a}}$, $S_{\mathrm{s.m}}$, and S_0, respectively.

A CA is an array of cells that interact with one another locally according to some neighborhood rules; in addition, the neighborhood relation is local and uniform. The temporal and spatial patterns of both *S. alterniflora and P. australis* on Jiuduansha Shoals have simple advancing wave fronts (Huang and Zhang 2007), which can be easily simulated by a Moore neighborhood (Sirakoulis et al. 2000). The Moore neighborhood is an enlargement of the von Neumann neighborhood (Fig. 7.11), containing diagonal cells

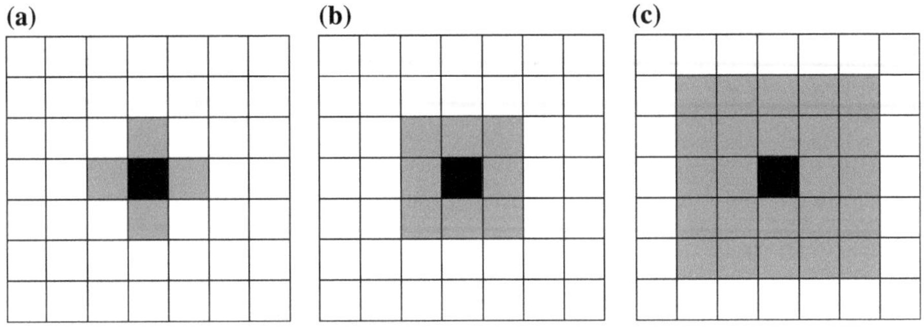

Fig. 7.11 Definitions of neighborhoods. **a** Neumann neighborhood, **b** Moore neighborhood, $r = 1$, and **c** Moore neighborhood $r = 2$. From Huang et al. (2008)

(Wallinga 1995). The rate of geographic spread for *S. alterniflora* and *P. australis* can be incorporated into the model by defining the radius $r = 0, 1, 2, \ldots n$.

The dynamics of the system takes place in discrete time steps, which are one year long in this case, and the transition function controls the cell state in the next time period. Each cell evolves according to local rules that depend only on the state of the cell and a finite number of neighboring cells. The spatial and temporal pattern of population and community is a product of the interactions among processes such as colonization, dynamics, renewal, competition, succession, and evolution. Salt marsh vegetation parameters in each cell take states of $S_{p.a}$, $S_{s.a}$, $S_{s.m}$, and S_O (representing the *P. australis*, *S. alterniflora*, and *S. mariqueter* communities, and bare mudflat, respectively) and are sorted spatially at the beginning of the simulation S_t. The cell states in the next time step S_{t+1} are then defined by salt marsh states and species interactions. *P. australis* and *S. alterniflora* are obstacles to each other and can only occupy the *S. mariqueter* cells, instead of mudflat with elevation ≤2 m asl. This can be summarized by the following rule:

$$S_{t+1} = \begin{cases} S_O, \text{ while } S_t = S_O \text{ (elevation} \le 2\text{m)} \\ S_{p.a}, \text{ while } S_t = S_{p.a} \\ S_{s.a}, \text{ while } St = S_{s.a} \\ S_{p.a}, \text{ while } N_{p.a} \ge 5 \text{ and } S_t = S_{s.m} \\ S_{s.a}, \text{ while } N_{s.a} > 3 \text{ and } S_t = S_{s.m} \\ S_{p.a}, \text{ while } N_{p.a} \ge 1 \text{ and } N_{s.a} = 0 \text{ and } S_t = S_{s.m} \\ S_{s.a}, \text{ while } N_{s.a} \ge 1 \text{ and } N_{p.a} = 0 \text{ and } S_t = S_{s.m} \\ S_{p.a}, \text{ while } N_{p.a} - N_{s.a} \ge 2 \text{ and } N_{s.a} < 3 \text{ and } S_t = S_{s.m} \\ S_{s.a}, \text{ while } N_{p.a} - N_{s.a} < 2 \text{ and } N_{p.a} < 5 \text{ and } S_t = S_{s.m} \end{cases} \quad (7.2)$$

where S_t and S_{t+1} represent the cell states, at times t and $t + 1$, respectively, and $N_{P.a}$ and $N_{S.a}$ are the numbers of *P. australis* and *S. alterniflora* cells around the center cell, respectively.

In this way, the interactions between species and habitat characteristics for the salt marsh vegetation on Jiuduansha Shoals (e.g., space preemption, succession, and inhibition) can be incorporated into the model. The rate of geographic spread of *S. alterniflora* and *P. australis* was incorporated into the model by adjusting the radius ($r = 0, 1, 2, \ldots n$) of the Moore neighborhood; i.e., one or more of the inputs was adjusted to determine the "fit" of the model to the observations. A series of radii were tested for *S. alterniflora* and *P. australis,* respectively, taking into account the accretion rate of the shoals. The tentatively simulated results were compared iteratively with the corresponding salt marsh classification maps until they reached an acceptable range.

Model accuracy assessment involves the derivation of accuracy metrics based on a comparison between the simulation results and the image classification in the corresponding year, using the classified salt marsh vegetation maps as a "checkpoint" (Messina and Walsh 2001). The accuracy (%) was defined as the derivation (or match) between the simulated results and the classified salt marsh vegetation maps for the whole area in corresponding years. The accuracy assessment was executed using ESRI Arc/GIS 8.1 software by overlaying corresponding classified salt marsh vegetation maps and the simulation images and calculating the percentage difference (Huang et al. 2008).

7.3.3 Population Expansion on the Middle Shoal

Using the results presented in Sect. 7.2.4 and Fig. 7.7, the rates of geographic spread (the radius) in 1997–2000, 2001, 2002, 2003, 2004, 2005, and 2008 for *S. alterniflora* and *P. australis* on the Middle Shoal were assigned to $r = 1, 3, 3, 1, 0, 0, 1$ and $r = 1, 1, 1, 2, 1, 0, 1$, respectively (Table 7.6). The simulation by the CA model for the Middle Shoal during 1997–2008 is presented in Fig. 7.12. The model accuracy assessment shows that accuracies ranged from 67 to 75 % (Table 7.6). In general, this model could satisfactorily simulate the broad-scale population dynamics of *S. alterniflora* and *P. australis* on the Middle Shoal.

Table 7.6 Expansion rate (r values) of salt marshes, and simulation accuracy of the CA model for the Middle Shoal (from Huang et al. 2008)

Year	1997–2000	2000–2001	2001–2002	2002–2003	2003–2004	2004–2005	2005–2008
Expansion rate of *S. alterniflora* (r)	1	3	3	1	0	0	1
Expansion rate of *P. australis* (r)	1	1	1	1	1	0	1
Accuracy	71.5 %	67.5 %	75 %	73 %	70 %	–	72.1 %

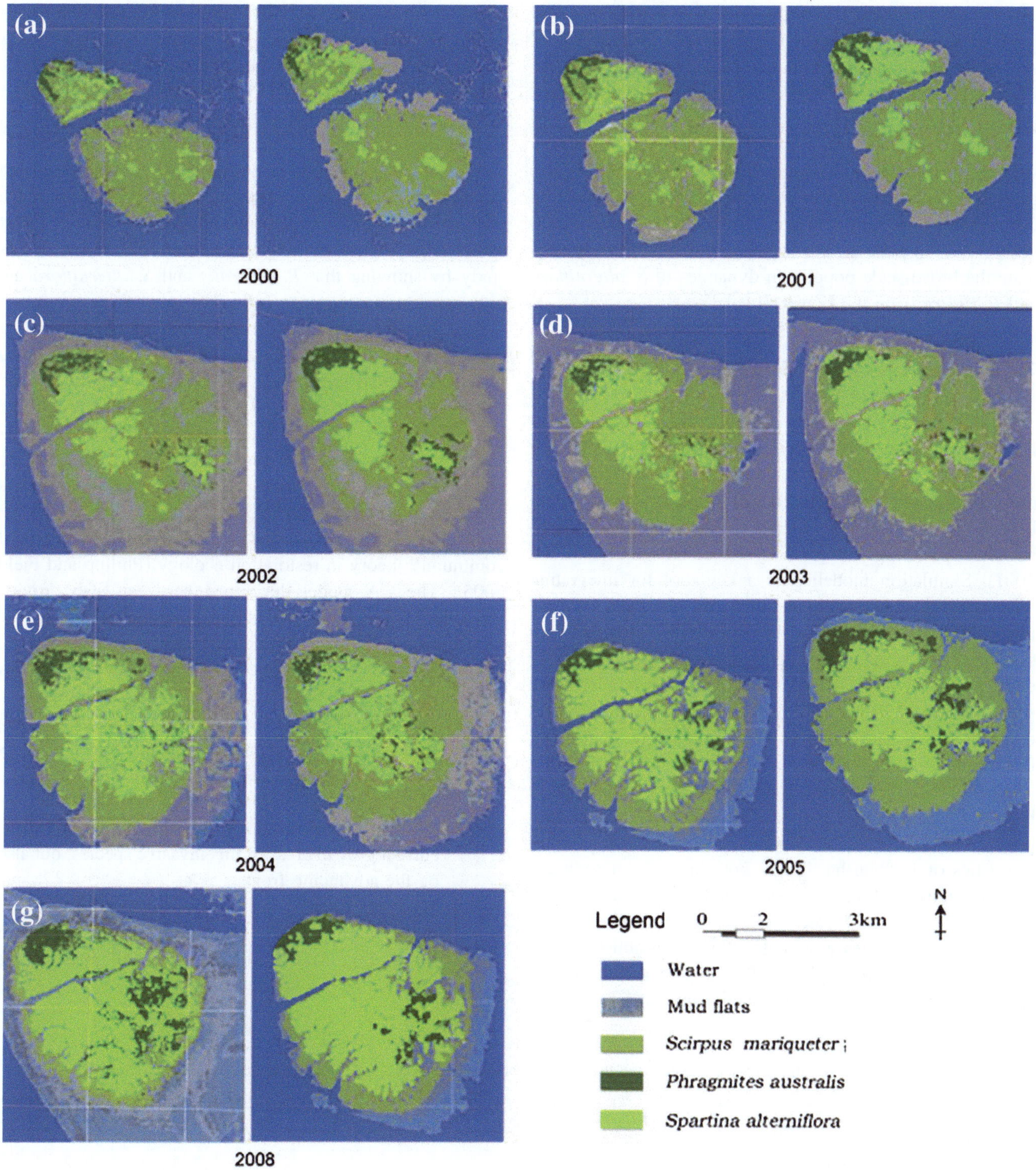

Fig. 7.12 Results of the CA simulation of *Phragmites australis* and *Spartina alterniflora* population dynamics on the Middle Shoal and the Lower Shoal of the Jiuduansha Shoals. Image classifications are shown in the *left panels,* and the simulations are shown in the *right panels* (from Huang 2009)

7.3.4 Population Expansion on the Lower Shoal

Compared with the Middle Shoal, the Lower Shoal is growing fast in a southeasterly direction (Fig. 7.7). Therefore, the rates of geographic spread (r value) in 1997–2000, 2001, 2002, 2003, 2004, 2005, and 2008 for *S. alterniflora* and *P. australis* were assigned to $r = 1, 4, 4, 4, 5, 3, 3$ and $r = 0, 0, 1, 1, 1, 2, 2$, respectively (Table 7.7). Figure 7.12 shows the results of the simulation by the CA model for the Lower Shoal during 1997–2008. It seems that the accuracy is relatively low (37 %) compared to post-2000, as the elevation was relatively low on the Lower Shoal, and there were no sufficient data on the location of *S. alterniflora* plantation and the origin of *P. australis*. The accuracies of CA simulations ranged from 57 to 71 % after 2000 (Table 7.7). In general, this model could satisfactorily simulate the broad-scale population dynamics of *S. alterniflora* and *P. australis* on the Lower Shoal.

7.3.5 Implications for Strategic Control of Invasive Plants

Like other complex biological processes, experimental or observational studies of population expansion are commonly difficult at larger scales, due to factors such as lack of replication in landscape ecological studies and interactions between spatial and temporal dynamics (Jenerette and Wu 2001). Simulation modeling is a key tool for integrating information and testing hypotheses and has been of great importance in understanding the complex dynamics of ecosystems. Simulation modeling is also the only means for extrapolating ecological relationships between research sites with different environmental driving functions and across a range of spatial and temporal scales over which controlling factors may vary significantly.

The Jiuduansha Shoals in the Changjiang Estuary have been undergoing rapid growth with relatively low anthropogenic disturbance. The dynamic, new, and natural characteristics of Jiuduansha Shoals, both in terms of sediment deposition, vegetation succession, and biological invasion, make it an ideal study area. *S. mariqueter* is the pioneer vegetation and creates new niches for the establishment of *P. australis* and *S. alterniflora*. All range expansions of these species depend on the accretion of dynamic shoals, i.e., the new niche regions. As the Middle Shoal reached a relatively stable state after 2002, the expansion of both species has slowed. Conversely, the Lower Shoal is growing rapidly, and the expansion of both species has remained high in this area.

The temporal and spatial expansion of both *S. alterniflora and P. australis* on the Jiuduansha Shoals exhibits simple advancing wave fronts (Huang and Zhang 2007), which could be simulated by the CA model by adjusting the radius of the Moore neighborhood. It is notable that *P. australis* had a similar pattern and range expansion process to *S. alterniflora*, although at a 3–5 times lower rate. This result indicates that *S. alterniflora* shares some traits (or niches) with *P. australis*. Once either of these two species occupied a new habitat, they could not out-compete and replace each other, which is clearly compatible with the classical space preemption model for plant community dynamics (Watt 1947). This hypothesis has been successfully tested in this study by showing that *P. australis* and *S. alterniflora* are only capable of occupying cells containing *S. mariqueter*.

The stronger capacity for competition and a broader niche are key factors for the successful invasion of *S. alterniflora* into the intertidal wetlands in the Changjiang Estuary. The CA model simulation showed that the range expansion rate of *S. alterniflora* was 3–5 times faster than that of *P. australis*. With the rapid development of Jiuduansha Shoals, the range expansion of *S. alterniflora* on the Lower Shoal is expected to continue well into the future.

Studies of invasive species are crucial for applications of community theory in restoration ecology (Philipp and Field 2005). The CA model developed here can also provide valuable insights into which strategic control efforts can be modeled and applied. The following conclusions on strategic control efforts can be obtained from the results of the CA model developed in this study.

(i) The invasion of *S. alterniflora* in the Changjiang Estuary is dependent on the increase of empty niches, and strategic control efforts should therefore be focused on accretive intertidal zones.
(ii) The control efforts should not only be located in the current growth areas of this invasive species, but also on the advancing fronts.
(iii) Measures, such as an isolation zone (barrier), should be taken to stop the rapid geographical spread of *S. alterniflora* and to control and manage the invasion of this species.

Table 7.7 Expansion rate (*r* values) of salt marshes and the accuracy of the CA model for the Lower Shoal (from Huang et al. 2008)

Year	1997–2000	2000–2001	2001–2002	2002–2003	2003–2004	2004–2005	2005–2008
Expansion rate of *S. alterniflora* (*r*)	1	4	4	4	5	3	3
Expansion rate of *P. australis* (*r*)	0	0	1	1	1	2	2
Accuracy (%)	37	57	71	67	57	68	70

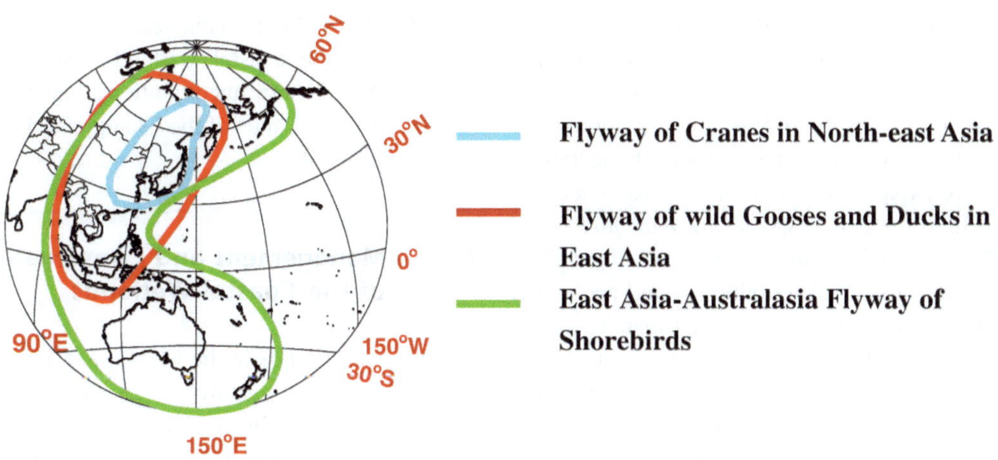

Fig. 7.13 Coastal wetlands in the Changjiang Estuary are an important destination and stopover point for migratory birds in the Asia–Pacific region (modified from Zhou and Xie 2012)

7.4 Management and Conservation of Coastal Wetlands in the Changjiang Estuary

Located at the junction of the Changjiang, the Yellow Sea, and the East China Sea, the coastal wetlands in the Changjiang Estuary form one of the largest estuarine alluvial wetlands in the world. Consisting of large areas of salt marshes, tidal creeks, and intertidal mudflat, the coastal wetlands are the north–south flyway of migratory birds (Fig. 7.13). Each year, millions of birds arrive for staging and wintering. The number of geese arriving for wintering can exceed 30,000, and four wintering species [the white stork (*Ciconia boyciana*), hooded crane (*Crus monacha*), black stork (*Ciconia nigra*), and white-tailed eagle (*Haliaeetus albicilla*)] are under state protection and listed in National first-class categories. More than 10 species of water birds that visit the wetlands exceed 1 % of the entire East Asian and Australian populations (Tian et al. 2008; Zhou and Xie 2012). The coastal wetlands in the Changjiang Estuary also provide channels for migratory fish and spawning and feeding grounds for many fish species, especially the Chinese sturgeon (Zhou and Xie 2012).

As the Changjiang Estuary is located adjacent to the megacity of Shanghai, human activities exert substantial pressures on the coastal wetlands and cause coastal ecosystem degeneration and consequent decreases or loss in ecosystem services. A stress is defined as a proximate activity or process that has caused or may cause the decrease or loss in ecosystem services in the coastal wetlands. The major stresses, mainly related to human activities in the coastal wetlands of the Changjiang Estuary, include urbanization, reclamation, invasive species, overfishing, and environmental pollution (Zhou and Xie 2012).

7.4.1 Stresses on the Coastal Wetlands in the Changjiang Estuary

Coastal areas are vital economic hubs in terms of settlement, industry, agriculture, trade, and tourism, as well as being of great environmental significance. The coastal wetlands in the Changjiang Estuary not only provide a vast number of ecological services, but also eco-security for the sustainable social and economic development of the city of Shanghai. Shanghai is now the fastest growing major Chinese city, with a population that increased from 5.2 million in 1949 to more than 20 million in 2010 (Zhou and Xie 2012). This megacity is accelerating the process of urbanization, which has imposed substantial pressures on the coastal wetlands, both in terms of space demand and environmental impacts. It is a great challenge to balance sustainable social–economic development with maintaining and preserving healthy coastal wetlands in the Changjiang Estuary.

As a fast growing megacity, Shanghai demands more land for social–economic development and accelerating urbanization. According to the Shanghai Water Authority (2010), a total 1040.9 km^2 of coastal wetlands has been reclaimed from 1949 to 2010 (Table 7.8). Large-scale reclamation therefore threatens the preservation of coastal wetlands in the Changjiang Estuary.

Table 7.8 Reclamation of coastal wetlands (km^2) in the Changjiang Estuary during 1950–2010 (data from Shanghai Water Authority 2010)

Period	1953–1959	1960–1969	1970–1979	1980–1989	1990–1999	2000–2005	2005–2010	Total
Reclamation	136.58	251.92	152.67	79.69	194.15	172.6	53.3	1040.9

The invasion of exotic species is a major threat to the biodiversity of the coastal wetlands in the Changjiang Estuary. *S. alterniflora*, a non-native plant species, was first introduced to this region in the 1990s, for the purpose of trapping sediment and encouraging salt marsh accretion. With its rapid range expansion, *S. alterniflora* expanded at a much faster rate than the native species, *S. mariqueter* and *P. australis*. The area of mono-dominant *S. alterniflora* increased rapidly from its initial introduction and accounted for almost one-third of the total intertidal salt marshes in the Changjiang Estuary in 2008 (Huang 2009). It has been reported that this exotic species may out-compete native plants, threaten the native ecosystems and coastal aquaculture, and cause a decline in the native species richness of coastal wetlands in the Changjiang Estuary (Li and Zhang, 2008).

Overfishing is another threat to the biodiversity of coastal wetlands in the Changjiang Estuary (see Chap. 8). Overfishing of specific fish or marine organisms could lead to some fish species failing to produce enough viable offspring to replace themselves. The species under threat of overfishing include the blue-spotted mudhopper (*Boleophthalmus pectinirostris*), so-iny mullet (*Liza haematocheila*), Asian freshwater goby (*Synechogobius ommaturus*), and the spotted sea bass (*Lateolabrax maculatus*). The highly affected macrobenthic species include *Moerella iridescens*, *Bullacta exarata*, *Helice tientsinensis*, and *Glaucomya chinensis*. Some fishing practices, such as benthos collection, adversely affect the benthic habitat and also disturb bird habitats. Inadequate fishing tools, such as netting that could result in incidental mortality through entanglement, or bycatch of migratory birds and rare species, are also a threat to estuarine ecosystems (Zhou and Xie 2012). Other low-ranked threats to the coastal wetlands in the Changjiang Estuary include illegal poaching and water pollution. All of these threats could greatly impact habitat quality, ecosystem services, and the biodiversity of the coastal wetlands (Zhou and Xie 2012).

The coastal wetlands in the Changjiang Estuary also face potential threats from the reduction in sediment supplied from upstream regions and rising sea levels. The large-scale engineering projects upstream, such as the Three Gorge Dam, have significantly reduced the amount of sediment transported downstream (Yang et al. 2011). On the other hand, the global climate system has experienced significant modification over the past century, due to the impacts of increased greenhouse gas concentrations, resulting in global warming and the associated melting of polar and alpine glaciers, and changing ocean currents (Wang et al. 2012). In recent decades, sea level in the Changjiang Estuary has been rising at a rate of about 2.6 mm per year, and this is predicted to accelerate, resulting in a sea-level increase of 40–50 cm by 2050 (SOA 2012). This sea-level increase may cause a reduction in the coastal wetland area in the Changjiang Estuary. All of these uncertainties about the future impacts on the coastal wetlands of the Changjiang Estuary need to be studied and verified.

7.4.2 Management and Conservation of the Coastal Wetlands

The coastal wetlands in the Changjiang Estuary are not only important natural resources, but also provide ecological security for sustainable social and economic development in this region. As drastic urbanization accelerates in this region, it is a great challenge to keep a dynamic balance between the preservation of coastal wetlands and sustainable social and economic development. The importance of the coastal wetlands in the Changjiang Estuary has been recognized widely both by local government and by Chinese citizens, resulting in the establishment of the Chongming Dongtan National Nature Reserve and the Jiuduansha Wetland Nature Reserve in the Changjiang Estuary. However, these two nature reserves are facing a great challenge due to the invasion of the exotic and invasive species, *S. alterniflora*.

By their very nature, the coastal wetlands in the Changjiang Estuary are dynamic, and their management relies on up-to-date and effective management plans. Control and eradication of *S. alterniflora* within the nature reserves are vital for the management and conservation of the coastal wetlands. Methods that have been suggested to control or eradicate *S. alterniflora* (Frid et al. 1999) include physical or mechanical control (Patten 2002) and biological control (Grevstad et al. 2003). Physical control methods include the application of manual labor or mechanical devices to remove, burn, prune, excavate, dredge, cover, flood, and drain the *S. alterniflora* plants. These control measures would restrict growth, hinder respiration and photosynthesis, and finally suppress or kill the plants. Chemical controls include the use of herbicides that may impact the soils or local ecosystems. Biological controls inevitably introduce natural enemies of the invasive species, but as this may bring another potential invasion, careful study is needed to validate the efficiency of this method (Li and Zhang 2008). As a result, physical control methods are usually the first and most preferred option.

A field experiment to examine physical controls on *S. alterniflora*, including digging and tillage, breaking of rhizomes, mowing, and biological substitution with *P. australis*, was conducted during 2005 and 2006 to find a means of controlling this invasive plant (Li and Zhang 2008). The growth parameters of plant density, coverage, and above-ground biomass were used to evaluate the efficiency of different treatments. The results showed that for all treatments, the plant density, coverage, and aboveground biomass of *S. alterniflora* were significantly lower in the treatment area than in the control area in the first growing season. However, in the second season, there was less difference between the treatment and control areas, and there were no significant differences by the end of the second growing season.

Breaking of rhizomes inhibited the growth of *S. alterniflora* significantly in the first growing season, and the inhibition increased with the depth of the treatment. However, the growth inhibition of *S. alterniflora* disappeared after two growing seasons, at which point there were no significant differences between the treatment and the control areas. Mowing significantly inhibited the growth of *S. alterniflora* in the first growing season, but by the end of the second growing season, the growth of *S. alterniflora* had recovered to some extent, and only the treatments during the summer significantly inhibited its growth. For this reason, July and August would be the most suitable time for controlling *S. alterniflora* via mowing. In the biological substitution treatment, the transplanted native plant, *P. australis*, survived quite well over both growing seasons, and both the plant height and fruiting percent increased considerably in the second growing season (Li and Zhang 2008).

A demonstration project was established at the experimental zone of the Chongming Dongtan National Nature Reserve. In this zone, controlled waterlogging and cutting were carried out to investigate effective measures for controlling *S. alterniflora* on salt marshes (Yuan et al. 2011). The results during 2007–2008 showed that although the managed waterlogging itself could significantly reduce biomass and seed production at an early stage, *S. alterniflora* adapted very quickly to long-term waterlogging stress. Thus, managed waterlogging alone could not effectively control or eradicate *S. alterniflora* (Fig. 7.12).

On the other hand, this plant could be effectively controlled and eradicated by managed waterlogging combined with cutting the aboveground part of *S. alterniflora* at a key stage (e.g., the florescence period in July) (Fig. 7.14). Using this technique, the aboveground and belowground parts of *S. alterniflora* were killed, the plants began to decompose after three months, and there was no re-growth of *S. alterniflora* in the following years (Fig. 7.15). Therefore, an integrated technique of cutting at the florescence period, combined with managed waterlogging for about three months, has been shown to be a realistic and effective technique for controlling and managing the invasion of *S. alterniflora* in the nature reserve (Yuan et al. 2011). However, once the cofferdam that impounded water for waterlogging was removed to restore the natural hydrodynamic regime of wetlands, the seeds and seedlings of *S. alterniflora* could reinvade the controlled site from the neighboring area and re-establish a new community (Xiao et al. 2012). Therefore, after eradication of *S. alterniflora*, proper measures should be taken to prevent its re-establishment. The results of this demonstration project provide a useful and effective approach for controlling and managing the large-scale invasion of *S. alterniflora* on coastal wetlands in the Changjiang Estuary.

Fig. 7.14 Effects of different treatments on the growth of *S. alterniflora* during 2007–2008 at the demonstration project site. **a** is in July 2007; **b** is in November 2007; **c** is in June 2008; and **d** is in October, 2008 (from Yuan et al. 2011)

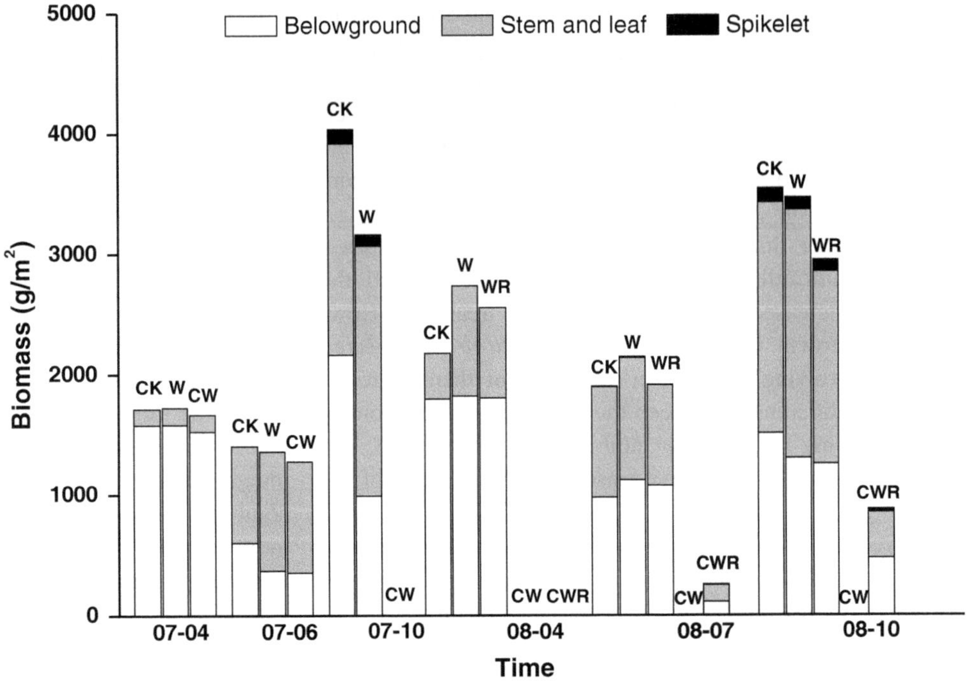

Fig. 7.15 Effects of different treatments on the biomass of *Spartina alterniflora* during 2007–2008 at the demonstration project site. *CW* is cutting *Spartina alterniflora* at the florescence period plus managed waterlogging; *W* is managed waterlogging alone; *CWR* and *WR* represent the cofferdam being broken to restore the natural hydrodynamic regime in 2008 after managed waterlogging in 2007; and *CK* is the control (from Yuan et al. 2011)

7.5 Summary and Conclusions

The enormous quantities of sediment produced by the Changjiang have created extensive coastal wetlands in the estuarine region, which have been colonized by various types of salt marsh vegetation. By their very nature, the coastal wetlands in the Changjiang Estuary are dynamic and provide high biodiversity and various ecosystem services.

This chapter described the succession of the coastal wetlands in the Changjiang Estuary. The spatial and temporal variations (1990–2008) of salt marshes were analyzed on the basis of RS mapping, combined with field inventories of sampling plots. The population expansion pattern of an exotic species, *S. alterniflora*, after its introduction in the 1990s to the coastal wetlands in the Changjiang Estuary, was described and analyzed.

The impacts of the invasion of *S. alterniflara* on benthic communities and the invasion history were examined along intertidal salt marsh gradients. The invasion of *S. alterniflora* had a profound impact on benthic assemblages, but unlike the native marshes, the effects decreased over time after the initial invasion. A CA model to simulate the *S. alterniflora* expansion was developed to study the interactions between the salt marsh vegetation spatial pattern and ecosystem processes. The stronger capacity for competition and a broader niche are key factors for the successful invasion of *S. alterniflora* into the intertidal wetlands of the Changjiang Estuary. The CA model developed here could also provide valuable insights into which strategic control efforts can be modeled and applied. Field experiments and a demonstration project were carried out in the Chongming Dongtan Nature Reserve to control the invasion of *S. alterniflora.* An integrated technique of cutting at the florescence period, combined with managed waterlogging, was shown to be a realistic and effective technique for controlling and managing the invasion of *S. alterniflora* in the Changjiang Estuary.

As the Changjiang Estuary is located adjacent to the most developed region in China, the intensified human activities such as urbanization, reclamation, invasive species, overfishing, and environmental pollution have exerted substantial pressures on the coastal wetlands. The coastal wetlands are not only important natural resources, but also provide ecological security for sustainable social and economic development in this region. The future development, biodiversity, management, and conservation of coastal wetlands in the Changjiang Estuary were discussed.

In conclusion, the coastal wetlands in the Changjiang Estuary provide high biodiversity and various ecosystem services. It is a great challenge to keep a dynamic balance between the preservation of coastal wetlands and sustainable social and economic development. A large number of scientific questions and management practices related to these issues need to be answered. At the same time, the feasibility and effectiveness of various management measures need to be studied and verified.

References

Callaway JC, Josselyn MN (1992) The introduction and spread of smooth cordgrass (*Spartina alterniflora*) in south San Francisco Bay. Estuaries 15:218–226

Chen ZY, Li B, Chen JK (2005) Some growth characteristics and relative competitive ability of invasive *Spartina alterniflora* and native *Scirpus mariqueter*. Biodivers Sci 13:130–136

Chen HL, Li B, Hu JB, Chen JK, Wu J (2007) Effects of *Spartina alterniflara* invasion on benthic nematode communities in the Yangtze Estuary. Mar Ecol Prog Ser 336:99–110

Chung CH (2006) Forty years of ecological engineering with *Spartina* plantations in China. Ecol Eng 27:49–57

Frid CLJ, Chandrasekara WU, Davey P (1999) The restoration of mud flats invaded by common cordgrass (*Spartina anglica* CE Hubbard) using mechanical disturbance and its effects on the macrobenthic fauna. Aquat Conserv Mar Freshw Ecosyst 9:47–61

Gao ZG, Zhang LQ (2006) Analyzing the multi-seasonal spectral characteristics of saltmarsh vegetation at the intertidal zone in Shanghai. Estuar Coast Shelf Scil 69:217–224

Grevstad FS, Strong DR, Garcia-Rossi D, Switzer RW, Weckere MS (2003) Biological control of *Spartina alterniflora* in Willapa Bay, Washington using the planthopper *Prokelisia marginata*: agent specificity and early results. Biol Control 27:32–42

Huang HM (2009) Spatial-temporal dynamics of saltmarsh vegetation at the intertidal zone in Shanghai based on remote sensing and GIS analysis (in Chinese). Doctor thesis, Eastern China of Normal University, Shanghai

Huang HM, Zhang LQ (2007) A Study on the population dynamics of *Spartina alterniflora* at Jiuduansha Shoals in Shanghai. Ecol Eng 29:164–172

Huang HM, Zhang LQ, Yuan L (2007) The spatio-temporal dynamics of saltmarsh vegetation for Chongming Dongtan National Nature Reserve, Shanghai (in Chinese). Acta Ecologica Sinica 27:4166–4172

Huang HM, Zhang LQ, Guan YJ, Wang DH (2008) A cellular automata model for population expansion of Spartina alterniflora at Jiuduansha Shoals, Shanghai, China. Estuar Coast Shelf Sci 77:47–55

Jenerette GD, Wu J (2001) Analysis and simulation of land-use change in the central Arizona- Phoenix region, USA. Landscape Ecol 16:611–626

Li HP, Zhang LQ (2008) An experimental study on physical controls of an exotic plant *Spartina alterniflora* in Shanghai, China. Ecol Eng 32:11–21

Li HP, Zhang LQ, Wang DH (2006) A study on the distribution of an exotic plant *Spartina alterniflora* in Shanghai (in Chinese). Biodiver Sci 14(2):114–120

Messina JP, Walsh SJ (2001) 2.5D morphogenesis: modeling landuse and landcover dynamics in the Ecuadorian Amazon. Plant Ecol 156:75–88

Mitsch WJ, Gosselink JG (2007) Wetlands, 4th edn. Wiley, New York

Mooney H, Cropper A, Reid A (2005) Confronting the human dilemma. Nature 434:561–562

Patten K (2002) Smooth cordgrass (*Spartina alterniflora*) control with imazapyr. Weed Technol 16(4):826–832

Philipp KR, Field RT (2005) *Phragmites australis* expansion in Delaware Bay saltmarshes. Ecol Eng 25:275–291

Quiroga E, Quinones R, Palma M, Sellanes J, Gallardo VA, Gerdes D, Rowe G (2005) Biomass size-spectra of macrobenthic communities in the oxygen minimum zone off Chile. Estuar Coast Shelf Sci 62:217–231

Sakai AK, Allendorf FW, Holt JS (2001) The population biology of invasive species. Annu Rev Ecol Syst 32:305–332

Shanghai Water Authority (2010) Shanghai foreshore resources report (in Chinese). Shanghai Water Authority, Shanghai

Sipper M (2002) Machine nature: the coming age of bio-inspired computing. McGraw-Hill, New York, p 242

Sirakoulis GCH, Karafyllidis I, Thanaikakis A (2000) A cellular automaton model the effects of population movement and vaccination on epidemic propagation. Ecol Model 133:209–229

SOA (State Oceanic Administration People's Republic of China) (2012) Bulletin of Chinese sea level rise. Available in http://www.coi.gov.cn/gongbao/haipingmian/201303/t20130308_26217.html

Tian B, Zhou YX, Zhang LQ, Yuan L (2008) Analyzing the habitat suitability for migratory birds at the Chongming Dongtan Nature Reserve in Shanghai, China. Estuar Coast Shelf Sci 80:296–302

Wallinga J (1995) The role of space in plant population dynamics: annual weeds as an example. Oikos 74:377–383

Wang RZ, Yuan L, Zhang LQ (2010) Impacts of *Spartina alterniflora* invasion on the benthic communities at saltmarshes in the Yangtze Estuary, China. Ecol Eng 36:799–806

Wang N, Zhang LQ, Yuan L, Cao HB (2012) Research into vulnerability assessment for coastal zones in the context of climate change (in Chinese). Acta Ecologica Sinica. 32(7):2248–2258

Watt AS (1947) Pattern and process in the plant community. J Ecol 35:1–22

Wolfram S (1984) Universality and complexity in cellular automata. Physica D 10:1–35

Xiao DR, Zhu ZC, Yuan L, Tian K (2012) Reinvasion of exotic plant species *Spartina alterniflora* in Chongming Dongtan Nature Reserve of Shanghai. Chin J Appl Ecol 23(11):2997–3002 (in Chinese with English abstract)

Xie ZF, He WS, Liu WL, Lu JJ (2008) Influence of *Spartina alterniflara* saltmarsh at its different development stages on macrobenthos (in Chinese). Chin J Ecol 27:63–67

Yang SL, Milliman JD, Li P, Xu K (2011) 50,000 dams later: erosion of the Yangtze River and its delta. Glob Planet Change 75:14–20

Yuan L, Zhang LQ, Xiao DR, Huang HM (2011) The application of cutting plus waterlogging to control *Spartina alterniflora* on saltmarshes in the Yangtze Estuary, China. Estuar Coast Shelf Sci 92:103–110

Yun CX (2004) The fundamental rules in the recent development of the Yangtze Estuary. Oceanic Press, Beijing (in Chinese)

Zhao YQ, Zeng JN, Chen Q (2009) Community pattern of macrozoobenthos in distinct habitat with different densities of *Spartina alterniflora* Loisel. J Nat Resour 24:630–639 (in Chinese)

Zhang LQ, Yong XK (1992) Studies on phenology and spatial distribution of *Scirpus mariqueter* population. Acta Phytoecologica Sinica. 16:43–51 (in Chinese with English abstract)

Zhu ZC, Zhang LQ, Wang N, Schwarz C, Ysebaert T (2012) Interactions between range expansion of saltmarsh vegetation and hydrodynamic regimes in the Yangtze Estuary, China. Estuar Coast Shelf Sci 96:273–279

Zhou YX, Xie YM (2012) Study on wetland resources survey and monitoring and assessment system, Shanghai (in Chinese). Shanghai Science and Technology Press, Shanghai

Marine Biology of the Changjiang (Yangtze River) Estuary and Adjacent East China Sea Shelf

8

Zhao Li Xu, Xiao Min Shen, and Qian Gao

Abstract

This chapter focuses on the biological components of the Changjiang (Yangtze River) Estuary and adjacent East China Sea shelf, including phytoplankton, zooplankton, benthic macro-invertebrate, and fish faunas. During recent decades, the marine ecosystems in this area have experienced marked disturbances induced by human activities and changing climate. An increase in the abundance of dinoflagellates constitutes the most important change in phytoplankton populations over the last 20 years. The accelerated eutrophication related to human activity, particularly increased anthropogenic inputs into the Changjiang, and the increasingly skewed ratio of nutrient species that has accompanied global warming, might be partially responsible for these changes. The Changjiang Estuary and the adjacent East China Sea shelf are situated in the subtropical–warm temperate biogeographic boundary zone of the NW Pacific Ocean. Warming of surface waters in this area has been occurring since the middle to late 1980s. Over this same period, the abundance of temperate and warm temperate zooplankton has significantly decreased, and the abundance and frequency of occurrence of subtropical and tropical zooplankton have significantly increased. The Changjiang Estuary and the adjacent East China Sea shelf are the habitats of many resident and migratory fish species (present in various life cycle stages) that are of commercial and ecological value. However, because of human-induced disturbances (such as over-fishing) and environmental changes, many of the larger and commercially important demersal species occupying higher trophic levels have been replaced by smaller commercially marginal pelagic species occupying lower trophic levels. Thus, the recent intensification of human-induced disturbances and those of climate change necessitate and demand an integrated estuarine, coastal, and marine management strategy for this region.

Keywords

Changjiang Estuary • East China Sea • Fish • HABs • Macrobenthos • Zooplankton • Phytoplankton

8.1 Introduction

The Changjiang (Yangtze River) Estuary and East China Sea comprise the largest estuary and adjacent marginal sea in the western Pacific Ocean. The area is characterized by dynamic interactions between water systems that include freshwater input from the Changjiang (Beardsley et al. 1985; Su 1998), warm and oligotrophic waters of the Kuroshio Current, and waters of the Taiwan Strait Current (Miao and Yu 1991). These regions are highly productive and are of great biological importance, insofar as they support diverse and ecologically important life forms, including commercially valuable resident and migratory fish species at various life cycle stages. The dominant primary consumers in the area are *Calanus sinicus* (Copepoda) and *Euphausia pacifica*

Z.L. Xu (✉) · X.M. Shen · Q. Gao
Chinese Academy of Fisheries Sciences, East China Sea Fisheries Research Institute, 300 Jungong Road, Shanghai 200090, China
e-mail: xiaomin1957@126.com

J. Zhang (ed.), *Ecological Continuum from the Changjiang (Yangtze River) Watersheds to the East China Sea Continental Margin*, Estuaries of the World,
DOI 10.1007/978-3-319-16339-0_8, © Springer International Publishing Switzerland 2015

Fig. 8.1 Conceptual model of the marine ecosystem of the Yellow Sea and East China Sea (modified from Cai et al. 2005)

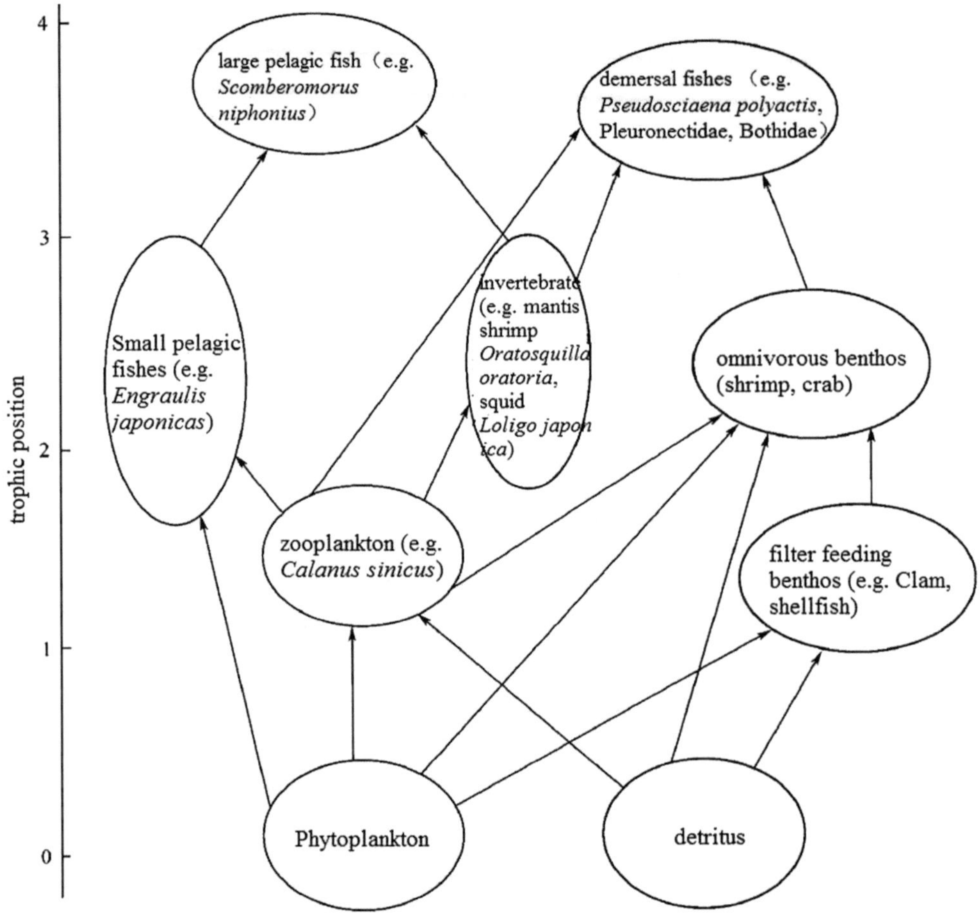

(Euphausiacea), and the most important secondary consumers are the invertebrate *Sagitta crassa* (Chaetognatha), the portunid crab *Charybdis bimaculata*, and the planktivorous Japanese anchovy *Engraulis japonicas*. Important tertiary consumers include commercially valuable fish species such as the largehead hairtail *Trichiurus lepturus*, the small yellow croaker *Larimichthys polyactis*, and the Japanese Spanish mackerel *Scomberomorus niphonius* (Cai et al. 2005; Fig. 8.1).

8.2 Materials and Methods

In this chapter, the dominance of species (Y) was calculated as follows (Zhao and Zhou 1984):

$$Y = n_i/N \times f_i \tag{8.1}$$

where n_i is the abundance of species i, f_i is the occurrence frequency of species i, and N is the total abundance of species. Dominant species are defined as those for which $Y > 0.02$ (Xu and Chen 1989).

The Shannon index (Shannon 1948) was used to measure diversity (H') according to:

$$H' = \sum (P_i \times \ln P_i) \tag{8.2}$$

where H' is the Shannon diversity index, P_i is the fraction of the entire population comprised of individuals of species i, S is the number of species encountered, and $\sum$ is the sum of species 1 to species S.

We used Pielou's measure of species evenness (J) which is:

$$J = H'/\ln(S) \tag{8.3}$$

where S is the total number of species in the sample across all samples in the dataset.

8.3 Phytoplankton in the Changjiang Estuary

The phytoplankton of the Changjiang Estuary and adjacent East China Sea (121°15′–123°15′E; 30°45′–32°00′N) are examined in light of investigations by Lin and colleagues during the winter (February), spring (May), summer (August), and autumn (November) of 2004, the results of which are presented by Lin et al. (2008).

8.3.1 Species Composition and Dominant Species

During their 2004 study, Lin et al. (2008) identified a total of 153 species of phytoplankton, of which 111 were Bacillariophyta and 42 were Dinophyta; 89 species (belonging to 26 genera of Bacillariophyta and 9 genera of Pyrrhophyta) were identified in the spring; 100 species (belonging to 27 genera of Bacillariophyta and 12 genera of Pyrrhophyta) were identified in the summer; 74 species (belonging to 22 genera of bacillariophyta and 6 genera of Pyrrhophyta) were identified in the autumn; and 54 species (belonging to 23 genera of bacillariophyta and 2 genera of Pyrrhophyta) were identified in the winter. Species numbers were higher in spring and summer than in the other seasons.

The cosmopolitan diatom *Skeletonema costatum* was the dominant species throughout the year (Table 8.1). In spring, *Skeletonema costatum* was the predominant species; in summer, the dominant species were the warm-temperate-adapted *Rhizosolenia alata* f. *gracillima*, the tropically adapted *Chaetoceros subtilis*, and the temperately adapted *Pseudo-nitzschia pungens*; in autumn, the widely distributed *Thalassiothrix nitzschioides*, *Coscinodiscus jonesianus*, and *Biddulphia sinensis* were the dominant species, and *Chaetoceros lorenzianus* was a subordinate species; in winter, the dominance of *Skeletonema costatum* decreased because of decreasing water temperatures, while *Coscinodiscus jonesianus* and *Biddulphia sinensis* were consistently abundant, and abundances of *Coscinodiscus asteromphalus* and *Melosira sulcata* increased markedly.

8.3.2 Phytoplankton Along Estuarine Gradients

In spring, four phytoplankton assemblages were identified in the four different areas of the Changjiang Estuary consisting of (1) inner waters; (2) southern neritic waters; (3) northern waters; and (4) southern oceanic waters. The inner waters (i.e., at the end of the riverine environment) contained mainly freshwater species, including *Melosirai slandica* and *Fragilaria* sp. In the northern waters, *Skeletonema costatum* was the predominant species, occurring at frequencies of as high as 83.2 %. In the southern neritic waters, *Coscinodiscus jonesianus* was the dominant species; its abundance was as much as six times greater than that of *Skeletonema costatum*. In the southern oceanic waters, *Prorocentrum dentatum* was the most abundant species, occurring at frequencies of as high as 98.8 %.

In summer, two phytoplankton communities were identified, one in the inner and one in the outer waters. The inner water communities were composed mainly of freshwater species, including *Melosirai slandica* and *Fragilaria* sp., while the outer waters were inhabited by a variety of different ecological groups.

In autumn, *Pediastrum* spp. were the most typical freshwater species, and the abundance of *Skeletonema costatum* was markedly decreased. The inner waters were dominated by *Pediastrum* spp., and *Skeletonema costatum* was rare. In the outer waters, however, *Skeletonema costatum* was the predominant species, and other polyhaline species were also common.

In winter, three distinctive phytoplankton communities were identified, in the inner waters, the southern neritic waters, and offshore waters. In the inner waters, *Thalassionema nitzschioides* and *Skeletonema costatum* predominated (and *Synedra* sp. was also common in freshwater areas). *Skeletonema costatum* was the predominant species in southern neritic waters. In the offshore waters, cosmopolitan estuarine species, including *Coscinodiscus jonesianus*, *Coscinodiscus asteromphalus*, and *Biddulphia sinensis*, were the most common. Additionally, the marine species *Rhizosolenia robusta* was found in offshore waters. In winter, water temperatures were low, and both the numbers

Table 8.1 Seasonal dominance (*Y*) and occurrence frequencies (*f*) of dominant phytoplankton in the Changjiang Estuary during 2004 (modified from Lin et al. 2008)

Dominant species	Spring		Summer		Autumn		Winter	
	Y	*f*	*Y*	*f*	*Y*	*f*	*Y*	*f*
Skeletonema costatum	0.658	66.7	0.688	89.7	0.658	83.3	0.035	31.0
Coscinodiscus asteromphalus							0.168	100.0
Biddulphia sinensis					0.034	93.3	0.155	100.0
Coscinodiscus jonesianus					0.034	100.0	0.133	96.6
Chaetoceros subtilis			0.025	100.0	0.024	93.3		
Melosira sulcata							0.104	75.7
Rhizosolenia alata f. *gracillima*			0.152	100.0				
Pseudo-nitzschia pungens			0.018	89.7				
Thalassionema nitzschioides					0.027	93.3		

and abundances of warm water species decreased; the abundance of *Skeletonema costatum* decreased, and the distribution of *Skeletonema costatum* retreated toward the inner waters. Concurrently, temperate neritic species, including *Melosira sultana, Chaetocero carstracanei, Coscinodiscus thorii, Ditylum brightwellii, Pleurosigma* sp., and *Corethron hystrix*, showed notable increases.

8.3.3 Phytoplankton Communities

Phytoplankton community parameters such as total abundance, species richness, and species diversity were taken as indicators of the health of the ecosystem. In 2004, total phytoplankton abundance was highest in spring (6.41×10^7 ind./m^3), intermediate in summer (8.72×10^6 ind./m^3) and autumn (2.91×10^5 ind./m^3), and lowest in winter (5.44×10^4 ind./m^3). The distribution of phytoplankton abundance also varied through the course of the year. In spring, phytoplankton species were most abundant in northern neritic waters. In summer and autumn, phytoplankton were most abundant in the southern neritic and southern oceanic waters of the area. During winter, species distributions were relatively uniform. Shannon diversity was highest in winter, intermediate in autumn, and lowest in spring and summer. The study indicates that rapid population growth of a few species led to the low diversity indices in spring and summer. In winter and spring, species richness values were highest in the southeastern part of the study area. In summer and autumn, species richness was highest in the eastern part of the study area. In the 1980s, *Skeletonema costatum* was the predominant species during spring, summer, and autumn (Guo and Yang 1992). However, by 2004, the dominance of *Skeletonema costatum* had significantly decreased (Wu et al. 2004). Additionally, as compared with the 1980s, the number of phytoplankton species, particularly of dinoflagellates, which are the main cause of red tides (Guo and Yang 1992; Wu et al. 2004; Lin et al. 2008), was obviously higher in the spring of 2004. In summer, autumn, and winter of 2004, the species diversity of phytoplankton was reduced, relative to that in the 1980s; however, the number of dinoflagellate species was greater than that in the 1980s. Generally, increases in dinoflagellates have constituted the major change in phytoplankton communities in the Changjiang Estuary over the last 20 years (Zhou et al. 2008). Accelerated eutrophication due to human activity, particularly due to increased anthropogenic inputs into the Changjiang, and the increasingly skewed ratio of nutrient species might be partially responsible for these changes (Anderson et al. 2002; Cai et al. 2006; Zhang this volume).

8.4 Zooplankton Communities of the Changjiang Estuary and the Shelf Waters of the East China Sea

8.4.1 Variations in Species Composition and Diversity Among Subregions of the Changjiang Estuary

Detailed investigations have been conducted on the zooplankton of the Changjiang Estuary, including those of the inner estuary, where turbidity levels are the highest in the area and in the waters outside of the estuary (Xu et al. 1995, 2009). Investigations of crustacean plankton in the north and south branches of the Changjiang Estuary during 1988, 1989, and 1990 (Zeng 1993) revealed 77 species belonging to a total of 17 families and 45 genera. Among these, a total of 7 families, 17 genera, and 34 species were cladocerans, and 10 families, 28 genera, and 43 species were copepods; these orders accounted for 44.2 and 55.8 % of the total numbers of species, respectively (Table 8.2). During the four investigations in 1988–1990, a total of 63 genera, which included 31 cladoceran and 32 copepod genera, were found in the south branch of the Changjiang Estuary (Zeng 1993); the major species were *Diaphanosoma sarsi, Daphnia hyalina, Moina brachiata, Moina micrura, Bosmina fatalis, Sinocalanus dorrii, Schmackeria forbesi, Neodiaptomus schmackeri, Limnoithona sinensis, Halicyclops latus, Mesocycolops leuckarti, Thermocyctops taihokuensis*, and *Thermocyclops hyalinus*. A total of 56 genera, which included 21 cladocerans and 35 copepods, were found in the north branch of the Changjiang Estuary; of these, *Diaphanosoma sarsi, Daphnia hyalina, Moina brachiata, Moina micrura, Tortanus vermiculus, Calanus sinicus, Sinocalanus dorrii, Schmackeria forbesi, Neodiaptomus schmackeri, Limnoithona sinensis, Limnoithona tertraspina, Halicyclops*

Table 8.2 The number of planktonic crustacean species during the investigations of 1988, 1989, and 1990 in the Changjiang Estuary (modified from Zeng 1993)

	1988–1990	1988.5	1989.5	1989.11	1990.7
No. of cladoceran species	34	14	27	12	13
No. of copepod species	43	29	33	21	18
Total no. of crustacean plankton species	77	43	60	33	31

latus, *Mesocycolops leuckarti*, *Thermocyctops taihokuensis*, and *Thermocyclops hyalinus* were the dominant species.

The above results indicate that the number of cladoceran species in the south branch was greater than that in the north branch and that the number of copepod species in the south branch was smaller than that in the north branch. Cladoceran species are suited to freshwater conditions, and the limitation of their distribution to the south branch is due to the high salinity of the waters in the north branch (Zhu et al. this volume). The presence of relatively high numbers of copepod species in the south branch was mainly related to the presence of brackish water species, such as *Tortanus vermiculus*, *Calanus sinicus*, *Schmackeria poplesis*, *Limnoithona tertraspinam*, *Euterpe* sp., *Diaphanosoma sarsi*, and *Moina micrura*.

The abundance and biomass of crustacean plankton also differed between the north and south branches of the Changjiang Estuary. The average abundance and biomass were higher in the north than in the south (Table 8.3). The average abundance and biomass of cladocerans were higher in the south than in the north, and the abundance of copepods was greater in the north than in the south (Table 8.3).

Xu et al. (2005a) found that in the north branch of Changjiang Estuary, the biomass of zooplankton in water near the north shore during the high water period was greater than that near the south shore, but was reversed during ebb tides. The tidal current on the north bank was stronger than that on the south bank during periods of tidal influx, due to the influence of the Coriolis force. The situation was reversed during fluvial outflux. The plankton was carried with the tidal current, and the Coriolis force accounted for the changes in the quantities of zooplankton in the north and south banks of the north branch via tidal movement.

Gao et al. (2008) compared changes in zooplankton communities off the east beach of Chongming Island as the zooplankton migrated with the freshwater from the north passage to the mesohaline water area of the north branch of the Changjiang Estuary. These researchers found that the dominant species in the north passage was *Sinocalanus sinensis*. Additionally, several species that were abundant in the north branch were also abundant during winter, including *Labidocera euchaeta, Schmackeria poplesis, Paracalanus parvus*, *Paracalanus aculeatus*, and *Calanus sinicus*. The abundance and biomass of mesozooplankton in the north passage were obviously lower than those in the north branch (Gao et al. 2008).

In waters of the Changjiang Estuary, the annual average total biomass was 170.7 mg/m^3, although the value varies seasonally. Biomass values during summer were highest, followed by values during the spring, then the autumn, and then the winter. Xu (2004) found that the distribution of zooplankton was heterogeneous. Zooplankton were most dense in outer sea areas (30°45′–31°15′N, 122°45′–123°15′E) and least dense (biomass, 50–100 mg/m^3) to the southeast of the Zhoushan Archipelago (from (30°00′N, from 122°15′E). The average total diet biomass of zooplankton (not including jellyfish or tunicates) in May 2002 was 196.0 mg/m^3 (range 55.5–496.1 mg/m^3). As shown in Fig. 8.2, the areas where the total diet of zooplankton was most dense (>250 mg/m^3) and most widely distributed are restricted to mainly offshore areas (30°15′–31°50′N and 122°15′E to the east of the Changjiang Estuary); most of these areas overlap with areas of the densest total biomass. Species such as *Calanus sinicus*, *Euphausia* larvae, *Euchaeta* larvae, *Euchaeta concinna*, and *Sagitta nagae* dominated the diet biomass in the investigated water areas.

8.4.2 Seasonal Succession of Zooplankton Populations

During 2002–2003, Xu (2005b) conducted an analysis of the zooplankton of the outer Changjiang Estuary, in the area of 29°–32°N to 122°–123°E. A total of 128 species of zooplankton were present in the investigated area, along with 16 planktonic and fish larvae that belonged to 5 families and 12 subtypes (Table 8.4). The copepods were the dominant taxon, followed by amphipods and the ostracods. The species distribution showed a tendency to decrease from the outer sea to nearshore areas. The outer area of the Changjiang Estuary has a simple species composition, an asymmetric interspecies distribution, and prominent species are present, especially in the north of the Changjiang Estuary (north of 31°30′N) and to the southeast of Dongshan Island (west of 123°E). These areas also showed low values of *H′*, *J′*, and *d* and characteristic species poverty. This low diversity might be an indication of

Table 8.3 The average abundance (ind./L) and biomass (mg/L) of crustacean plankton in the north and south branches of the Changjiang Estuary (modified from Zeng 1993)

Groups	South branch		North branch	
	Abundance	Biomass	Abundance	Biomass
Cladoceran species	2.91	0.054	0.50	0.008
Copepod species	15.25	0.385	21.54	0.592
Total crustacean plankton	18.16	0.439	22.04	0.600

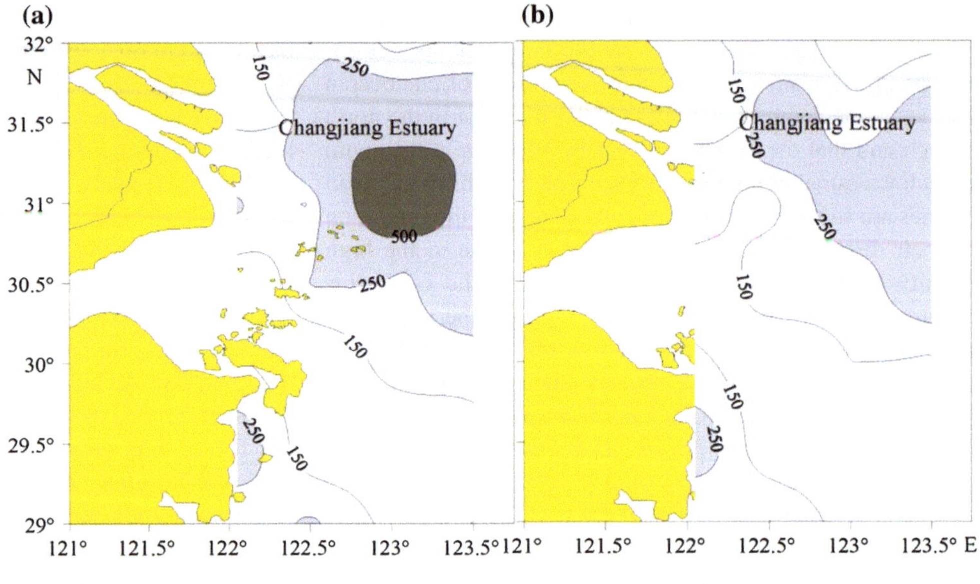

Fig. 8.2 Distribution of total biomass and diet biomass (mg/m^3) of zooplankton investigated in May 2002: **a** total biomass and **b** diet biomass

Table 8.4 Species distributions and average abundances of zooplankton in the Changjiang Estuary

Group	Species richness	Percentage (%)	Abundance (ind./m^3)	Percentage (%)
Polychaeta	1	0.78	0.04	0.02
Heteropoda	3	2.34	0.17	0.08
Pteropoda	5	3.91	2.19	1.05
Copepoda	40	31.25	154.97	74.26
Amphipoda	22	17.19	1.35	0.65
Euphausiacea	5	3.91	2.29	1.1
Decapoda	4	3.13	0.46	0.22
Mysidacea	8	6.25	0.8	0.39
Cumacea	1	0.78	0.05	0.02
Ostracoda	16	12.5	4.85	2.32
Chaetognatha	6	4.69	4.29	2.06
Appendiculata	1	0.78	0.06	0.03
Pelagic larvae	16	12.5	36.85	17.66
Fish eggs			0.08	0.04
Fish larvae			0.24	0.11
Total	128		208.69	

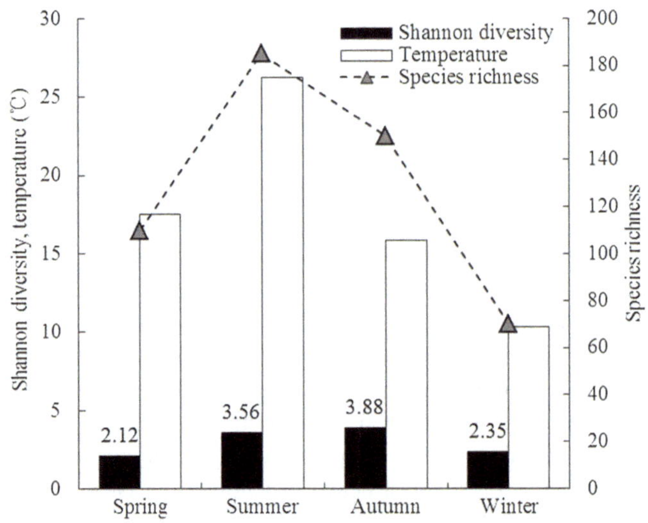

Fig. 8.3 Zooplankton Shannon diversities and species richness in the Changjiang Estuary during May 2002, August 2002, November 2002, and February 2003

unstable zooplankton communities (Ives and Carpenter 2007). The characteristics of the distribution of the diversity index were as follows: The values of H' and d were lowest in the spring and highest in the autumn, the values of C were highest in the summer, and the seasonal values of H' were ordered as autumn (3.88) > summer (3.56) > winter (2.35) > spring (2.12) (Fig. 8.3). In spring, values of H' less than 2 accounted for approximately 37.03 % of the total diversity, and values of greater than 3 accounted for 11.11 % of the total diversity.

As shown in Table 8.5, 18 species were dominant throughout the year (based on a value of Y, which represents the relative abundance of a species multiplied by the occurrence frequency of that species, set to not less than 0.02 for the dominant species of zooplankton in the investigated area). Among these species, only *Calanus sinicus* was dominant during all four seasons. The numbers of dominant species were 6, 11, and 7 for summer, autumn, and winter, respectively. The abundance of *Calanus sinicus* was as high as 0.68, and the average abundance of this species accounted

Table 8.5 Seasonal dominance (*Y*) of the dominant species of zooplankton in the Changjiang Estuary and the adjacent East China Sea during 2002/2003 ($Y \geq 0.02$)

Dominant species	Dominance (*Y*)			
	Spring	Summer	Autumn	Winter
Calanus sinicus	0.68	0.50	0.03	0.10
Muggiaea atlantica	0.14			0.04
Temora discaudata		0.02		
Acartia pacifica		0.04	0.03	
Lucifer intermedius		0.03		
Sagitta enflata		0.05	0.03	0.03
Diphyes chamissonis			0.04	0.03
Acrocalanus gibber			0.02	
Nannocalanus minor			0.02	
Oncaea venusta			0.02	
Canthocalanus pauper			0.03	
Scolecithricella longispinosa			0.04	
Euchaeta concinna			0.27	0.02
Sagitta bedoti			0.03	
Euchaeta plana				0.02
Scolecithrix nicobarica				0.03
Sagitta nagae				0.07

for 68.09 % of the total abundance of zooplankton during spring, which was far greater than that of any other species.

8.4.3 Zooplankton Along the Salinity Gradients

The border between the outer and inner areas of the Changjiang Estuary is a zone of maximum turbidity. In terms of adaptations to the different environments in the Changjiang Estuary, the zooplankton can be divided into the following ecological groups:

1. **Freshwater species**, which include *Brachionus falcatus, Brachionus calycif lorus, Sinocalanus dorrii,* and *Mesocycolops leuckarti*. Freshwater species live mainly in areas with salinities less than 2.
2. **Brackish water species**, which include *Tortanus vermiculus, Schmackeria poplesis*, and *Monoculodes limnophilus*. These species usually live in brackish water areas of the estuary where salinities are from 2 to 10. The ranges of these species often change with changes in rainfall intensity, and their distributions during flood season are usually further toward the outer estuary as compared with their distributions during the dry season. For example, *Tortanus vermiculus* can be found in the area of 122.5°E during flood season, whereas it is found only in areas west of 122.5°E during the dry period.
3. **Marine species,** which include most of the species that live in the plume front region of the Changjiang Estuary. Generally, these species include *Labidocera euchaeta, Sagitta nagae, Pseudeuphausia sinica,* and *Acanthomysis sinensis*. The ranges of these species reflect the extent and intensity of runoff of the Changjiang.
4. **Offshore sea species,** which are limited in this area due to the limited influence of water currents in the study area. Offshore species mainly consist of *Sagitta enflata* and *Calanus sinicus*. The abundances and distributions of these species reflect the intensity of the Taiwan Warm Current and the influence of the intensity of the dilution by Changjiang water (Zhu et al. this volume).

8.4.4 Changes in Zooplankton in the Changjiang Estuary and the Adjacent East China Sea

The Changjiang Estuary and the adjacent East China Sea are situated in the subtropical–warm temperate biogeographic boundary zone of the NW Pacific Ocean (Fig. 8.4). Zooplankton living in this region are mainly warm temperate and temperate species in spring and winter and subtropical and tropical species in summer and autumn (Chen 1992). In recent decades, the studied area has experienced environmental changes and increases in sea-surface temperatures (Belkin 2009). Warming of the surface waters in the Changjiang Estuary and the adjacent East China Sea has been occurring since the middle to late 1980s. Sea-surface temperatures (SSTs) in the East China Sea rose by 2.2 °C between 1982 and 1998 (Belkin 2009). Warming has been especially rapid (up to 1 °C/decade between 1982 and 2003) in the area encompassing the Changjiang Estuary (Ho et al. 2004). Over this period, the abundance of temperate and warm temperate zooplankton significantly decreased in the Changjiang Estuary and the adjacent waters (Table 8.6; Li et al. 2009; Xu and Gao 2009; Zhou et al. 2009; Xu et al. 2013). Temperate species, such as *Themisto gaudichaudii* (Amphipoda), were previously the dominant species during the winter, but these species have retreated from the East China Sea (Zhou et al. 2009). The warm temperate species *Calanus sinicus* (Copepoda), which is a key zooplankton species, was most abundant in the study area during spring. In the spring of 2005, the timing of the occurrence and reduction in abundance of *Calanus sinicus* advanced by about one month, which accompanied the increase in SSTs (Xu et al. 2011). In 1959, abundance of *Calanus sinicus* peaked in June and decreased in July, whereas, in 2005, it peaked in May and was sharply attenuated by early June (Fig. 8.5). Both the abundances and the frequencies of occurrence of subtropical and tropical species have significantly increased. For example, the abundances of the

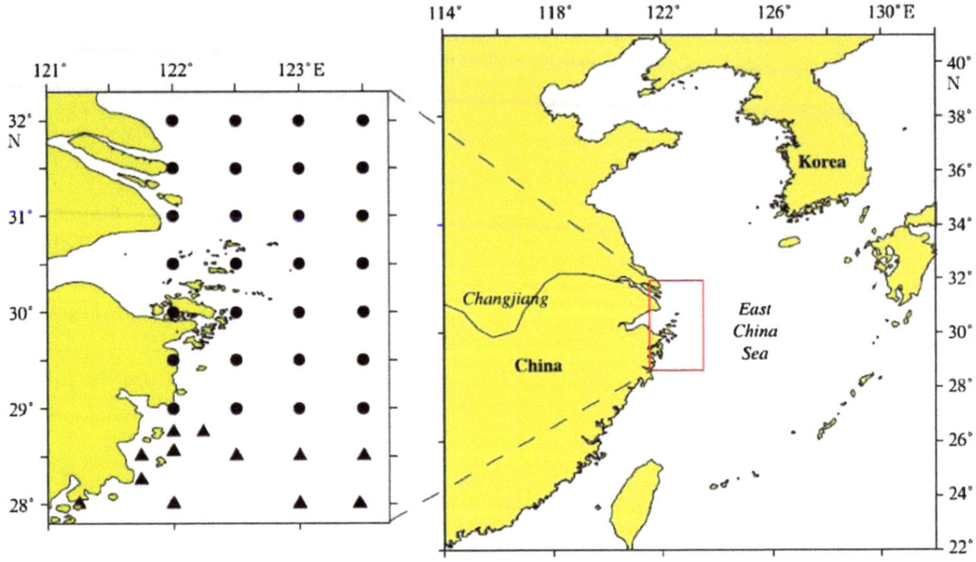

Fig. 8.4 Locations of sampling stations during the cruises. The symbol *filled circle* denotes the sites sampled in 1959 and 2002, and *filled triangle* denotes additional sites sampled in 1959

Table 8.6 Seasonal average abundance (AA), patchiness, and frequency of the warm temperate species *Euchaeta plana* in the Changjiang Estuary and the adjacent East China Sea in 1959 and 2002/2003 (modified from Xu et al. 2013)

Seasons	Year	Temperature (°C)		*Euchaeta plana*		
		Surface water	Bottom water	AA (ind./m^3)	Patchiness	Frequency (%)
Spring	1959	17.86	17.21	14.1	26.1	96.3
	2002	19.53	19.08	0.6	0.3	51.9
Summer	1959	26.82	22.76	2.2	6.1	59.3
	2002	27.26	23.08	1.4	2.5	74.1
Autumn	1959	19.14	19.43	1.3	44.9	72.4
	2002	18.84	20.06	4.3	6.0	51.7
Winter	1959	9.07	9.60	4.1	95.3	58.6
	2003	10.36	10.99	0.6	0.3	75.9

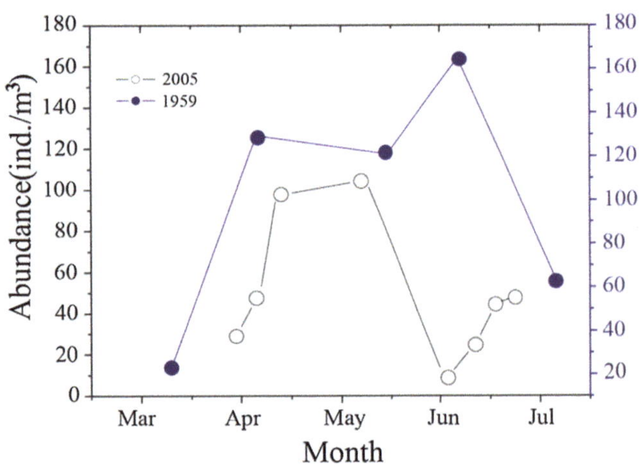

Fig. 8.5 Changes in the mean abundances of *Calanus sinicus* in the Changjiang Estuary and adjacent waters during the spring of 1959 and 2005

subtropical species *Lucifer intermedius* (Decapoda) and *Pseudeuphausia sinica* (Euphausiacea) have significantly increased in spring (Ma et al. 2009; Gao and Xu 2011).

8.5 The Benthic Macro-Invertebrate Community of the Changjiang Estuary

Liu et al. (2012) studied the seasonal variations of macrobenthos in the waters of the Changjiang Estuary based on investigations during May–July 2005, September 2005, November 2005, and June of 2006. A total of 330 species of macrobenthos were identified, including 122 of Mollusca, 83 of Polychaeta, 67 of Crustacea, 23 of Echinodermata, 28 of Pisces, and 7 unidentified species. Based on geographical characteristics, the study area could be divided into three

Table 8.7 Community indices of different forms of macrobenthos in different zones

Region	Biomass (g/m^2)	Abundance (ind./m^2)	Shannon diversity	Species richness	Evenness index
Inner water	3.2 ± 2.5	25.5 ± 7.3	0.67 ± 0.17	0.28 ± 0.08	0.44 ± 0.10
Outer water	19.9 ± 5.2	173.7 ± 40.2	1.73 ± 0.25	1.19 ± 0.27	0.65 ± 0.07
Zhoushan Archipelago	14.8 ± 5.4	128.4 ± 35.5	1.74 ± 0.25	1.17 ± 0.27	0.73 ± 0.08

distinct subareas that included the inner waters, outer waters, and waters off the Zhoushan Archipelago.

In the inner waters, a total of 24 species were identified at 21 stations. Eight species of macrobenthos were caught using Agassiz trawling nets, and on average, 3.8 species were found at each station (Liu et al. 2012); the common species included *Mcrobrachium nipponense*, *Corbicula largillierti*, *Eriocheir leptognathus*, and *Exopalaemon annandalei*, with occurrence frequencies of 83.3, 66.7, 66.7, and 58.3 %, respectively. Eighteen species were caught using van Veen grabs, and on average, 1.7 species were found at each station; *Nephtys polybranchia*, *Notomastus latericeus*, and *Corbicula largillierti* were common, with occurrence frequencies of 47.6, 28.6, and 28.6 %, respectively.

In the outer waters, a total of 149 benthic species were found at 25 sampling stations. Sixty-four species of macrobenthos were caught using Agassiz trawling nets, and on average, 14 species were found at each station; *Palaemon gravieri*, *Exopalaemon annandalei*, and *Oratosquilla oratoria* were the common species, with occurrence frequencies of 81.8, 63.6, and 45.5 %, respectively. One hundred species were caught using the van Veen grab, and on average, 7.4 species were found at each station; *Sternaspis scutata*, *Heteromastus filiforms*, *Nassarius festivus,* and *Mediomastus californiensis* were the common species, with occurrence frequencies of 32, 28, 20, and 16 %, respectively.

Off the Zhoushan Archipelago, a total of 126 benthic species were found at 22 sampling stations. Fifty-six species were caught at 13 stations using Agassiz trawling nets; common species included *Acetes chinensis*, *Leptochela gracilis*, *Palaemon gravieri,* and *Alpheus japonicus*, with occurrence frequencies of 84.6, 61.5, 53.8, and 46.2 %, respectively. Seventy-six species were caught at 22 stations using van Veen grabs, and on average, 7 species were found at each station; *Eocylichna cylindrella*, *Sternaspls scutata*, and *Heteromastus filiformis* were the most common species, with occurrence frequencies of 31.8, 27.3, and 22.7 %, respectively.

In the waters of the Changjiang Estuary, the average abundances of macrobenthos were lowest in the inner waters (25.5 ind./m^2), highest in the outer waters (173.7 ind./m^2), and intermediate in the waters off the Zhoushan Archipelago (128.4 ind./m^2). The distribution pattern of macrobenthos was distinct. In the western part of the area, species richness was low and the biomass and abundance of macrobenthos were low, indicating that the benthic community was unstable. However, in the eastern part (including in the outer waters and the Zhoushan Archipelago), the biomass, abundance, and species diversity of macrobenthos were high, which may indicate that the community was stable (Table 8.7).

8.6 Fish Communities of the Changjiang Estuary and the Adjacent East China Sea

8.6.1 Species Composition

Fish populations in the Changjiang Estuary and adjacent waters have been investigated by Zhuang and colleagues. During their investigation, a total of 332 fish species were identified in the Changjiang Estuary (Table 8.8). Most of these species belonged to the Osteichethyes (298 species), while the remaining 34 species belonged to the Chondrichthyes. Perciformes were the most diverse (107 species), followed by Cypriniformes (53 species), Tetraodontiformes (22 species), Pleuronectiformes (21 species), Scorpaeniformes (16 species), and Clupeiformes (16 species). Among the Chondrichthyes, Rajiformes were the most diverse (19 species) (Zhuang et al. 2006).

8.6.2 Temporal–Spatial Distributions of Fishes

Freshwater species predominated in the waters of the Changjiang Estuary with salinities <5. From upstream of Jiangyin to the mouth of the estuary, 76 freshwater species were present, including members of the Cypriniformes (53 species), Perciformes, and Siluriformes.

The fish species present in waters with salinities of 5–20 consisted mainly of estuarine species; the 53 estuarine species identified belonged to 9 orders and 17 families. Perciformes were the most diverse (29 species), followed by Osmeriformes (7 species) and Pleuronectiformes (4 species). Some estuarine species, such as the tank goby *Glossogobius giuris*, mudskipper (Periophthalmini), tongue sole *Cynoglossus semilaevis*, and yellowfin puffer *Takifugu xanthopterus*, were also found in freshwater. Some species may have migrated, but they would not have migrated far, and these migrations would have generally been limited to migrations in neritic waters.

Table 8.8 Numbers of fish species in different families and environments in the Changjiang Estuary (Zhuang et al. 2006)

Order	Family	Species	Freshwater species	Marine species	Estuary species	Migration species
Carcharhiniformes	3	9		9		
Lamniformes	2	2		2		
Squaliformes	1	3		3		
Squatiniformes	1	1		1		
Rajiformes	9	19		19		
Acipenseriformes	2	2	1			1
Anguilliformes	5	10		9		1
Clupeiformes	2	16		11	2	3
Gonorhynchiformes	1	1		1		
Cypriniformes	3	53	53			
Siluriformes	3	11	9	1	1	
Osmerforems	1	8			7	1
Aulopiformes	1	4		4		
Myctophiformes	1	1		1		
Lampridiformes	1	1		1		
Gadiformes	2	2		2		
Lophiiformes	2	3		3		
Mugiliformes	1	5		2	3	
Atheriniformes	1	1		1		
Beloniformes	3	5	1	1	3	
Cyprinodontiformes	1	1	1			
Beryciformes	1	1		1		
Zeifromes	1	1		1		
Gasterosteiformes	2	4		4		
Synbranchiformes	2	2	2			
Scorpaeniformes	7	16		14	1	1
Perciformes	38	107	9	69	29	
Pleuronectiformes	4	21		17	4	
Tetraodontiformes	5	22		18	3	1
Total	106	332	76	195	53	8

The waters located to the west of 123°23′E contained mainly marine species; 34 species belonged to the Chondrichthyes, among which Rajiformes were the most diverse group. Most of the marine species belonged to the Osteichethyes (161 species), among which Perciformes (69 species) were the most diverse, followed by Tetraodontiformes (18 species), Pleuronectiformes (17 species), Scorpaeniformes (14 species), and Clupeiformes (11 species). The Perciformes included 13 species of Sciaenidae. The marine fish species living in the Changjiang Estuary were mainly associated with the East Asia subregion of the Pacific Boreal Region (Zhu et al. 1963). The common marine species included the rayfish *Raja porosa*, the Chinese herring *Ilisha elongata*, the scaly hairfin anchovy *Setipinna taty*, the Bombay duck *Harpadon nehereus*, hoki *Johnius belengerii*, the big-head croaker *Collichthys lucidus*, and the silver pomfret *Pampus argenteus*. These species are euryhaline and were mainly distributed in waters with salinities of 30. Most of the marine fishes completed their life cycles (including spawning, nursery stages, and foraging) in the Changjiang Estuary and the adjacent neritic waters of the East China Sea; thus, these species were found in this area year-round (Zhuang et al. 2006). Other marine species, however, were found in the Changjiang Estuary and adjacent neritic waters of the East China Sea only during specific seasons. These species were found mainly in offshore areas and migrated to the Changjiang Estuary and adjacent neritic waters of the East China Sea for spawning or foraging (Zhuang et al. 2006). For example, each year, the large yellow croaker *Larimichthys crocea*, the small yellow

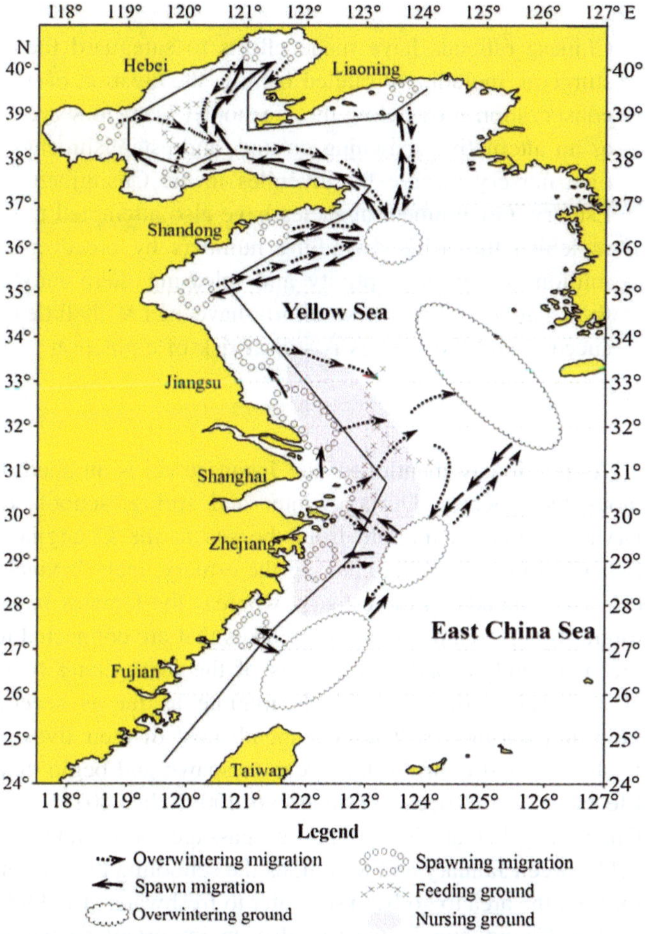

Fig. 8.6 Migration routes of the small yellow croaker *Larimichthys polyactis* in the Bohai Sea, Yellow Sea, and East China Sea (modified from Xu and Chen 2009)

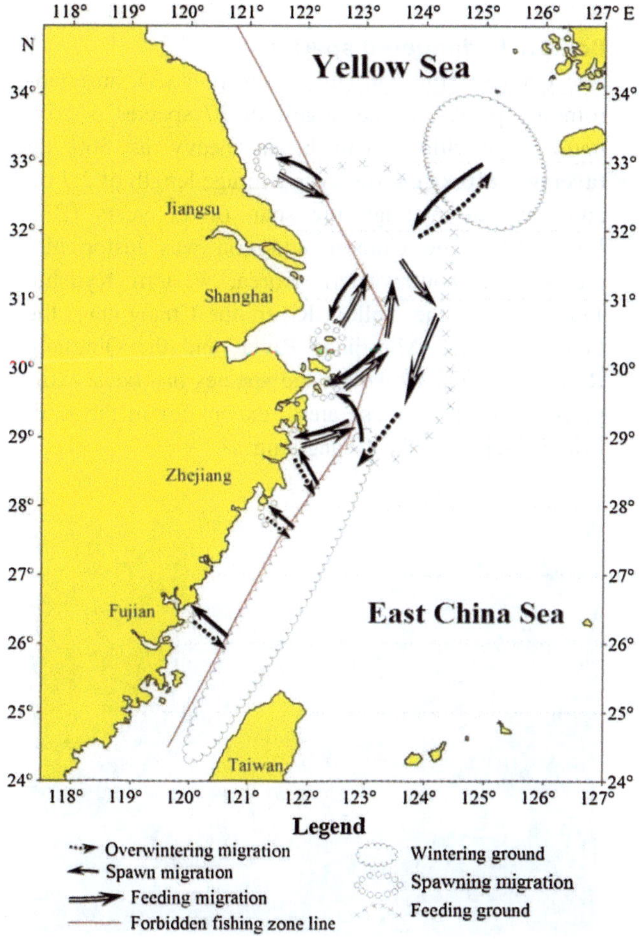

Fig. 8.7 Migration routes of the large yellow croaker *Larimichthys crocea* in the Yellow Sea and East China Sea (modified from Xu and Chen 2011)

croaker, the largehead hairtail and the chub mackerel *Scomber japonicus* migrated to the Changjiang Estuary to spawn (Figs. 8.6 and 8.7). The Japanese anchovy *Engraulis japonicas* and the Japanese Spanish mackerel *Scomberomorus niphonius* foraged in the adjacent waters of the Changjiang Estuary during the summer.

8.6.3 Fish Migration Patterns

The Changjiang Estuary is an important passage for actively migrating fish and passively migrating ichthyoplankton juveniles. Environmental factors such as water temperature, salinity, runoff, tides, and food availability may influence the migration of these species.

Many anadromous fish species must migrate to freshwater to reproduce, including populations of Acipenseridae and the Japanese grenadier anchovy *Coilia nasus* which migrate into the Changjiang to spawn (Fig. 8.8). Conversely, the Japanese eel *Anguilla japonica* and a number of other facultative catadromous species, such as the roughskin sculpin *Trachidermus fasciatus*, grow in estuaries or inland freshwater until they become sexually mature, at which time they migrate back to their oceanic or estuarine spawning grounds (Fig. 8.9). Approximately 300 species of fish live in the Changjiang, and one-third of these species, mostly migratory fishes, inhabit the lower reaches of the river (Ke et al. 1993). Human influences, including overfishing, water pollution, and hydropower and dam construction, have sharply reduced the natural populations of these fishes; some of these species, such as the Chinese sturgeon *Acipenser sinensis*, the Chinese paddlefish *Psephurus gladius*, the finless porpoise *Neophocaena phocaenoides*, the Chinese sucker *Myxocyprinus asiaticus*), Reeves shad *Tenualosa reevesii*, and the roughskin sculpin, are even facing extinction (Box. 8.1), which would result in a loss of biological diversity (Zhuang et al. 2006). Within this chapter, the migration patterns of the Japanese eel are included in the following section.

Box 8.1 Endangered species

The Chinese sturgeon (*Acipenser sinensis*), an anadromous species, is the largest of 27 species of *Acipenser*. Individuals can be as heavy as 500 kg (average, 216 kg), grow to an average length of 2.7 m, and have an average life span of 21 years (Box Fig. 8.1). The Chinese sturgeon was historically recorded in southwestern Korea, western Kyushu, Japan, and in the Yellow River, the Changjiang, the Pear River, the Mingjiang River, and the Qingtang River in China. However, the species has been extirpated from all of these areas, except for in the mid-lower reaches of the Changjiang.

Box Fig. 8.1 The Chinese sturgeon (*Acipenser sinensis*)

The Chinese sturgeon has a habit of upstream migration; they dwell along China's eastern coast areas and migrate upstream in the Changjiang to propagate upon reaching sexual maturity. The species has been overfished. In the 1970s, the total spawning population of this species was estimated at 10,000 individuals, but in 2005–2007, the total spawning stock of Chinese sturgeon was estimated to be 203–257 individuals (Wei et al. 2005). These data indicate a 97.5 % reduction in the total spawning population over a 37-year period. For this reason, the Chinese sturgeon has been assessed as a Critically Endangered Animal since the 1970s.

Because of their long reproductive cycles, long migrations, and sensitivity to environmental conditions, the Chinese sturgeon is under severe threat from the destruction of natural habitats and spawning sites, pollution, and overfishing of the wild population. The sturgeon is also highly sensitive to increased noise on the river caused by growing river traffic, as well as being vulnerable to death or injury by boat propellers. Chinese officials have made efforts to safeguard the sturgeon, including curtailed fishing, the creation of a conservation area below the Gezhouba Dam to serve as an alternative spawning ground, and establishment of a nursery reserve for juveniles in the Changjiang Estuary. Government agencies have also attempted to replenish the fish's dwindling numbers by breeding millions of fry in captivity and releasing them into their native rivers. These efforts have met with little success, and the species is still at risk of extinction.

As previously mentioned, the Japanese eel is an anadromous fish species. During winter and spring, schools of larvae of this eel migrate from the sea to the Changjiang Estuary (Fig. 8.9). After entering the estuary, male Japanese eels grow mainly in the estuary, whereas the females begin migrating upstream, living in the lakes that are connected to the main and secondary branches of the Changjiang (Guo et al. 2011). Some individuals migrate as far as several thousand kilometers. The eels spend most of their lives in freshwater and return to the ocean to spawn and begin their life cycle in the ocean (Tsukamoto 2006; Han 2011). The Changjiang Estuary is an important passage for the Japanese eel between January and March, as the schooling elvers pass through the area to go from seawater to freshwater (Li 1998).

The Changjiang Estuary is also an important spawning ground for many fishes, and spawning times and places intersect across species (Yang et al. 1990; Jiang et al. 2010). Generally, fish species spawn during the first half of the year and nurse during the latter half of the year. In the spring, particularly in May of each year, fishes, such as the Japanese grenadier anchovy, the small yellow croaker, the Japanese Spanish mackerel, Osbeck's grenadier anchovy *Coilia mystus*, and the silver pomfret, spawn in the Changjiang Estuary (Figs. 8.6, 8.8 and 8.10). The diversity and density of fish larvae reach their highest levels during June–August of each year; these larvae include those of the Japanese grenadier anchovy, Osbeck's grenadier anchovy, the large yellow croaker, the hairtail, the Japanese Spanish mackerel, the silver pomfret, the hoki, the big-head croaker, cyprinid fishes, and Cyonglossidae (e.g., Figs. 8.7, 8.8 and 8.10). The larvae are mainly distributed in the neritic waters of the south branch and to the southeast of Dasha. In autumn and winter, both the abundances and diversities of ichthyoplankton are low. The eggs or larvae of *Stolephorus chinensis*, the Japanese sea bass *Lateolabrax japonicus*, the skinnycheek lantern fish *Myctophum pterotum*, and the Pacific Tarpon *Megalops cyprinoides* have been found in this area (Fig. 8.11).

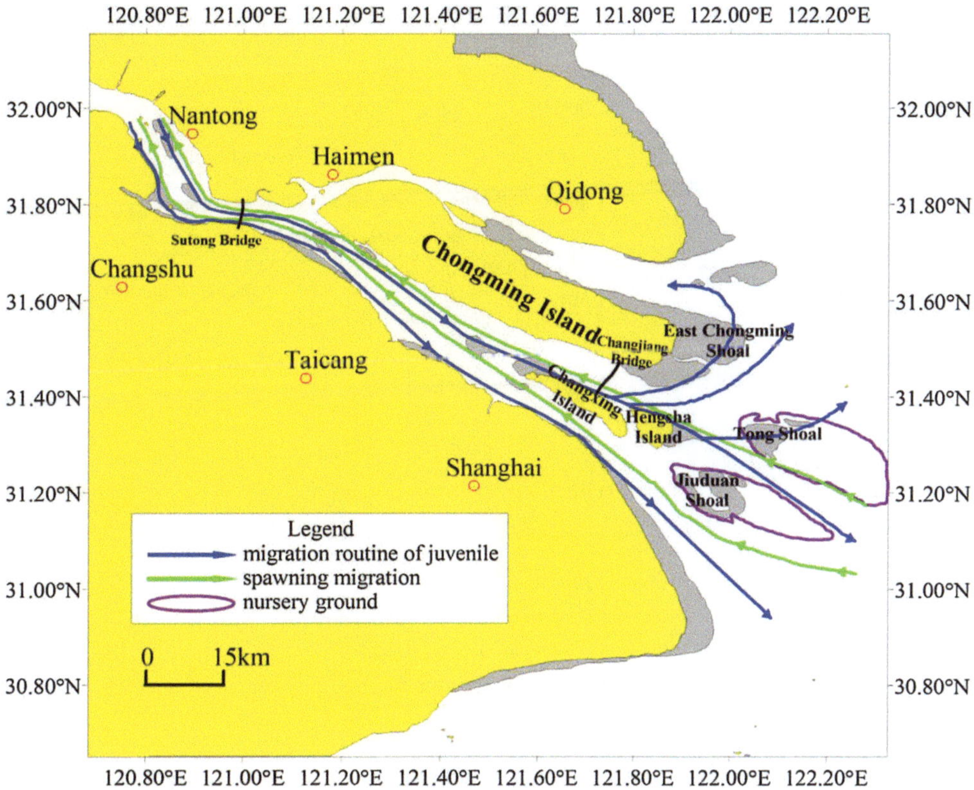

Fig. 8.8 Migration routes of the Japanese grenadier anchovy *Coilia nasus* in the Changjiang Estuary

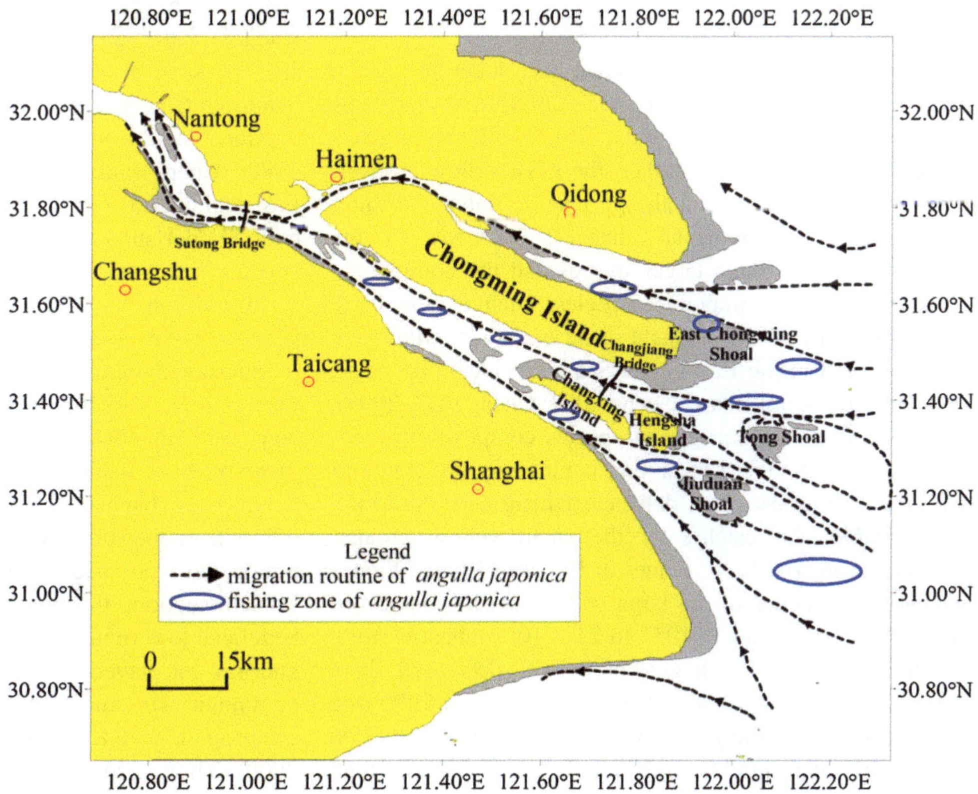

Fig. 8.9 Distribution and migration of the elvers of the Japanese eel *Anguilla japonica* in the Changjiang Estuary

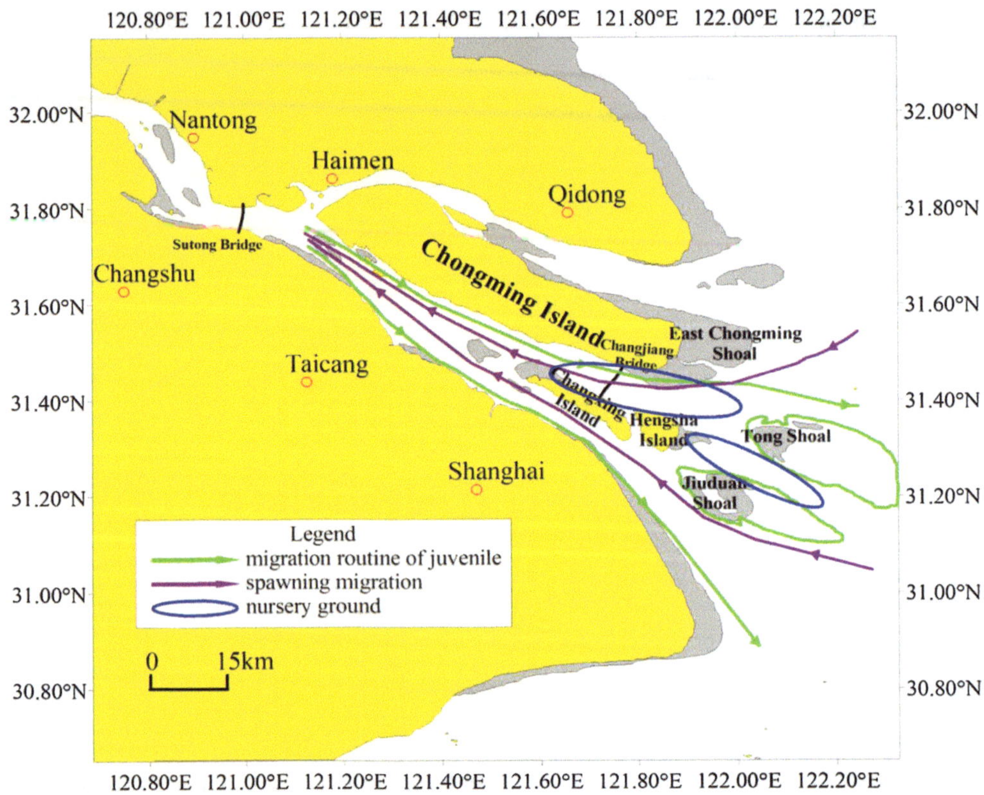

Fig. 8.10 Migration of Osbeck's grenadier anchovy *Coilia mystus* in the Changjing Estuary

8.6.4 Fishery Resources in the East China Sea

Approximately 730 fish species have been identified in the East China Sea, along with 91 species of shrimp and crabs and 64 species of cephalopods. The traditional catches include the small yellow croaker, the large yellow croaker, mackerel, and largehead hairtail. Due to intense fishing pressure and environmental changes, decreases in the abundances of the main target species and the commercial extinction of some important species, such as the large yellow croaker and the elongate ilisha *Ilisha elongate*, have been observed in the East China Sea since the 1970s (e.g., Tang and Su 2000; Lin and Cheng 2004; Cheng et al. 2006). Moreover, dramatic changes in species composition, composition of dominant species, and community structures of the fishery have also been detected (Cheng and Yu 2004). For example, the catches of the small yellow croaker decreased from 163 × 10^3 tonnes in 1957 to 23 × 10^3 tonnes in 1990; the catches of the large yellow croaker decreased from 178 × 10^3 tonnes in 1957 to 25 × 10^3 tonnes in 1990; and the yields of the largehead hairtail decreased from 577 × 10^3 tonnes in 1974 to 498 × 10^3 tonnes in 1990 (the latter being likely due to an increase in fishing effort) (Qu et al. 2005). Over time, the trends in the major fisheries tend to be catches of smaller size, lower value, younger fish, and fish at earlier stages of maturation; such trends have been observed in the largehead hairtail fisheries. In addition, smaller, faster growing, shorter lived, and lower valued species, such as the glowbelly *Acropoma japonicum*, the benttooth *Champsodon snyderi*, and the cardinalfish *Apogon lineatus*, significantly increased in abundance during the 1980s (Cheng et al. 2006; Lin et al. 2007). As a result, in the East China Sea, larger, higher trophic level, and commercially important demersal species have been replaced by smaller, lower trophic level, pelagic, less-valuable species.

In addition, to effectively protect fishery resources, the Chinese government implemented a new seasonal fishing moratorium system (annually from 16 June to 15 September) beginning in 1998 for areas north of 26°N in the Yellow Sea and East China Sea. During the annual moratorium, the two main types of fishing gear (bottom trawling and stow net fishing) are banned. This moratorium has shifted fishing efforts from the summer to other seasons, and annual fishing efforts have not been effectively reduced. Ten years after implementation, this summer moratorium has only been beneficial to certain stocks, such as the small yellow croaker, and has not played a positive role in the restoration of fish community structures and functions in the study area or in ensuring the maintenance of a healthy ecosystem (Jiang et al. 2009). Thus, a comprehensive set of restrictions on fishing efforts must be imposed for the fishery in this area.

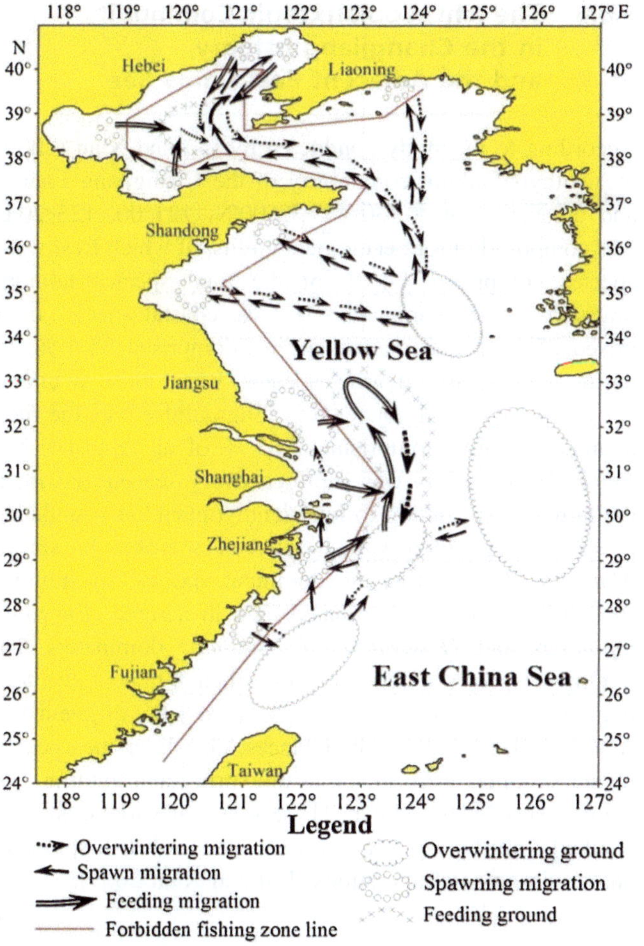

Fig. 8.11 Migration of the white croaker in the Yellow Sea and East China Sea (modified from Xu and Chen 2010)

8.6.5 Fish Migratory Routes in the East China Sea

8.6.5.1 Large Yellow Croaker

The migratory route of the large yellow croaker in the East China and Yellow Seas is presented in Fig. 8.8. Only one population, the East China Sea–Yellow Sea population (ECYSP) was found in the sampled areas. Two wintering grounds for the population were located. The offshore wintering ground lies in the waters of 30°00′–32°00′N, 124°00′–126°00′E, and the nearshore wintering ground lies near the periphery of the forbidden fishing zone limit in the mid-southern East China Sea. The offshore group began their migration for spawning toward the waters of the west Zhoushan Archipelago and the Changjiang Estuary during March and April; in these waters, they merged with fish assembles that came from the nearshore group from the mid-southern East China Sea. Some of these fish continued to migrate northward into the waters of the Lvsi Fishing Ground and some continued westward to the waters of the Daiquyang, Huangzeyang, and Damuyang near the Zhoushan Archipelago for spawning in May. After spawning, these fish groups moved northeast to the nearshore waters of the south Yellow Sea for feeding and growing. These fish species finally came to the area of 34°00′N, located near the forbidden fishing zone, in September. Because of the influence of the cold northwest monsoon after October, the population returned to the Changjiang Estuary. From there, these fishes split into two major groups: one group swam eastward back to their wintering ground in offshore waters and the other moved back to their wintering ground in the nearshore waters of the mid-southern East China Sea. The group that wintered in the waters of the mid-southern East China Sea moved back and forth between the wintering grounds and nearby spawning grounds, such as the waters of Maotouyang in Sanmen Bay, Dongtouyang at Oujiang Estuary, and Guanjingyang in Sansha Bay. Thus, the groups grew up in waters near the spawning ground (Xu and Chen 2011).

8.6.5.2 Small Yellow Croaker

The migrations of the small yellow croaker in the East China Sea, Yellow Sea, and Bohai Sea are presented in Fig. 8.6. The main wintering grounds of this species are influenced by warm currents and are located west of Jeju Island and in the middle and mid-southern parts of the East China Sea, respectively. Their migration for spawning first passes through the waters of 32.00°N, 123.3°–124.3°E, in the middle of the East China Sea toward the Zhoushan Fishing Ground, where they merge with another spawning group moving northward from the southern nearshore, and then the mixed group that had partially spawned in the Zhoushan Archipelago. Another portion of the small yellow croakers moved to the Changjiang Estuary and the Lvsi Fishing Ground to spawn. After spawning, the fish groups continued to move to the Dasha Fishing Ground for feeding. After October, the majority of the fish in the group migrated eastward back toward the offshore wintering ground, and some others moved south along the outside of the forbidden fishing zone to the mid-southern nearshore wintering ground. The routines of migration for spawning and wintering were different. In addition to the above major migration and spawning groups, parts of the groups were located between the outside of the forbidden fishing line and the coastal spawning ground. These groups migrated to the closest islands, estuaries, and gulfs to spawn and then returned to where they had left from for feeding and wintering (Xu and Chen 2009).

8.6.5.3 White Croaker

The populations and migratory routines of the white croaker *Argyrosomus argentatus* in China's seas, including the Bohai Sea, the Yellow Sea, and the East China Sea, are presented in Fig. 8.10 (Xu and Chen 2010). Two populations, the Yellow Sea–Bohai Sea population (YSBSP) and the Yellow Sea–East China Sea population (YSECSP), were identified in the study area. The peak abundances of the YSBP occurred in October and November when the fish school was located in the mid-Bohai Sea and the southern waters of the Yellow Sea off the Shandong Peninsula. The YSECSP was largest during June–August in the waters of the Changjiang Estuary, which is the location of their feeding grounds. The wintering grounds of the two populations did not overlap. The distinct geographic distribution patterns of the two populations suggest that the two populations differ in terms of thermal adaptations. The YSBSP is a characteristically warm temperate water population, while the YSECSP is more likely composed of warm water individuals. Moreover, the routines of their spawning migrations were also different. The YSBSP migrated toward the northwest, while the YSECSP migrated toward the southwest. The YSBSP overwintered in the northern waters in the middle Yellow Sea near 34°00′N. Afterward, this population migrated into the coastal waters of the Bohai Sea, the Yalvjiang Estuary, the Shandong Peninsula, and the Haizhou Gulf for spawning in March and April. After spawning, this population grew up during May–September in waters near to the spawning ground and aggregated in October to November to migrate toward the wintering ground, which they arrived at in January.

The offshore waters of the East China Sea and the nearshore waters of the mid-southern East China Sea were the two major overwintering grounds for the YSECSP. The offshore group began their spawning migration during March and April and moved toward the west Zhoushan Fishing Ground, where they merged with the fish that came from the nearshore group and continued to migrate northward into the waters of the Zhoushan Archipelago and the Changjiang Estuary for spawning in May. After spawning, the fish moved north to the nearshore waters of the south Yellow Sea and matured there from May to September. Due to the influence of the cold northwest monsoon after October, the population returned to the Changjiang Estuary. From there, the fish split into two major groups: one of which migrated eastward to a wintering ground in offshore waters and the other moved to a wintering ground in nearshore waters of the mid-southern East China Sea. In addition, part of the nearshore group never migrated to the Yellow Sea, but matured in waters near to the spawning ground.

8.6.6 The Ichthyoplanktonic Community in the Changjiang Estuary and the Adjacent East China Sea

According to the study conducted by Liu and Xian (2009), the ichthyoplankton community of the Changjiang Estuary and adjacent waters (30°45′–32°00′N, 121°00′–123°20′E) was composed of four ecological guilds, of which freshwater species comprised 4.4 % of the total species number, brackish species comprised 31.1 %, coastal species comprised 22.2 %, and marine species comprised 35.6 %. Of these species, the marine species were most abundant (74.6 % of the total abundance). Engraulidae was the most abundant family, comprising 76.5 % of all teleost fishes. *Engraulis japonicus* was captured every season and made the largest contribution to total ichthyoplankton abundance. Species composition varied between seasons: In spring, *Allanetta bleekeri* was the dominant species; in summer, *Engraulis japonicus* dominated; and in autumn, *Engraulis japonicus* and *Hemisalanx prognathous* dominated. The dominant species comprised more than 89 % of the total number of ichthyoplankton species across the different seasons. Species numbers, abundances, and diversities peaked in the summer, were lower in the spring and autumn, and were lowest in the winter. These seasonal variations likely resulted from both migration associated with fish spawning and environmental alterations (Liu and Xian 2009).

8.7 A General View of Ecocatastrophe in the Marine Ecosystems of the Changjiang Estuary and the Adjacent East China Sea

The Changjiang Estuary and the adjacent shelf of the East China Sea are high-frequency occurrence areas for ecological disasters, which include those related to harmful algal blooms (HABs), jellyfish outbreaks, alien species invasions, benthic habitat desertification, hypoxia, and commercial extinction of some important fishery species (Wang 2002; Cheng et al. 2004; Jiang et al. 2009; Wang et al. 2012). Among these threats, HABs have become the principal marine ecological disaster. According to the annual report on the Chinese marine environment, HABs of over 1000 km^2 in size occurred 10 times in 2005 in China. In terms of both area and frequency, the Changjiang Estuary accounted for 60 % of these occurrences nationwide. Thus, this area has become the area most seriously affected by red tides in the world State Oceanic Administration (1990–2005).

Since the 1980s, red tides have occurred frequently in the Changjiang Estuary and nearby areas. Red tides usually occur between May and August. Prior to the 1980s, a total of 23 red tide incidents had been recorded in the waters off the Changjiang Estuary. Of these red tides, 13 incidents were caused by *Noctiluca scintillans*, 6 were caused by *Skeletonema costatum*, and 1 each were caused by *Eucampia zoodiacus*, *Ceratium trichoceros*, *Prorocentrum*, and *Trichodesmium* (Wang 2002). In the 1990s, a total of 13 red tide incidents were recorded; of these, 5 were caused by *Noctiluca scintillans* and 4 were caused by *Skeletonema costatum*. Other reported red tide species include *Prorocentrum donghaiense*, *Prorocentrum micans* APBM, *Mesodinium* spp., *Pseudo-nitzschia pungens*, and *Peridinium bipes* (Wang 2002). In the 2000s, the dinoflagellates *Prorocentrum* and *Karenia* played dominant causal roles in the red tides, along with *Skeletonema costatum* and *Noctiluca scintillans*. These patterns reveal an obvious trend from diatoms to dinoflagellates (Wang 2002; Chen et al. 2003). Concurrently, phytoplankton cell size declined and the numbers of toxic and noxious phytoplankton species increased (Ning et al. 2004); these factors may be associated with the highly skewed ration nutrient species, decreased silicon concentrations, and higher N/P ratios (Zhang this volume). Additionally, the predominance and low functional diversity of phytoplankton are given as reasons (Chen et al. 2003; Ning et al. 2004) to partially explain why HABs tend to occur during spring and summer.

Why is the frequency of red tides in the Changjiang Estuary so high? Based on their times of occurrence and the vnational communiqués on oceanic disasters, red tides frequently occur in the Changjiang Estuary during the transition from spring to summer (State Oceanic Administration 1990–2005). Eutrophication has been thought to be the main cause of red tide outbreaks in Changjiang Estuary.

Zooplankton, which can exert top-down control of phytoplankton, are also closely associated with the HABs in the Changjiang Estuary. Xu (2004) found that in spring, the diatom grazer *Calanus sinicus* was the predominant zooplankton species and numerically contributed to over 90 % of the zooplankton. Zooplanktonic communities were heterogeneous in terms of species compositions; thus, the communities are thought to be unstable. The asynchrony and asymmetry between biomass and community compositions and the frangibility of community structures of zooplankton together form the biological and environmental characteristics of the red tides that occur in the East China Sea in the spring. However, it was reported that in 1959, in addition to *Calanus sinicus*, several other dominant species, such as *Euchaeta plana* and *Labidocera euchaeta*, were present in the spring. This succession of dominant species might be associated with regional warming (Xu and Gao 2009; Xu et al. 2011). However, the causes of red tide outbreaks are still an important issue requiring further research, and investigations into the causes and processes of red tide outbreaks in the Changjiang Estuary are still in progress.

During the end of spring and the beginning of summer, the seawater temperature increased continuously and the Taiwan Warm Current strengthened in the waters off the Changjiang Estuary. As shown in Fig. 8.12, the number of

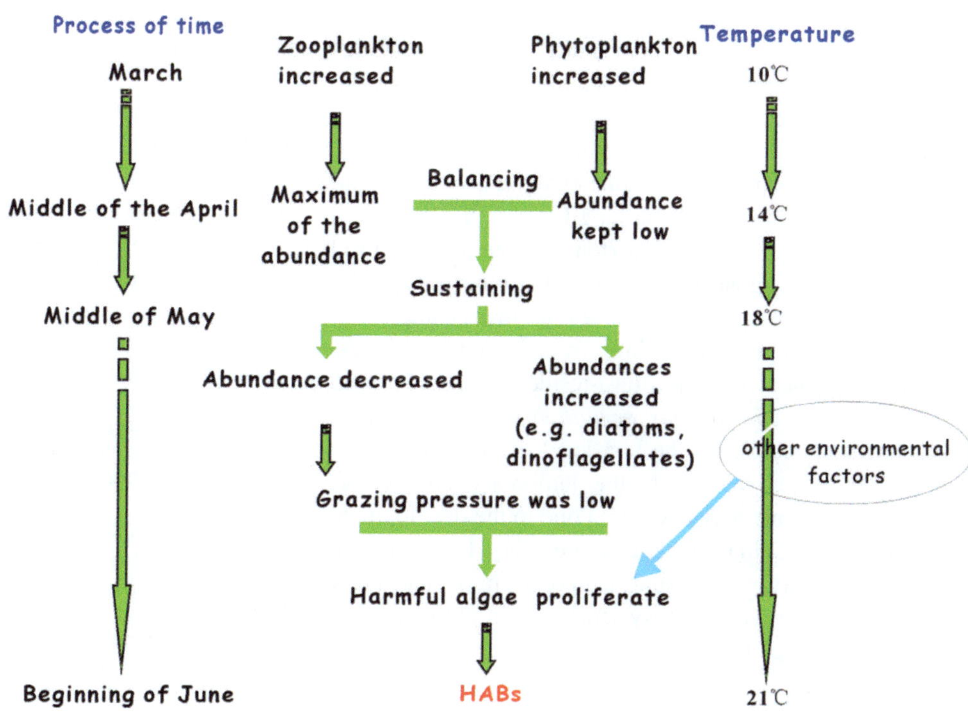

Fig. 8.12 Conceptual model of the dynamics of zooplankton during spring HABs in the Changjiang Estuary and the adjacent East China Sea

warm water zooplankton species increased. Furthermore, with the continuously changing hydrological environment, alterations between warm temperate species and warm water species were observed. The former zooplankton communities, which consist mainly of warm temperate species, gradually decreased. Meanwhile, the zooplankton community consisting mainly of warm water species had not yet become abundant. The decreasing numbers of zooplankton may have affected the grazing pressure on the phytoplankton at the level of the single dominant species, which would ultimately influence the red tide generation process. The influence of global warming on the top-down control of phytoplankton becomes weaker during the alternations between warm temperate zooplankton and warm water zooplankton. This may represent a rational explanation for the observation that the red tides have often occurred in the Changjiang Estuary when spring was transitioning into summer (Xu 2004). Moreover, recent global warming has intensified the situation.

Hypoxia is one of the important ecological disasters occurring in the waters of the Changjiang Estuary. Summer hypoxia appeared in this area in the early 1980s (Limeburner et al. 1983) and has become increasingly severe in recent decades (Li et al. 2002). Hypoxia decreases in the autumn and disappears in winter (Wang et al. 2012). In the 1980s, the lowest levels of dissolved oxygen (DO) were 2–3 mg/L (Wei et al. 2007). In 2000, the lowest DO values were <1 mg/L, and the hypoxic area extended to an area of 10,000 km^2. In July 2006, the hypoxic area expanded to 13,740 km^2, indicating that the area of hypoxia has been steadily increasing (Wei et al. 2007). Based on investigations of the macrobenthic communities in the Changjiang Estuary across four seasons, Liu et al. (2008) found that species numbers and biomasses of macrobenthic communities were lowest in the summer when hypoxia levels were greatest and that hypoxia may have limited the movements of mollusks, polychaetes, and echinoderms. Jiang et al. (2014) found that, in waters adjacent to the Changjiang Estuary, the minimum species diversities of demersal nekton communities occurred in the summer (August–September). This time is typically associated with increased diversities in many other subtropical coastal regions in China. To a certain extent, the temporary absence of some quasi-residents (e.g., *Parapenaeopsis hardwickii*, *Octopus ocellatus*) in the study area was likely responsible for the low level of diversity during August–September. However, the temporary absence of these quasi-residents may have been due to the development of summer hypoxia in the Changjiang Estuary. Extensions of the hypoxia (both spatial and temporal) may negatively impact local and regional ecosystems, and in particular, the fishery resources of the region.

8.8 Overall Evaluation

The Changjiang Estuary and the adjacent East China Sea are characterized by high productivity due to the combined influence of freshwater runoff from the Changjiang, waters from Chinese coastal areas, waters of the Taiwan Warm Current flowing from the Taiwan Strait, and waters from a branch of the Kuroshio Current that extends into the area from the northeast of Taiwan (Su 1998). Additionally, seasonal upwellings frequently occur in the vicinity of the Changjiang Estuary and in adjacent waters (Zhao 1993). Like most estuaries worldwide, the Changjiang Estuary is a highly productive and resource-rich aquatic ecosystem and is one of the most important fishing grounds in this area of China.

The Changjiang Estuary and adjacent waters are the habitats of mainly warm temperate communities in winter and spring, and subtropical communities in summer and autumn. The succession from warm temperate to subtropical communities begins in late spring and lasts through early summer, which makes the area ideal for studying the impacts of climate changes on marine ecosystems. In addition, community succession is characterized by instability in the marine ecosystem and is increasingly influenced by eutrophication; consequently, one of world largest HABs occurs frequently in this area.

Acknowledgments This review was based on the data collected in the mid-1990s with financial support from the Natural Science Foundation of China (Nos 41176131 and 90511005), the National Basic Research Program of China (Nos 2001CB409700 and 2010CB428705), and the Ministry of Agriculture of China. We are indebted to our colleagues and former students from the East China Sea Fisheries Research Institute for their support in the field campaigns and cruises, sample collection, and laboratory analyses. Special thanks are owed to Jiajie Chen for data processing.

References

Anderson DM, Glibert PM, Burkholder JM (2002) Harmful algal blooms and eutrophication: nutrient sources, composition, and consequences. Estuaries 25:704–726

Belkin IM (2009) Rapid warming of large marine ecosystems. Prog Oceanogr 81:207–213

Beardsley RC, Limeburner R, Yu H, Cannon GA (1985) Discharge of the Changjiang (Yangtze River) into the East China Sea. Cont Shelf Res 4:57–76

Cai DL, Li HY, Tang QS (2005) Establishment of trophic continuum in the food web of the yellow sea and East sea ecosystem: insight from carbon and nitrogen stable isotopes. Sci China B Life Sci 35:123–130

Cai C, Yu Z, Song X, Cao X (2006) The status and characteristics of eutrophication in the Yangtze River (Changjiang) Estuary and the adjacent East China Sea, China. Hydrobiologia 563:313–328

Chen CS, Zhu JR, Beardsley RC, Franks PJS (2003) Physical-biological sources for dense algal blooms near the Changjiang River. Geophys Res Lett 30:1515

Chen QC (1992) Zooplankton of China Seas (1). Science Press, Beijing and New York

Cheng J, Ding F, Li S, Yan L, Li J, Liang Z (2006) Changes of fish community structure in the coastal zone of the northern part of East China Sea in summer. J Nat Resour 21:775–781 (In Chinese)

Cheng J, Li S, Ding F, Yan L (2004) Primary analysis on the jellyfish blooms and its causes in the East China Sea and the Yellow Sea. Mod Fish Inf 19:10–12 (In Chinese)

Cheng J, Yu L (2004) The change of structure and diversity of demersal fish communities in the Yellow Sea and East China Sea in winter. J Fish Sci China 28:29–34 (In Chinese)

Gao Q, Xu ZL (2011) Effect of regional warming on the abundance of *Pseudeuphausia sinica* Wang et Chen (Euphausiacea) off the Changjiang River (Yangtze River) Estuary. Acta Oceanol Sin 30:122–128

Gao Q, Xu ZL, Zhang JP (2008) Comparison of mesozooplankton communities in north channel and north branch of Yangtze River Estuary. Chin J Appl Ecol 19:2049–2055 (In Chinese)

Guo HY, Zheng Y, Tang WQ, Shen H, Wei K, Xie ZL, Tsukamoto K (2011) Behavioral migration diversity of the Yangtze River Japanese Eel, *Anguilla japonica*, based on otolith Sr/Ca ratios. Zool Res 32(4):442–450 (In Chinese)

Guo Y, Yang Z (1992) Quantitative variation and ecological analysis of phytoplankton in the estuarine area of the Changjiang river. Stud Mar Sin 33:167–189 (In Chinese)

Han YS (2011) Temperature-dependent recruitment delay of the Japanese glass eel *Anguilla japonica* in East Asia. Mar Biol 158 (10):2349–2358

Ho CR, Lin CY, Huang SJ, Kuo NJ, Zheng Q (2004) Sea surface temperature variations in the China seas. Paper presented at the proceedings of the 13th workshop of OMISAR (WOM-13) on validation and application of satellite data for marine resources conservation, Bali, Indonesia, 5–9 Oct 2004

Ives AR, Carpenter SR (2007) Stability and diversity of ecosystem. Science 317:58–62

Jiang RJ, Zhong JS, Zhou YD (2010) The spatio-temporal distribution of fish larvae and juveniles in the surf zone of the Yangtze River estuary. J Shanghai Ocean Univ 19(6):828–835 (In Chinese)

Jiang Y, Cheng J, Li S (2009) Temporal changes in the fish community resulting from a summer fishing moratorium in the northern East China Sea. Mar Ecol Prog Ser 387:265–273

Jiang Y, Ling J, Li J, Yang L, Li S (2014) Seasonal changes in the demersal nekton community off the Changjiang River estuary. Chin J Oceanol Limnol 32:278–289

Ke F, Wei Q, Luo J, Yang W (1993) Effects of the Yangtze Gorges project on fishery resources and remedy measurement. Freshw Fish 23:15–18 (In Chinese)

Li CH (1998) Interrelationship among daily age, body length, birth date and sampling location at recruitment of *Anguilla japonica* elvers in the Chinese coast. Acta Oceanol Sin 20:107–113

Li D, Zhang J, Huang D, Wu Y, Liang J (2002) Oxygen depletion off the Changjiang (Yangtze River) Estuary. Sci China D Earth Sci 45:1137–1146

Li Y, Xu ZL, Gao Q (2009) Effects of global warming on *Sagitta crassa* and *Sagitta enflata* (Chaetognatha) in the Changjiang Estuary during different years. Acta Ecol Sin 29:4773–4780 (In Chinese)

Limeburner R, Beardsley RC, Zhao J (1983) Water masses and circulation in the East China Sea. Proceedings of international symposium on sedimentation on the continental shelf, with special Reference to the East China Sea, April 12–16, Hangzhou, China, vol 1. China Ocean Press, Beijing, pp 285–294

Lin L, Cheng J (2004) An analysis of the current situation of fishery biology of small yellow croaker in the East China Sea. Periodical Ocean Univ China 34:565–570 (In Chinese)

Lin L, Cheng J, Ling J (2007) Analysis on recent status of the bottom trawl fishery resources in the East China Sea region. Mar Fish 29 (4):371–374 (In Chinese)

Lin ZF, Wu YL, Yu HC, Xian WW (2008) Phytoplankton community structure in the Changjiang Estuary and its adjacent waters in 2004. Oceanol Limnol Sin 39(4):401–410 (In Chinese)

Liu LS, Zheng BH, Li BQ, Cai WJ, Han WX, Lin KX (2012) Long-term trends of macrobenthos in Changjiang Estuary, China in relation to environmental changes. Acta Oceanol Sin 34(3):134–145 (In Chinese)

Liu SD, Xian WW (2009) Temporal and spatial patterns of the ichthyoplankton community in the Yangtze Estuary and its adjacent waters. Biodivers Sci 17(2):151–159 (In Chinese)

Liu Y, Xian WW, Sun SC, Wu YQ (2008) Primary studies on the biomass abundance and secondary production of macrobenthos in Changjiang Estuary. Periodical Ocean Univ China 38(5):749–756

Ma ZL, Xu ZL, Zhou J (2009) Effect of global warming on the distribution of *Lucifer intermedium* and *L. hanseni* (Decapoda) in the Changjiang estuary. Prog Nat Sci 19:1389–1395

Miao YT, Yu HH (1991) Spatial and temporal variations of water type mixing characteristic in the East China Sea. In: Su JL, Chen ZS, Yu GH (eds) Transactions of scientific survey on Kuroshio current, vol 3. Ocean Press, Beijing, China, pp 193–203 (In Chinese)

Ning XR, Shi JX, Cai YM, Liu CG (2004) Biological productivity front in the Changjiang Estuary and the Hangzhou Bay and its ecological effects. Acta Oceanol Sin 26(6):96–106

Qu J, Xu Z, Long Q, Wang L, Shen X, Zhang J, Cai Y (2005) East China Sea, GIWA regional assessment 36. University of Kalmar, Kalmar

Shannon CE (1948) A mathematical theory of communication. Bell Syst Tech J 27:379–423

State Oceanic Administration (1990–2005) Year Report of China Marine Environment (In Chinese)

Su J (1998) Circulation dynamics of the China Seas North of 18°N. In: Robinson AR, Brink KH (eds) The Sea, vol 11. Wiley, New York, pp 483–505

Tang QS, Su JL (eds) (2000) Aspects of marine ecosystem dynamics in China I: scientific problems and research strategy. Science Press, Beijing (In Chinese)

Tsukamoto K (2006) Oceanic biology: spawning of eels near a seamount. Nature 439:929

Wang B, Wei Q, Chen J, Xie L (2012) Annual cycle of hypoxia off the Changjiang (Yangtze River) Estuary. Mar Environ Res 77:1–5

Wang JH (2002) HAB alga nearby Changjiang Estuary. Mar Environ Sci 21(2):37–41 (In Chinese)

Wei H, He YC, Li QL, Liu ZY, Wang HT (2007) Summer hypoxia adjacent to the Changjiang Estuary. J Mar Syst 67(3–4):292–303

Wei QW, Chen XH, Yang DG, Liu JY, Zhu YJ, Zheng WD (2005) Variations in spawning stock structure of *Acipenser sinensis* within 24 year since damming of Gezhou Dam. J Fish Sci China 12 (4):452–457 (In Chinese)

Wu YL, Fu YN, Zhang YS, Pu XM, Zhou CX (2004) Phytoplankton distribution and its relation to the runoff in the Changjiang (Yangtze) Estuary. Oceanol Limnol Sin 35(3):246–251 (In Chinese)

Xu ZL (2004) Relationship between red tide occurrence and zooplankton communities structure in the coastal sea of East China Sea in spring. China Environ Sci 24(3):257–260 (In Chinese)

Xu ZL (2005a) Zooplankton in north branch waters of Changjiang Estuary. Chin J Appl Ecol 16(7):1341–1345 (In Chinese)

Xu ZL (2005b) Character of zooplankton community and its variation in the water near the Yangtze River estuary. Chin J Ecol 24(7):780–784 (In Chinese)

Xu ZL, Chen JJ (2011) Analysis of migratory route of *Larimichthys crocea* in the East China Sea and Yellow Sea. J Fish China 35:429–437 (in Chinese)

Xu ZL, Chen JJ (2009) Analysis on migratory routine of *Larimichthys polyactis*. J Fish Sci China 16(6):931–940 (In Chinese)

Xu ZL, Chen JJ (2010) Analysis to population division and migratory routine of populations and migratory routines of *Argyrosomus argentatus* in the north China waters. Acta Ecol Sin 30:6442–6450 (In Chinese)

Xu ZL, Chen YQ (1989) Aggregated intensity of dominant species of zooplankton in autumn in the East China Sea and Yellow Sea. J Ecol 8(4):13–15 (In Chinese)

Xu ZL, Gao Q (2009) *Labidocera euchaeta*: its distribution in Yangtze River estuary and responses to global warming. Chin J Appl Ecol 20(5):1196–1201 (In Chinese)

Xu ZL, Gao Q, Kang W, Zhou J (2013) Regional warming and decline in abundance of *Euchaeta plana* (Copepoda, Calanoida) in the nearshore waters of the East China Sea. J Crustac Biol 33:323–331

Xu ZL, Ma Z, Wu Y (2011) Peaked abundance of *Calanus sinicus* earlier shifted in the Changjiang River (Yangtze River) Estuary: a comparable study between 1959, 2002 and 2005. Acta Oceanol Sin 30:84–91

Xu R, Li YH, Li ZE, Wang JH (2009) Quantitative comparison of zooplankton in different habitats of the Changjiang Estuary. Acta Ecol Sin 29:1688–1696 (In Chinese)

Xu ZL, Wang YL, Chen YQ (1995) An ecological study on zooplankton in maximum turbid zone of estuarine area of Changjing River. J Fish Sci China 2(1):39–48 (In Chinese)

Yang DL, Wu GZ, Sun JR (1990) The investigation of pelagic eggs, larvae and juveniles of fishes at the mouth of the Changjiang river and adjacent areas. Oceanol Limnol Sin 21(4):346–355 (In Chinese)

Zeng Q (1993) Survey of crustacean plankton in south and north branch of Yangtze River. Freshw Fish 23(1):33–35 (In Chinese)

Zhao BR (1993) Upwelling phenomenon off Changjiang estuary. Acta Oceanol Sin 15:106–114 (In Chinese)

Zhao ZM, Zhou XY (1984) Introduction to ecology. Scientific and Technical Documents Publishing House, Chongqing, pp 108–119 (In Chinese)

Zhou J, Xu ZL, Ma Z (2009) Effect of global warming on abundance variation of *Parathemisto gaudichardi* (Amphipoda) in the Changjiang Estuary. Acta Ecol Sin 29:5758–5765 (In Chinese)

Zhou MJ, Shen ZL, Yu RC (2008) Responses of a coastal phytoplankton community to increased nutrient input from the Changjiang (Yangtze) River. Cont Shelf Res 28:1483–1489

Zhu YD, Zhang CL, Cheng QT (1963) Fish fauna of the East China Sea. Science Press, Beijing (In Chinese)

Zhuang P, Wang YK, Li SF, Deng SM, Li CS, Ni Y (2006) Fisheries of the Yangtze Estuary. Shanghai Scientific and Technical Publishers, Shanghai, p 497 (In Chinese)

Socioeconomic Dimensions of the Coastal Environment of the East China Sea

9

Qian Long, and Jing Zhang

Abstract

The four coastal provinces (Jiangsu, Zhejiang, Fujian, and Taiwan) and autonomous city (Shanghai) adjacent to the East China Sea are among the most developed and the densely populated areas in China. Over the last 30 years, these areas have experienced rapid urbanization and industrialization. With increasing incomes and improved living standards in China, so tourism, traffic, and related industries have also developed rapidly. The rapid urbanization of these provinces along the coast of the East China Sea has adversely impacted the marine environment. Domestic sewage discharge per capita has increased significantly, especially in metropolitan cities such as Shanghai. Illegal reclamation and dumping have also had adverse effects on the marine environment. Increased use of fertilizer to promote higher agricultural crop yields has directly impacted water quality in the East China Sea, leading to widespread eutrophication. Incident such as oil spills has had serious impacts on the marine ecosystem. River inputs represent the largest source of water pollution in the East China Sea, but sewage outfalls, mariculture, engineering activities, agriculture, and nonpoint source pollutants are other key contributors. At the present time, water quality in the East China Sea meets national standards, but it has been deteriorating for the last 10 years. Measures must be implemented to prevent further deterioration in water quality as a result of human activities.

Keywords

Socioeconomic • East China Sea • Urbanization • Marine environment • Pollution

9.1 Introduction

The four coastal provinces (Jiangsu, Zhejiang, Fujian, and Taiwan) and autonomous city (Shanghai) adjacent to the East China Sea are among the most developed and densely populated areas of China (Fig. 9.1). Over the last thirty years, these areas have experienced rapid urbanization and industrialization (BSC 1949–2012a; BSF 1950–2012; BSJ 1957–2012; BSZ 1952–2012; BSS 1952–2012; CEPDRC 2001–2010). The level of economic development and living standards have increased, but urban expansion and population growth have put great stress on the environment. Industrial production processes and domestic activities generate high volumes of liquid and solid waste, and the construction of sewage and solid waste treatment systems has not kept pace with the rate of urbanization. Large amounts of wastewater and solid waste are directly discharged into rivers and the coastal environment. Consequently, anthropogenic pollution in the marine environment has also affected human well-being. For instance, along the coast of Zhejiang, marine litter and industrial wastewater discharge have had serious negative effects on fisheries and

Q. Long (✉)
Central Library, East China Normal University,
500 Dongchuan Road, Shanghai 200241, China
e-mail: xlong@library.ecnu.edu.cn

J. Zhang
State Key Laboratory of Estuarine and Coastal Research,
East China Normal University, 3663 Zhongshan Road North,
Shanghai 200062, China
e-mail: jzhang@sklec.ecnu.edu.cn

J. Zhang (ed.), *Ecological Continuum from the Changjiang (Yangtze River) Watersheds to the East China Sea Continental Margin*, Estuaries of the World,
DOI 10.1007/978-3-319-16339-0_9,

Fig. 9.1 Sketch map of the coastal East China Sea and the study area, including Jiangsu, Shanghai, Zhejiang, Fujian, and Taiwan (the *shadow areas*)

aquaculture (Data source: http://jingji.cntv.cn/2013/06/10/VIDE1370875919287918.shtml). Large-scale red tide outbreaks (harmful algal blooms) have damaged coastal marine aquaculture (ECSOA 2012a). Environmental issues associated with urbanization along the coast of the East China Sea have received increasing attention, and there is a need to restrain development if we are to protect the environment.

9.2 Urbanization and Environmental Quality

9.2.1 Temporal Variations in the Urbanization Rate in Each Region

China experienced rapid urbanization after reform policies were implemented in 1978. As shown in Fig. 9.2, urbanization rates in all of the subregions studied here were higher than the Chinese average, especially over the last 10 years (BSC 1949–2012a; BSC 1980–2012b; BSF 1950–2012; BSJ 1957–2012; BSZ 1952–2012; BSS 1952–2012; CEPDRC 2001–2010).

Due to widespread migration into cities, urban areas have also expanded rapidly. The total population of Shanghai, for example, increased by 42 % in 30 years, from 10.9 million in 1978 to 18.9 million in 2008 (BSS 1978–2008). The area of Shanghai also expanded by 6.8 % over a similar period, from 5910 km^2 in 1978 to 6340 km^2 in 2008 (BSS 1978–2008) (Box 9.1).

Box 9.1 Urbanization rates in the study region compared with other developed countries

Urbanization rates in the Changjiang Delta: Urbanization rates, or the percentage of the total population living in urban areas, in all subregions of this study, are higher than the Chinese average. The Changjiang Delta (including Zhejiang, Jiangsu, and Shanghai) is the most developed area in China and has the highest urbanization rate in the region. The economy of the Changjiang Delta has maintained rapid growth, with an annual GDP increase of approximately 10 % in 2012. The GDP of this region accounted for 21.4 % of the total GDP of China in 2009 (BSC 2009a; BSF 2009–2012; BSJ 2009–2012; BSZ 2009–2012; BSS 2009–2012).

Urbanization rate in Jiangsu and Fujian: The urbanization rates in Jiangsu and Fujian were 60.6 and 57.1 % in 2010, respectively, which is lower than for the other three regions in the study area. These rates are slightly lower than for Zhejiang, but far lower than that of Shanghai.

Urbanization rate in Zhejiang: The urbanization rate in Zhejiang also shows rapid growth after 2000. The urbanization rate increased 28 % between 2000 and 2011, from 48.7 to 62.3 %.

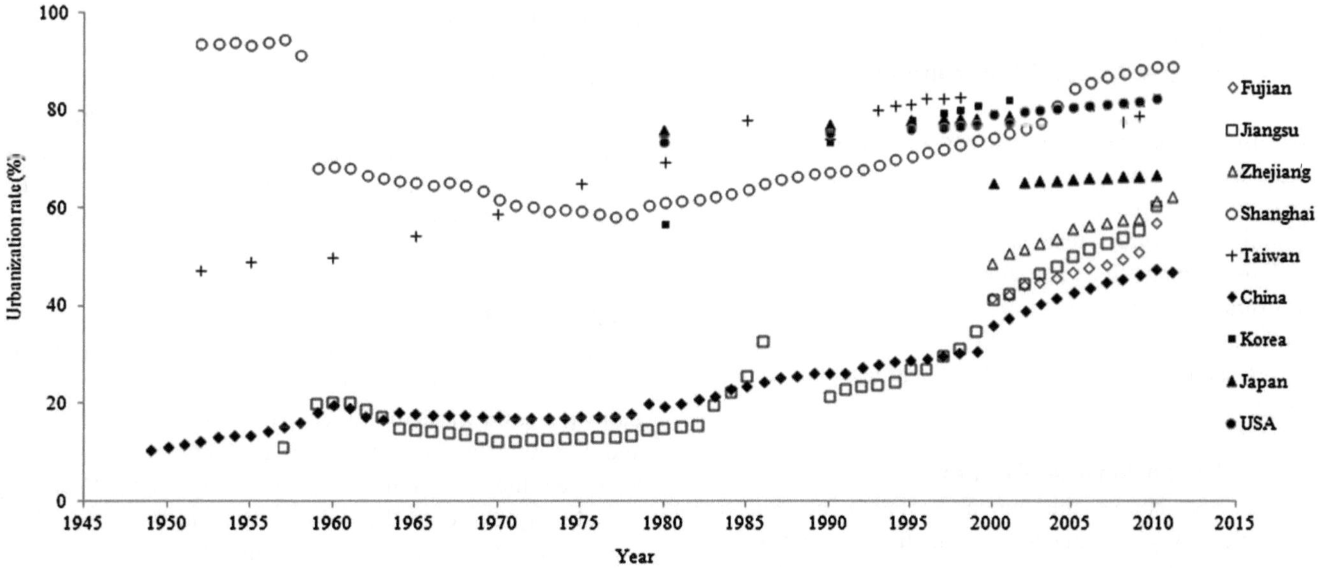

Fig. 9.2 Temporal variation in urbanization rates in the study area and in China, Korea, Japan, and the USA between 1949 and 2011 (BSC 1949–2012a; BSC 1980–2012b; BSF 2000–2012; BSJ 1957–2012; BSZ 2000–2012; BSS 1952–2012; CEPDRC 2001–2010)

Urbanization rate in Shanghai: The urbanization rate in Shanghai decreased until 1978. After this time, Shanghai experienced increased urbanization, particularly in the past 10 years, due to its role as China's financial center. In 2010, the urbanization rate in Shanghai was 88.9 %, which is the highest of all coastal cities in mainland China.

Urbanization rate in Taiwan: As a result of its early development, Taiwan experienced relatively high urbanization rates in the latter part of the twentieth century. The urbanization rate in Taiwan increased steadily from 1950 to 1998 and reached 83 % in 1998. After this time, there has been a slight decrease in urbanization rates in Taiwan.

Urbanization rates in China, Korea, Japan, and the USA: As a developing country, the urbanization rate in China is much lower than in developed countries such as Korea, Japan, and the USA. Shanghai is the exception, with urbanization rates that are higher than in both Japan and Korea over the past five years, making it one of the most rapidly developing cities in the world. For example, the urbanization rates in Korea, Japan, and the USA were 83, 67, and 82.3 % in 2010, respectively, but the urbanization rate in Shanghai was 88.9 % (BSC 1949–2012a, 1980–2012b; BSF 1950–2012; BSJ 1957–2012; BSZ 1952–2012; BSS 1952–2012; CEPDRC 2001–2010).

9.2.2 Impact of Urbanization on the Coastal Environment

Because of population growth and the expansion of urban areas, heavily polluting industries have moved from urban areas to the suburbs, and farmland has been replaced by buildings, factories, residential areas, and parkland. This land use change has introduced anthropogenic pollutants into rural soils (Sowana et al. 2011). High population densities have also brought pressure on suburban areas due to factors such as traffic jams and garbage leaching (Xia et al. 2011).

9.2.2.1 Urban Sewage Discharge

Figure 9.3a shows the annual variation in per capita domestic sewage discharge in each subregion (BSF 1998–2012; BSJ 1998–2012; BSZ 1998–2012; BSS 1998–2012). The annual discharge of domestic sewage in each subregion has increased gradually. Shanghai has the highest per capita domestic sewage discharge, with volumes that are three times that in Zhejiang, Jiangsu, and Fujian. Significant positive correlations are found between time and per capita volume of domestic sewage discharge in all subregions, especially in Zhejiang, Jiangsu, and Fujian. For example, the correlation coefficient (R^2) of per capita volume of domestic sewage discharge in Fujian with time, for the period 1998–2010, is 0.953. Figure 9.3b shows the variation in volume of domestic sewage discharge with per capita gross domestic product (GDP) for 1998–2010. The volume of domestic sewage discharge per capita has a significant positive correlation with per capita GDP in all subregions, which illustrates the effect of economic development on wastewater discharge. As a metropolitan and economic center in China, Shanghai has a higher temporary population than other regions for reasons such as travel, eviction, and hospitalization. It also therefore has a higher volume of domestic sewage discharge than other subregions.

9.2.2.2 Industrial Wastewater Discharge

The volume of industrial wastewater discharge has increased gradually in the last 30 years. The ratio of industrial to total wastewater in Fig. 9.4a shows a significant decrease between 1985 and 2012 in all subregions. With growing individual and governmental environmental consciousness, the investment in environmental protection has increased every year. Industrial wastewater has increasingly been treated and/or recycled before discharge, which has resulted in reduced industrial wastewater discharge into coastal environments. Figure 9.4b shows the variation in the ratio of industrial to total wastewater with per capita GDP in each subregion from 1985 to 2012. This ratio decreases gradually with increased GDP in Shanghai, Fujian, and Zhejiang, but was stable in Jiangsu (BSF 1985–2012; BSJ 1985–2012; BSZ 1985–2012; BSS 1985–2012).

Table 9.1 lists the per capita volume of wastewater discharge into the East China Sea between 2003 and 2010. The volume of wastewater per capita has increased significantly in Fujian and Zhejiang over the last 10 years, but has decreased gradually in Shanghai due to wastewater treatment and recycling.

9.2.2.3 Solid Waste

Figure 9.5 shows changes in the volume of industrial solid waste generated between 1986 and 2010. The volume of industrial solid waste discharge decreased rapidly after 1986, and the recycling rate has increased gradually in coastal areas near the East China Sea (BSC 1985–2010a). Recycling can sharply decrease the volume of solid waste and limit adverse affects on the local environment.

9.2.2.4 Public Transportation and Private Car Use

The per capita road area in urban centers in each subregion has not increased significantly since 1995, but the number of private cars has increased sharply, especially after 2005 (BSF

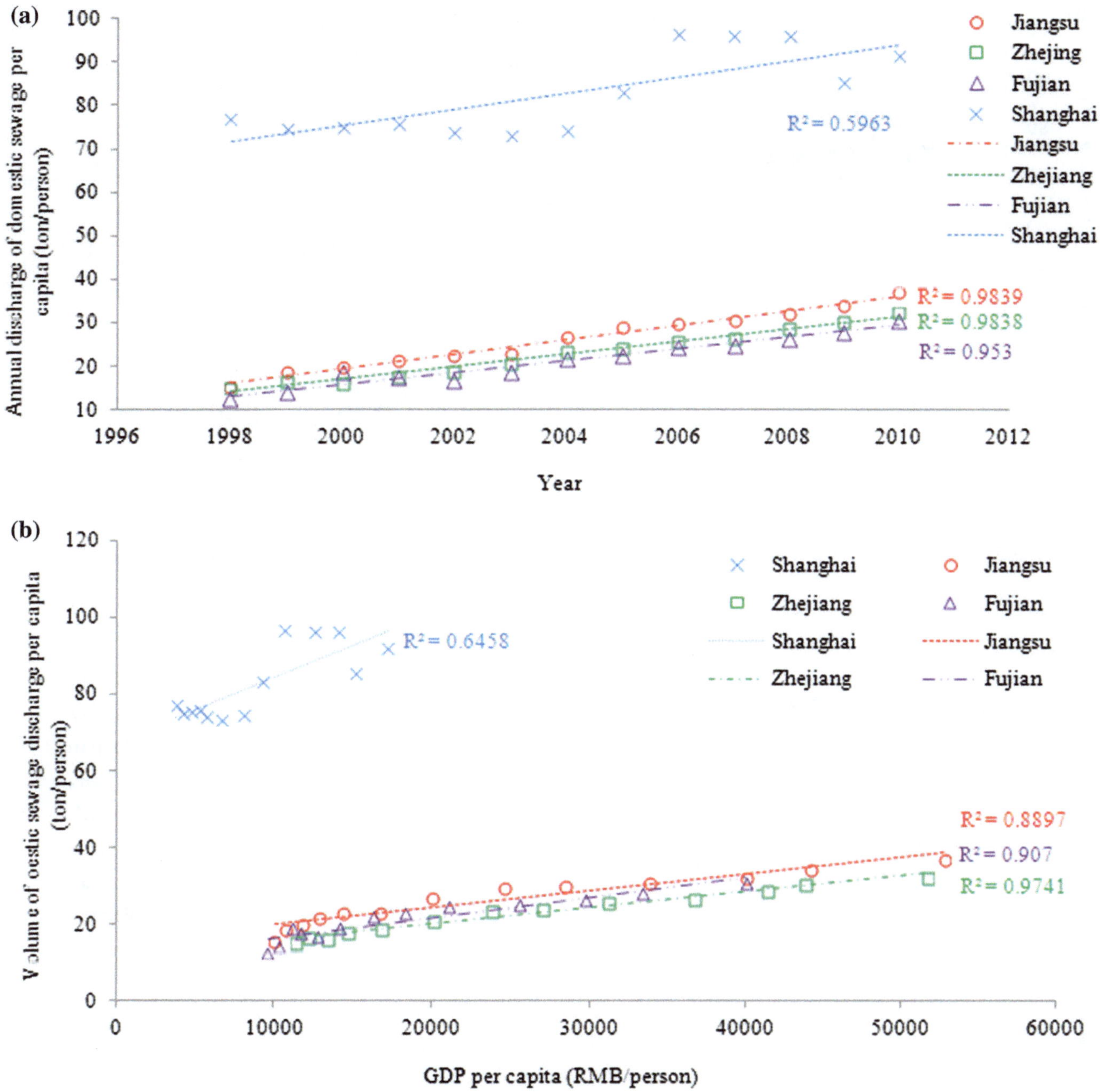

Fig. 9.3 Domestic sewage discharge in the study area, including **a** annual discharge of domestic sewage per capita and **b** variations in domestic sewage discharge with per capita GDP (the per capita GDP in Shanghai is 10^9 RMB) during 1998–2010 (BSF 1998–2010; BSJ 1998–2010; BSS 1998–2010; BSZ 1998–2010)

1985–2012; BSJ 1985–2012; BSZ 1985–2012; BSS 1985–2012). Due to urbanization and population expansion, farmland and the outer suburbs have been developed into higher density residential areas. As public transit has not been developed to accommodate these suburban populations, many people use cars to travel long distances to and from work. For instance, the number of vehicles in 2006 in Shanghai was 10 times greater than in 1990 (BSS 1990–2006). This dramatic increase in the number of cars has brought environmental pollution to suburban areas.

In contrast, the increase in the number of agricultural vehicles is lower than the increase in private cars. Table 9.2 shows the ratio between the number of private and agricultural vehicles in each subregion. This shows that the increase in private car ownership has been increasing faster than agricultural vehicles, especially in Shanghai.

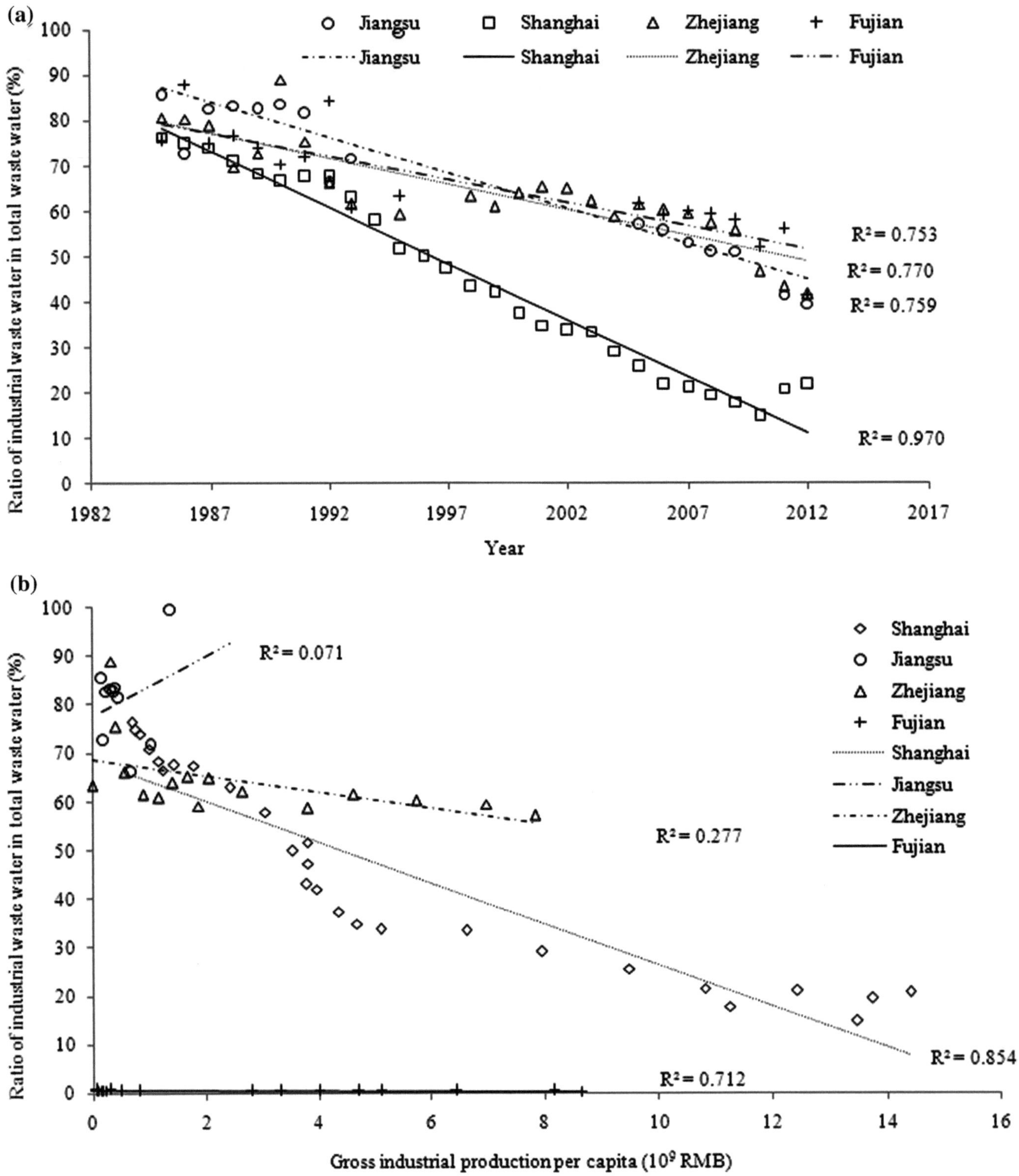

Fig. 9.4 The ratio of industrial to total wastewater discharge in the study area from 1985 to 2012, including **a** the ratio of industrial to total wastewater discharge and **b** the variation in the industrial to total wastewater ratio with per capita gross industrial production (BSF 1985–2012; BSJ 1985–2012; BSS 1985–2012; BSZ 1985–2012)

Table 9.1 The volume of wastewater discharge into the sea per capita in all subregions (tons/person) (BSC 2003–2010a)

	2003	2004	2005	2006	2007	2008	2009	2010
Shanghai	8.40	9.65	9.25	7.44	7.93	5.15	1.15	0.87
Jiangsu	0.11	0.28	0.07	0.09	0.12	0.47	0.21	0.09
Zhejiang	1.06	0.90	1.16	1.80	2.46	2.48	2.21	2.27
Fujian	12.09	15.22	17.62	15.79	17.58	18.56	20.13	16.04

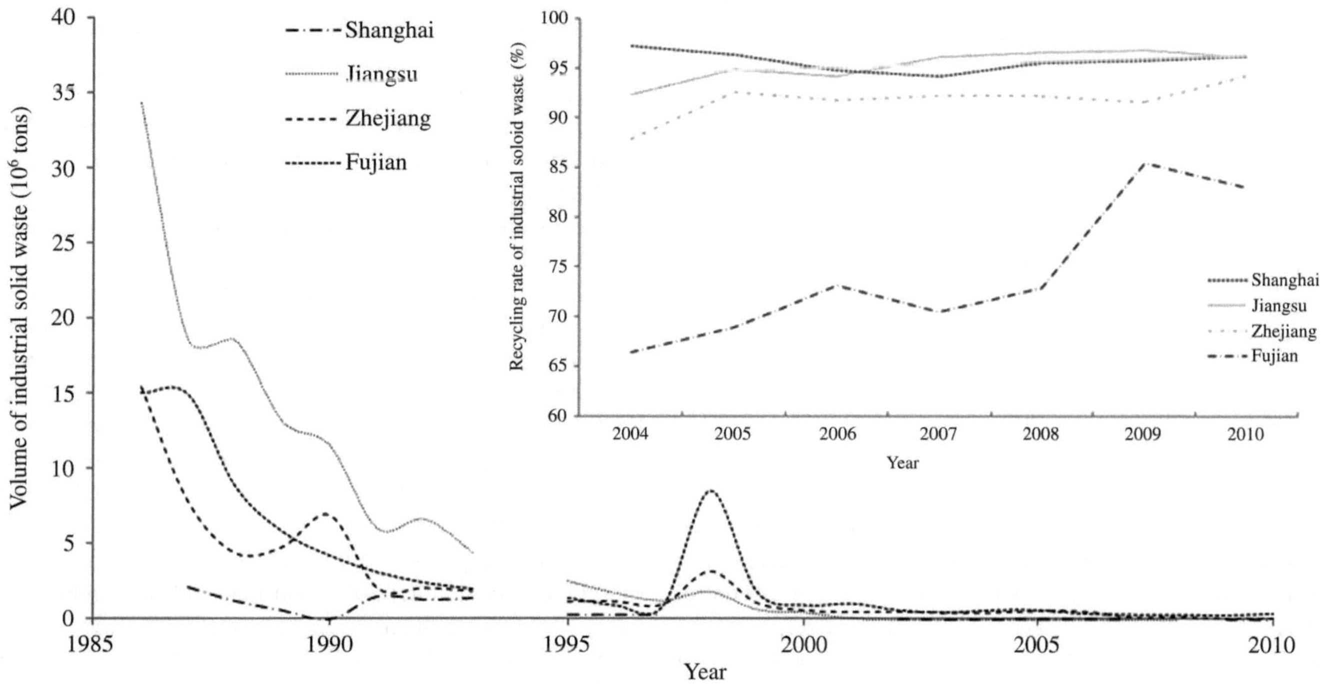

Fig. 9.5 Industrial solid waste discharge in each subregion. The *inset* map shows the recycling rate for solid waste in all areas, 1986–2010 (BSC 1986–2010a)

9.2.2.5 Reclamation and the Quality of the Coastal Environment

As urbanization has continued, large areas of coastal wetland have been reclaimed for the development of residential, industrial, and aquaculture areas. Figure 9.6 shows the total area of reclamation in each subregion from 1954 to 2012. The reclamation trends for Shanghai and Fujian follow a u-shaped profile, showing an initial decrease in reclamation area until 2001 and 2006, respectively, and then an ongoing increase in both subregions after this time.

Reclamation of coastal wetlands has caused serious damage to coastal ecosystems (Chen and Xu 2004; Li et al. 2006, Liu et al. 2007). The main consequences of reclamation have been the loss of wetland area, mudflat ecosystem degradation, a decrease in biodiversity, habitat reduction, and pollution from waste discharge. There have also been instances of illegal reclamation near the East China Sea coast, resulting in the loss of coastal beaches and wetlands, and changing coastal topography (CCICED 2011). More effort is needed to supervise and manage development along this sensitive coastal margin.

9.2.3 Environmental Investigation

Environmental pollution has received increasing attention as the Chinese economy has continued to develop. The Chinese government has taken measures to reduce the effect of waste on the environment, and there has been an increase in pollution control in all regions. However, the ratio of investment in environmental pollution control to total GDP has not increased significantly (BSJ 2000–2010; BSF 2000–2010; BSZ 2000–2010; BSS 2000–2010).

Environmental pollution management funds are mainly used for urban environmental infrastructure, industrial pollution source control, and environmental management

Table 9.2 The ratio of the number of private cars to agricultural vehicles in each subregion (BSF 1987–2008; BSJ 1987–2008; BSS 1987–2008; BSZ 1987–2008)

Year	Shanghai	Jiangsu	Zhejiang	Fujian
1987		5.36	0.66	0.89
1988		4.60	0.66	1.17
1995		0.59	1.33	1.00
1996		0.64	1.55	1.70
1997		0.55	1.72	1.70
1998		0.57	1.37	2.02
1999		0.69	1.48	3.62
2000		1.03	2.73	3.83
2001		1.45	2.91	4.95
2002		1.98	3.42	10.02
2003		2.71	4.93	15.11
2004	212	3.54	5.78	20.98
2005	241	4.47	6.28	28.32
2006	283	6.45	7.95	36.76
2007	323	8.23	9.94	44.53
2008	379	9.97	11.35	56.82

during construction. Most funds are used for the control of wastewater and gas. In Jiangsu, in 2009, the investment in wastewater, waste gas, solid waste, and noise control was 53.8, 38.7, 1.3, and 0.1 %, respectively, of the total investment in environmental pollution control (BSJ 2009).

9.3 Economy and Environmental Quality

9.3.1 GDP

The Chinese economy entered a period of rapid development after 1978. The Chinese GDP maintained high growth after 1990, with an annual growth rate of between 7 and 14 % (BSC 1978–2012a). The gross value of agricultural and industrial production increased rapidly, and the ratios of gross industrial production and gross agricultural production to total GDP increased after 1950 and accelerated after 1995 in each subregion (BSF 1950–2012; BSJ 1957–2012; BSS 1952–2012; BSZ 1952–2012).

Urbanization and industrialization have sharply increased the GDP of each subregion over the last 15 years. The ratio of primary-industry GDP to total GDP decreased sharply, but the ratio of secondary-industry GDP to total GDP increased gradually in Fujian, Zhejiang, and Jiangsu provinces (BSF 1950–2012; BSJ 1957–2012; BSS 1952–2012; BSZ 1952–2012). In Shanghai, the ratios of primary- and secondary-industry GDP to total GDP decreased sharply from 1980 onward (BSF 1950–2012; BSJ 1957–2012; BSS 1952–2012; BSZ 1952–2012).

The port cargo throughput in all subregions increased sharply, especially after 2000, and Jiangsu had a higher port cargo throughput than Shanghai, Zhejiang, and Fujian (BSF 1950–2012; BSJ 1957–2012; BSS 1952–2012; BSZ 1952–2012).

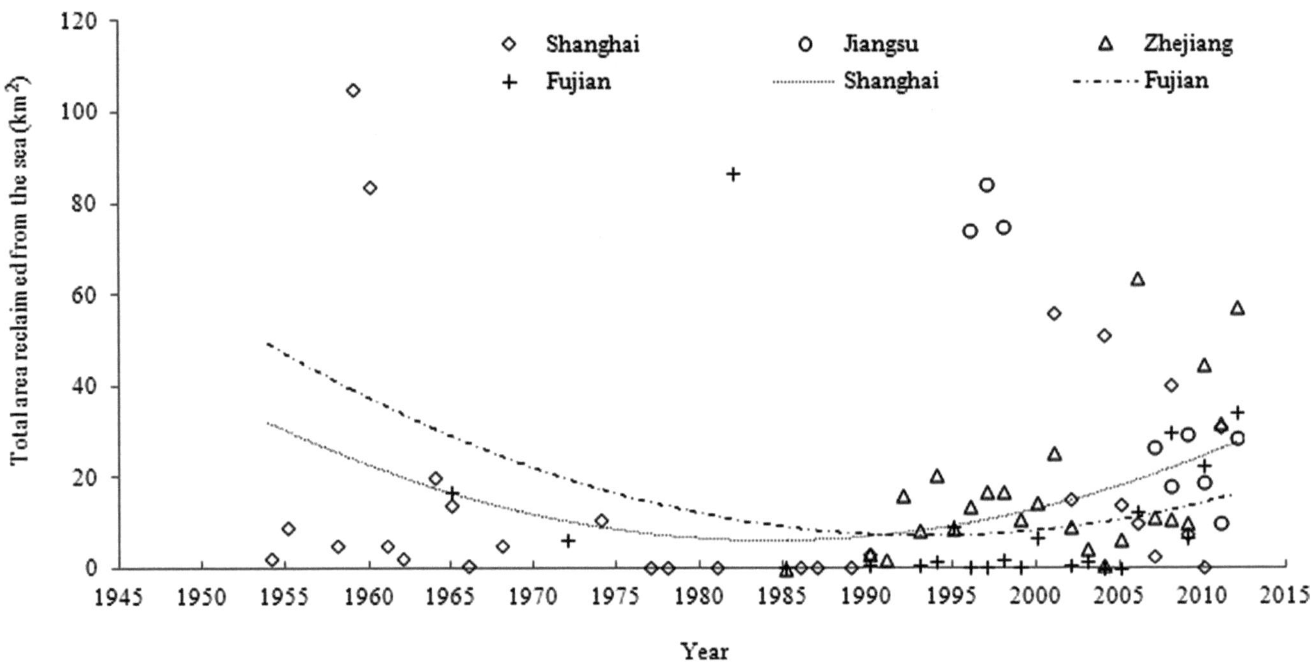

Fig. 9.6 Total area reclaimed from the sea in each region, 1954–2012 (BSF 1965–2012; BSJ 1996–2012; BSS 1954–2012; BSZ 1985–2012)

9.3.1.1 Value of Agricultural Planting and Crop Production

The main agricultural products of all subregions are rice, wheat, maize, soybeans, peanuts, rape grape, cotton, and vegetables. Per capita grain production has increased gradually since 1990 in Shanghai and Jiangsu, but has decreased in Fujian and Zhejiang (BSF 1950–2012; BSJ 1957–2012; BSS 1952–2012; BSZ 1952–2012). Per capita vegetable production in all subregions increased gradually since 1990, especially in Shanghai and Jiangsu (BSF 1950–2012; BSJ 1957–2012; BSS 1952–2012; BSZ 1952–2012). Planting efficiency can be investigated using the ratio of per capita grain to vegetable production in different subregions. Figure 9.7 shows that the ratio between grain and vegetable production in all subregions has decreased since 1990 and illustrates the declining reliance on grain as the main crop in the study area. These changing planting patterns may lead to the release of different pollutants into watercourses in these subregions.

9.3.1.2 Animal Husbandry

The value of animal husbandry in each region has varied sharply over the last 50 years (BSF 1950–2012; BSJ 1957–2012; BSZ 1952–2012; BSS 1952–2012; CEPDRC 2001–2010). Shanghai and Taiwan have the lowest values, and the rate of increase was either very low, or even decreased, in the last 10 years (BSS 1952–2012; CEPDRC 2001–2010). In Jiangsu, Zhejiang, and Fujian, the value of animal husbandry increased significantly over the past 50 years (BSF 1950–2012; BSJ 1957–2012; BSZ 1952–2012). Shanghai and Taiwan have experienced particularly high urbanization rates in the last 20 years and are more industrialized than other provinces, which may explain the low rate of growth in animal husbandry.

9.3.1.3 Fisheries and Marine Aquaculture

Compared with agricultural production, fisheries production has increased sharply since 1990 (Fig. 9.8a). However, the ratio of fisheries output to GDP has decreased gradually since 2000 (Fig. 9.8b). The ratio between aquaculture and marine fisheries production (Fig. 9.9) shows that the value of aquaculture has increased sharply since 1980, especially in Jiangsu.

9.3.1.4 Industrial GDP Structure

Different urbanization and industrialization rates have caused variations in the ratios of primary-, secondary-, and tertiary-industry production to GDP in the five subregions.

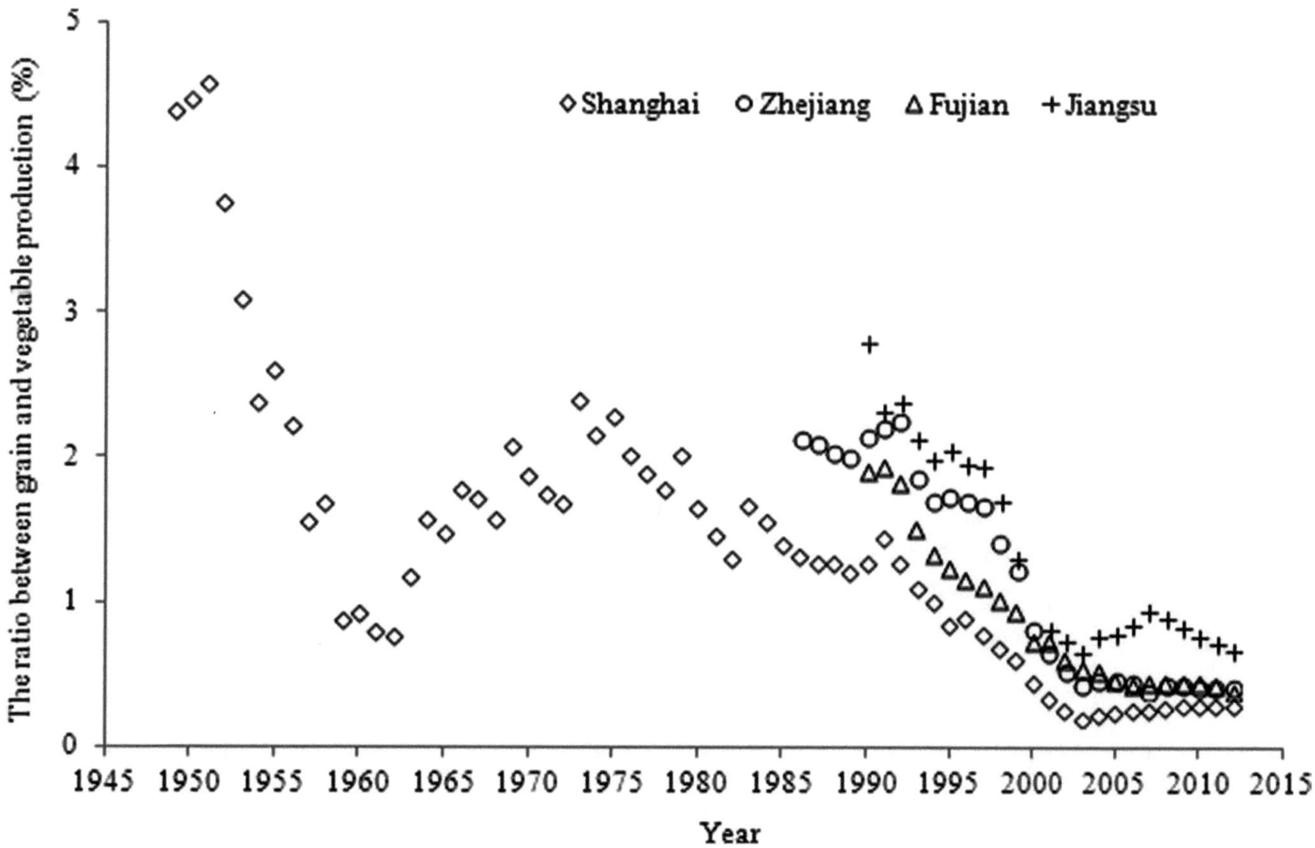

Fig. 9.7 The ratio of grain production to vegetation in each subregion (BSF 1990–2012; BSJ 1990–2012; BSS 1949–2012; BSZ 1985–2012)

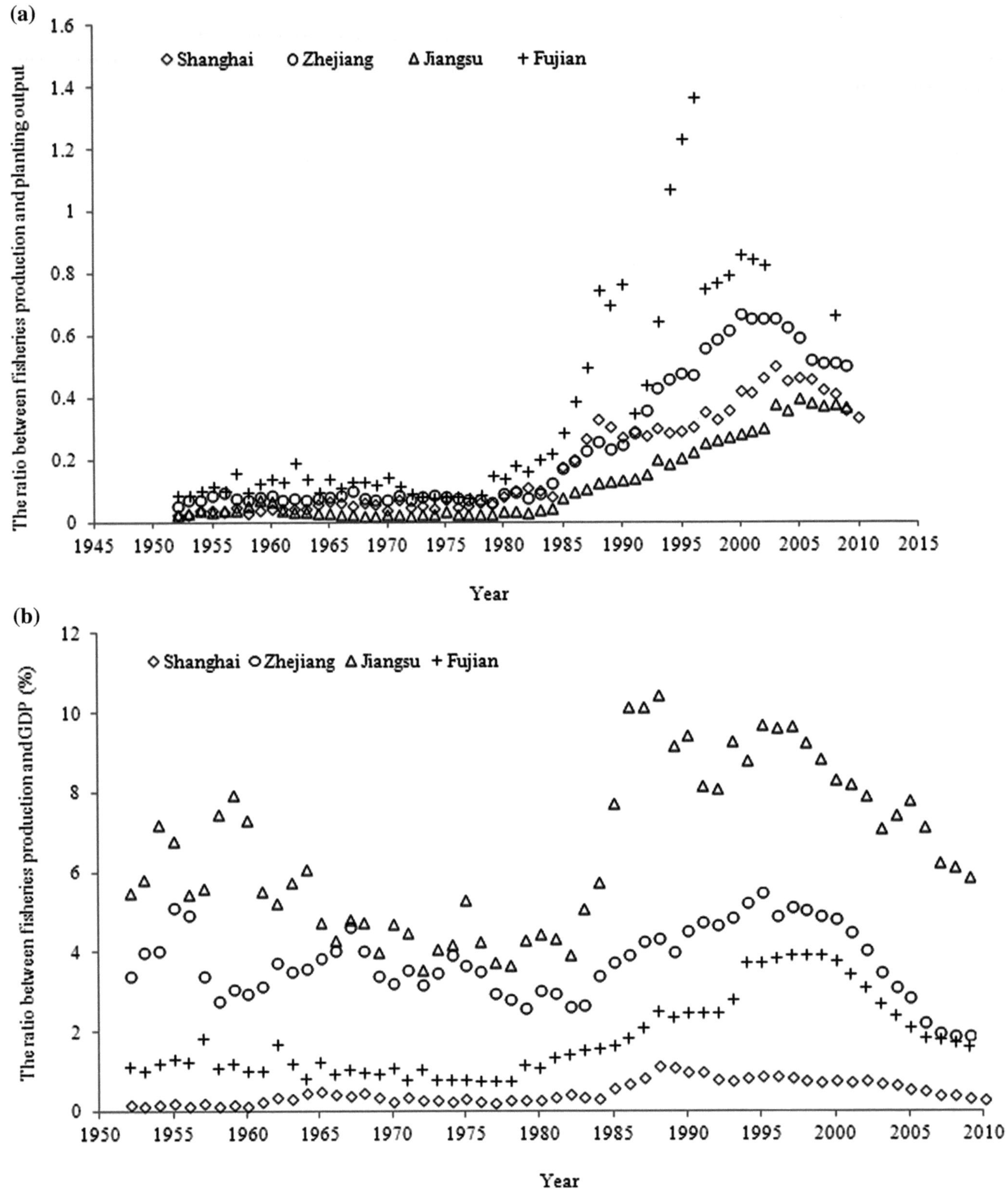

Fig. 9.8 **a** The ratio of fisheries production to planting output and **b** the ratio of fisheries production to GDP from 1952 to 2010 (BSC 1949–2012a)

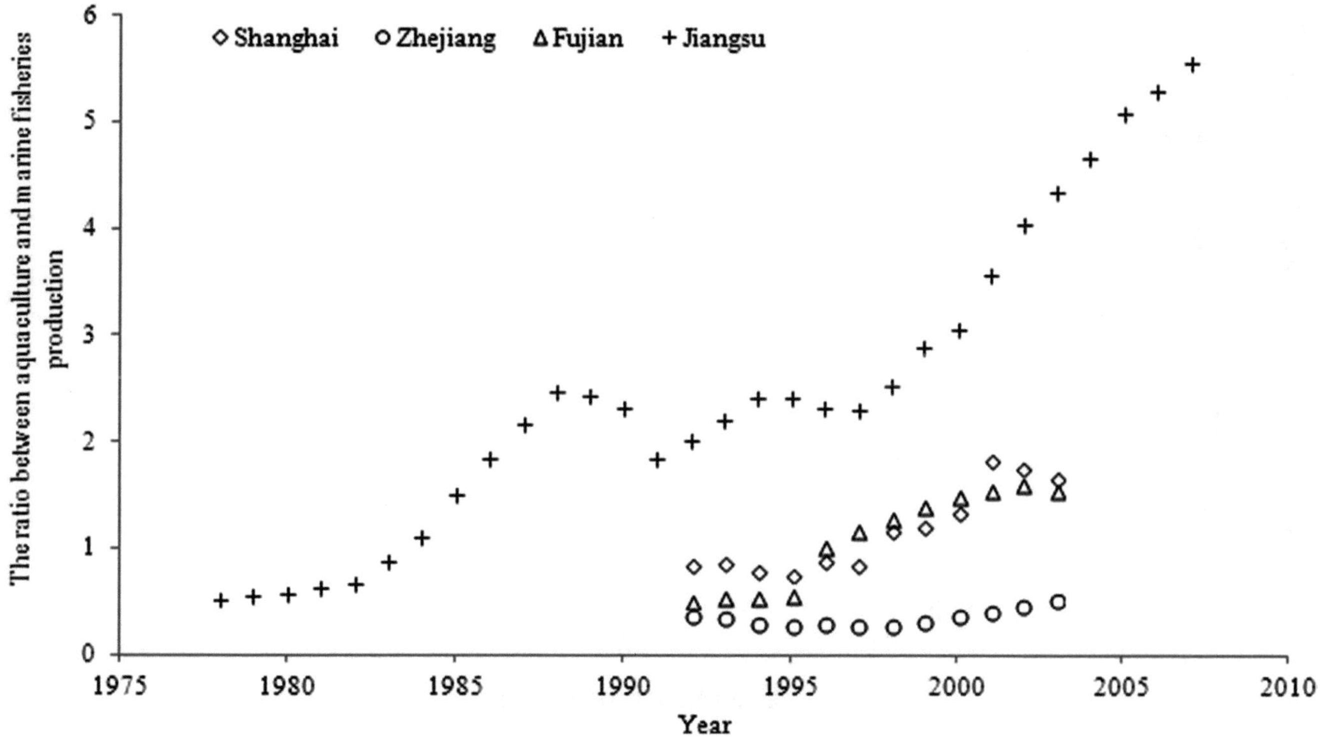

Fig. 9.9 The ratio of aquaculture to marine fisheries production, 1957–2012 (BSF 1957–2012; BSJ 1957–2012; BSZ 1957–2012; BSS 1957–2012)

The ratio of primary-industry production to GDP decreased gradually in each region alongside urbanization. Shanghai, the most developed city in all subregions, had the lowest ratio of primary-industry production to GDP (BSF 1950–2012; BSJ 1957–2012; BSZ 1952–2012; BSS 1952–2012). The ratio of secondary-industry production to GDP remained stable in each subregion, aside from a significantly decreasing trend in Shanghai (BSF 1950–2012; BSJ 1957–2012; BSZ 1952–2012; BSS 1952–2012). Urbanization causes heavily polluting industries to move outside city limits or into undeveloped provinces. However, the ratio of tertiary-industry production to GDP has increased significantly since 1990, especially in Shanghai, which had the highest value of all regions (BSF 1950–2012; BSJ 1957–2012; BSZ 1952–2012; BSS 1952–2012).

Tourism is one of the fastest growing tertiary industries in China. The value of international tourism in 2010 was US $45.8 billion, or 175 times greater than in 1978 (BSC 1978–2010a). As a large international city, the income from tourism in Shanghai is higher than in other regions, with a net value of up to US$6.341 billion in 2010 (BSS 2010). The value of international tourism in Jiangsu, Zhejiang, and Fujian was US$4.78, US$3.93, and US$2.98 billion dollars in 2010, respectively (BSF 2010; BSJ 2010; BSZ 2010). Meanwhile, with the rapid increase in personal income in China, the domestic tourism industry has also expanded rapidly. The value of domestic tourism in 2010 was US$184 billion, or 14.7 times that of 1993 (BSC 2010a). The rapid growth in tourism has led to the development of transportation, shopping, accommodation, and entertainment. After 1990, in Shanghai, domestic tourism income increased gradually alongside urbanization (Fig. 9.10a), and domestic wastewater discharge also increased sharply with increased domestic tourism income (Fig. 9.10b). This illustrates that the development of tourism has also caused stress on the local environment (Box 9.1).

9.3.2 Impact of the Economy on the Environment

9.3.2.1 Changing Land Use

During the period 1950–2012, there was more arable land in Jiangsu than in Shanghai, Fujian, and Zhejiang. As the most developed city in the study area, Shanghai had the lowest area of arable land, and this has continued to decrease over the past six decades. In the other three provinces, however, agriculture was still an important source of income for local government (BSF 1950–2012; BSJ 1957–2012; BSZ 1952–2012; BSS 1952–2012).

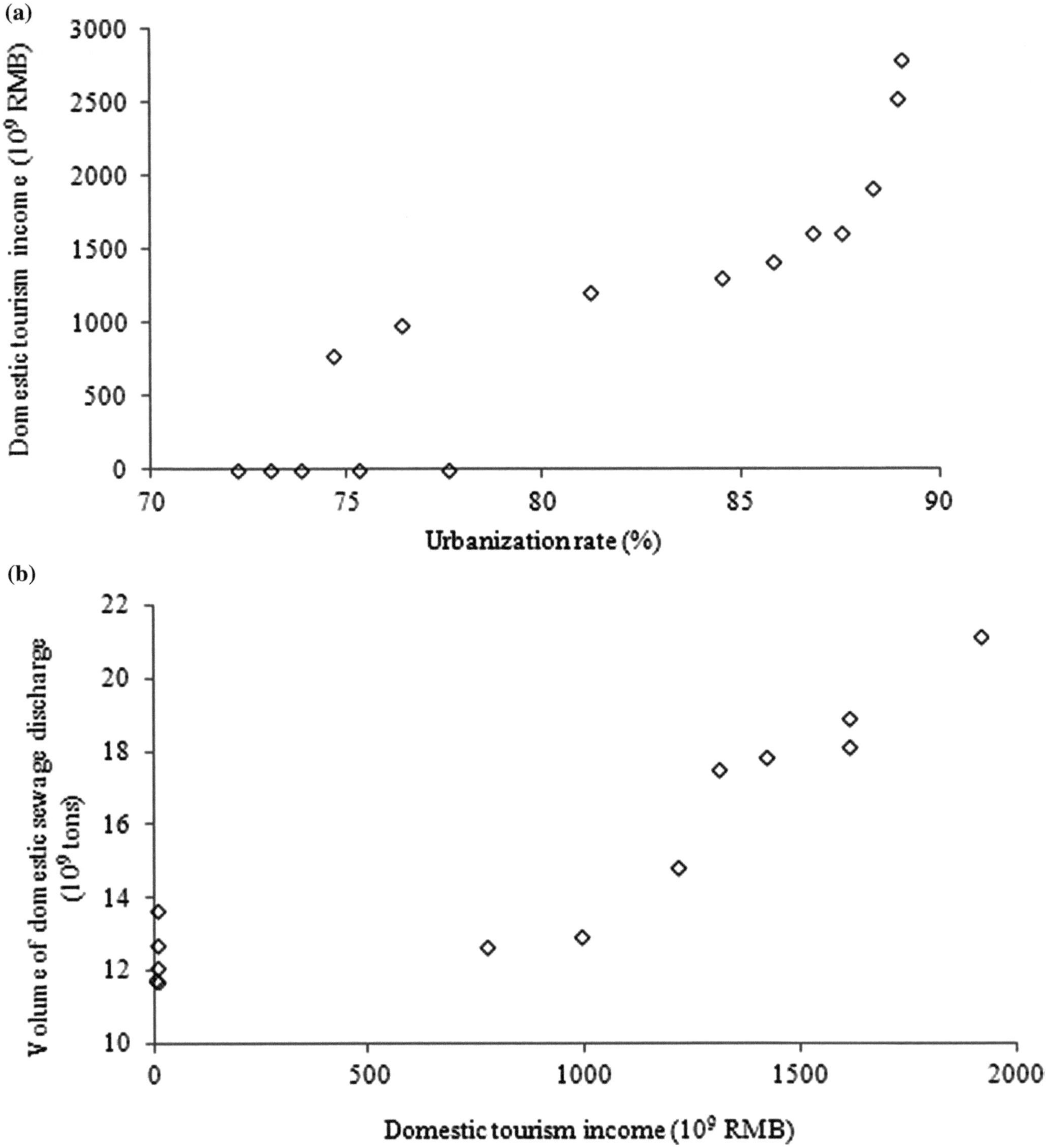

Fig. 9.10 **a** The relationship between domestic tourism income and urbanization rates, and **b** the relationship between domestic sewage discharge and domestic tourism income, 1990–2011 (BSS 1990–2011)

9.3.2.2 Fertilizer Application

Fertilizer consumption in cultivated areas in Fujian, Jiangsu, and Zhejiang has increased gradually since 1949 and has caused significant agricultural pollution. In Shanghai, fertilizer consumption increased after 1979 and fell after 2001. Pesticide consumption in cultivated areas was stable from 1990 onward in all subregions (BSF 1950–2012; BSJ 1957–2012; BSZ 1952–2012; BSS 1952–2012). The Shanghai government has controlled fertilizer use for many years and has obtained effective results. In

2009, the volume of fertilizer use was only 23 % of that in 1980. However, the consumption of chemical fertilizer has increased sharply in Fujian, Jiangsu, and Zhejiang over the last 50 years and especially after 1979 (BSF 1950–2012; BSJ 1957–2012; BSZ 1952–2012; BSS 1952–2012).

Since 2003, the ratio of the area of grain crops (including wheat, rice, corn, oats, rye, and barley) to economic crops (including vegetables, fruits, and flowers) in each subregion has decreased gradually in Fujian and Jiangsu and increased gradually in Shanghai (BSF 1950–2012; BSJ 1957–2012; BSZ 1952–2012; BSS 1952–2012). In 2012, the main crop types in Zhejiang, Shanghai, and Fujian were vegetables, but grain was still the dominant crop type in Jiangsu (BSF 1950–2012; BSJ 1957–2012; BSZ 1952–2012; BSS 1952–2012).

9.3.2.3 Marine Aquaculture

Figure 9.11 shows the increase in the mariculture area in Jiangsu, Zhejiang, and Fujian in the last 50 years. The main mariculture products in Zhejiang, Jiangsu, and Fujian are shellfish. Shrimps and crabs are the main mariculture products in Shanghai (BSF 2010; BSJ 2010; BSZ 2010; BSS 2010), but with more stringent environmental regulations in this area, the mariculture area in Shanghai has been reduced significantly in recent years (Fig. 9.11).

9.4 Environmental Quality in the East China Sea

The *East China Sea Marine Environment Quality Bulletin for 2012* (ECSOA 2012a) reports good overall seawater quality in the East China Sea. Approximately 83.3 % of the waters met Grade I Chinese water quality standards. However, inorganic nitrogen and active phosphate are major pollutants in the East China Sea, and as shown in Fig. 9.12, the area of Grade IV and inferior Grade IV water has increased over the past 10 years (ECSOA 2012a). Box 9.2 explains the seawater grading system.

Box 9.2 China's national standard—seawater quality standards (GB 3097–1997)

Seawater quality is divided into four classes:

Grade I: Seawater is suitable for marine fisheries, marine sanctuaries, and protected areas for rare and endangered marine organisms.

Grade II: Seawater is suitable for aquaculture, bathing beaches, entertainment districts, and industrial water associated with human food processing.

Grade III: Seawater is suitable for industrial water and coastal scenic areas.

Grade IV: Seawater is suitable for ocean ports and engineered structures.

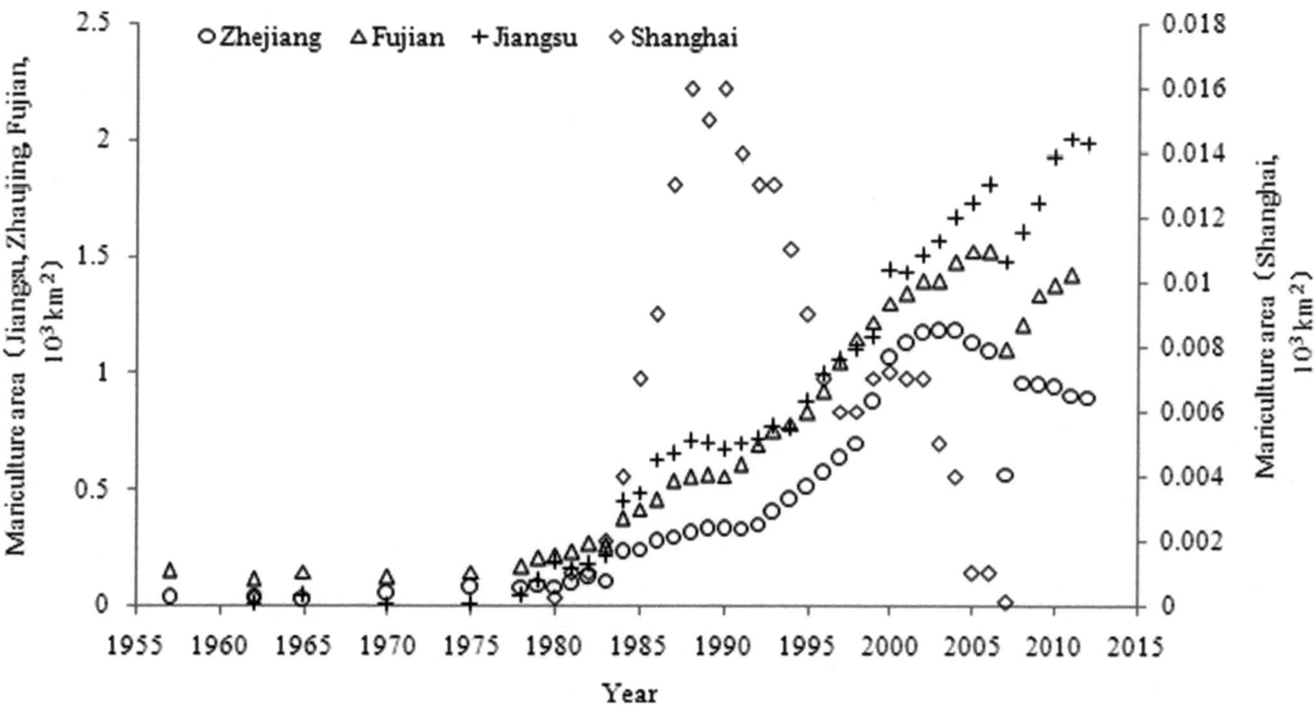

Fig. 9.11 Mariculture areas in each subregion, 1957–2012 (BSF 1950–2012; BSJ 1957–2012; BSZ 1952–2012; BSS 1952–2012)

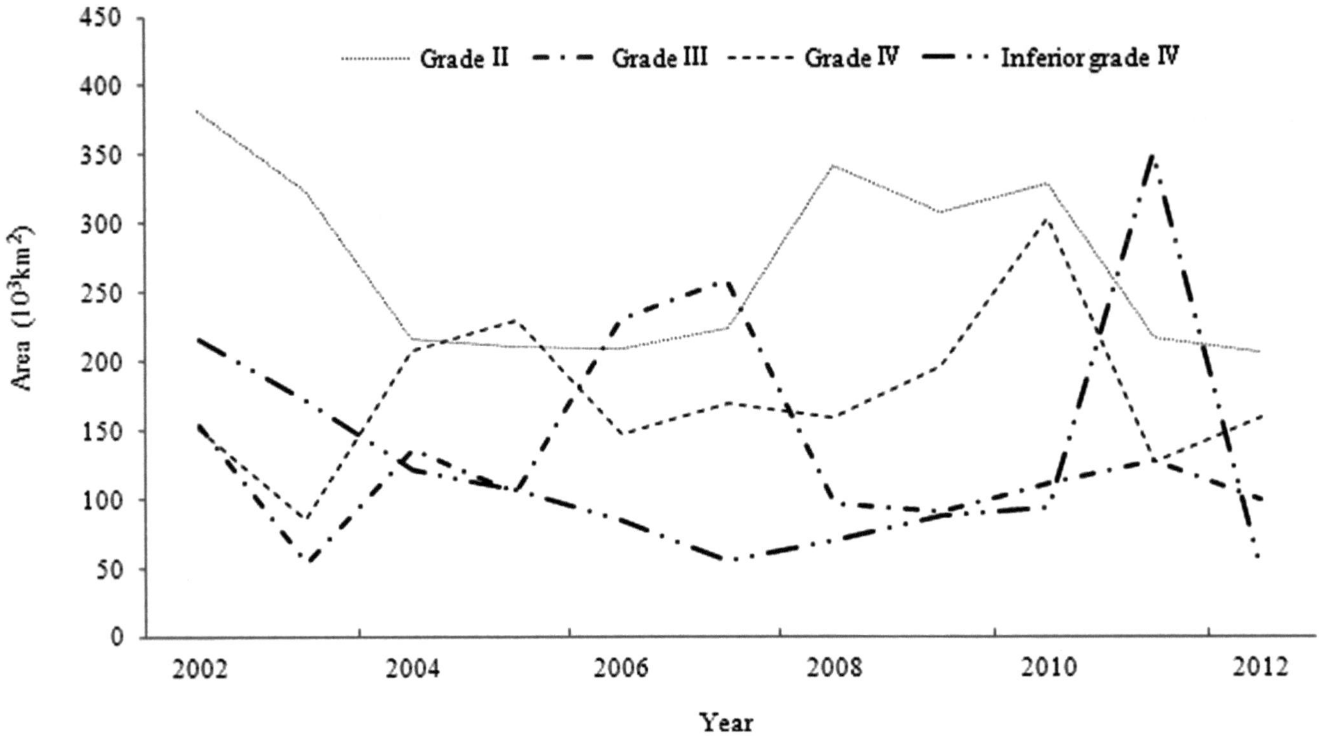

Fig. 9.12 Temporal variations in water quality in the East China Sea from 2001 to 2012 (ECSOA 2001–2012a)

The southern Yancheng coast (in Jiangsu), Changjiang Estuary, Hangzhou Bay, Xiangshan Harbor, Oujiang Estuary, and Minjiang Estuary are all areas with poor water quality (ECSOA 2012a). Most pollutants in these areas come from coastal river inputs and sewage outfalls. In 2012, the 37 major rivers flowing into the East China Sea carried tons of pollutants, including COD_{Cr} (11.65 million tons), petroleum (63.8 thousand tons), nutrients (0.45 million tons), heavy metals (Cu, Pb, Zn, Cd, Hg; 40.2 thousand tons), and As (2.85 thousand tons) (ECSOA 2012a). The rivers with the highest pollutant load were Changjiang, Qiantangjiang, Minjiang, and Oujiang. Moreover, the components that most exceeded the maximum allowable concentrations in sewage outfall discharged into coastal East China Sea were fecal coliform, total phosphorus, COD_{Cr}, and suspended matter, which exceeded the recommended levels by 60, 35, 23, and 21 %, respectively (ECSOA 2012a).

Algae blooms, known as red and green tides, have adversely affected the marine environment and economy of the East China Sea coast. In 2012, 42 red tides, including 22 toxic red tides, were recorded in the East China Sea, which affected areas of up to 2515 km^2 and caused the large-scale death of local cultured marine organisms (ECSOA 2012a). About 60 % of red tide outbreaks in coastal China have been located in the coastal East China Sea. On average, 41 red tides have been recorded in this region every year from 2008 to 2012, and these tides have affected an average annual area of 6591 km^2 of ocean (ECSOA 2012a).

Oil and gas exploration in the East China Sea has had no adverse impact on the environment, but sea-closure engineering, power plants, port and waterway construction, and bridge construction projects have had significant adverse impacts on marine organisms (ECSOA 2012a).

9.4.1 Microbiological Pollution

A study by Lu et al. (2007) found bacterial contamination of oysters being sold in Shanghai, with total bacterial and *Escherichia coli* counts above national standards, especially in summer and autumn. Bacterial detection tests on shellfish from coastal areas in Zhejiang showed that most samples were contaminated (Shen et al. 2005). The most prevalent species identified in these samples was *E. coli*, followed by *Vibrio parahaemolyticus*, *Salmonella*, and *Listeria*. *E. coli* mainly originate from sewage and industrial wastewater, which is discharged directly into marine environments and, in turn, threatens human health.

Table 9.3 The occurrence of red tides between 2003 and 2012 (ECSOA 2012a)

Year	Number of outbreaks	Affected area (km^2)	Main red tide organisms
2003	86	12,990	*Prorocentrum donghaiense, Skeletonema costatum*
2004	53	17,880	*Prorocentrum donghaiense, Noctiluca scintillans*
2005	49	15,237	*Prorocentrum donghaiense, Karenia mikimotoi*
2006	57	14,500	*Prorocentrum donghaiense, Karenia mikimotoi*
2007	65	9955	*Skeletonema costatum, Karenia Mikimotoi*
2008	50	14,592	*Prorocentrum donghaiense, Skeletonema costatum*
2009	48	7244	*Karenia Mikimotoi, Skeletonema costatum*
2010	42	6994	*Prorocentrum donghaiense, Skeletonema costatum*
2011	24	1612	*Prorocentrum donghaiense, Skeletonema costatum*
2012	42	2515	*Karenia Mikimotoi, Prorocentrum donghaiense*

9.4.2 Eutrophication

Eutrophication in the Changjiang Estuary and adjacent marine areas has been increasing since the 1960s as a result of increased inputs of anthropogenically derived nutrients. The use of chemical fertilizers, the expansion of aquaculture, and the discharge of wastewater have all contributed to this problem. With rapid urbanization and economic development, nutrient inputs into the East China Sea will continue to increase.

The main pollutants in the East China Sea are inorganic nitrogen and phosphorus, which are both found at concentrations much higher than national standards. The main source of nitrogen is runoff from agriculture, while most phosphorus comes from domestic sewage and industrial wastewater discharge. The introduction of these nutrients has resulted in the increase in phytoplankton biomass and an increased frequency of harmful algal blooms in the East China Sea. The high frequency of red tide outbreaks, higher chlorophyll-a concentrations, and hypoxia of bottom water are the main symptoms of eutrophication in these areas.

Table 9.3 shows the number and areas of red tide outbreaks for the period 2003–2012, along with the dominant organisms found in the algae blooms. The outbreak frequency and the affected surface areas have decreased gradually over the last 10 years. The dominant organisms in the red tides were *Prorocentrum donghaiense*, *Skeletonema costatum*, and *Karenia Mikimotoi* (ECSOA 2012a).

Figure 9.13 shows the red tide outbreak frequency and affected surface area in each subregion between 1997 and 2012. Zhejiang and Fujian, which have longer coastlines than the other provinces, had higher red tide outbreak frequencies, with 60 and 31 %, respectively, of the total number of outbreaks in the study region. Zhejiang was more frequently affected by red tides than other areas, but the highest economic loss associated with red tides occurred in Fujian.

The outbreak of red tides has a close relationship with seawater temperature. As shown in Table 9.4, 70 % of red tide outbreaks occurred from May to July in 2012 in the East China Sea (SOA 2012b). Most red tide outbreaks occurred close to the coast and caused significant economic loss in the mariculture industry. In Fujian, the area affected by red tides was only 0.2 % of the total East China Sea area in 2012, but the direct economic loss from red tides amounted to US $31.6 million (FPDOF 2012).

Table 9.5 lists the pollutant load per basin area carried by large rivers that drain into the East China Sea. The riverine input of pollutants into the East China Sea is derived mainly from the Changjiang, Qiantangjiang, and Minjiang. Among these rivers, the Changjiang carries the largest pollutant load and contributes the highest nutrient input, but the pollutant load per basin area is the lowest. The Qiantangjiang, which has the smallest drainage area, delivers the highest concentration of COD_{Cr}, ammonia, nitrate nitrogen, and nitrite nitrogen. This nutrient input is the main reason for the high frequency of green and red tides in Zhejiang. The Changjiang basin is China's major agricultural region, and large quantities of fertilizer are used in Jiangsu, Zhejiang. Agricultural nonpoint source pollution will cause ongoing eutrophication in the East China Sea if no effective measures are taken by government to reduce nutrient input to coastal oceans.

9.4.3 Chemical Pollution

Of the four coastal areas around China (Bohai, Yellow Sea, East China Sea, and South China Sea), the East China Sea has the poorest water quality (MEP 2012). Nearly 90 % of the water in the Changjiang Estuary is Grade IV or below in terms of the Chinese national standards. The main pollutants in this region are inorganic nitrogen and active phosphate, followed by heavy metals and oil. Figures 9.14 and 9.15

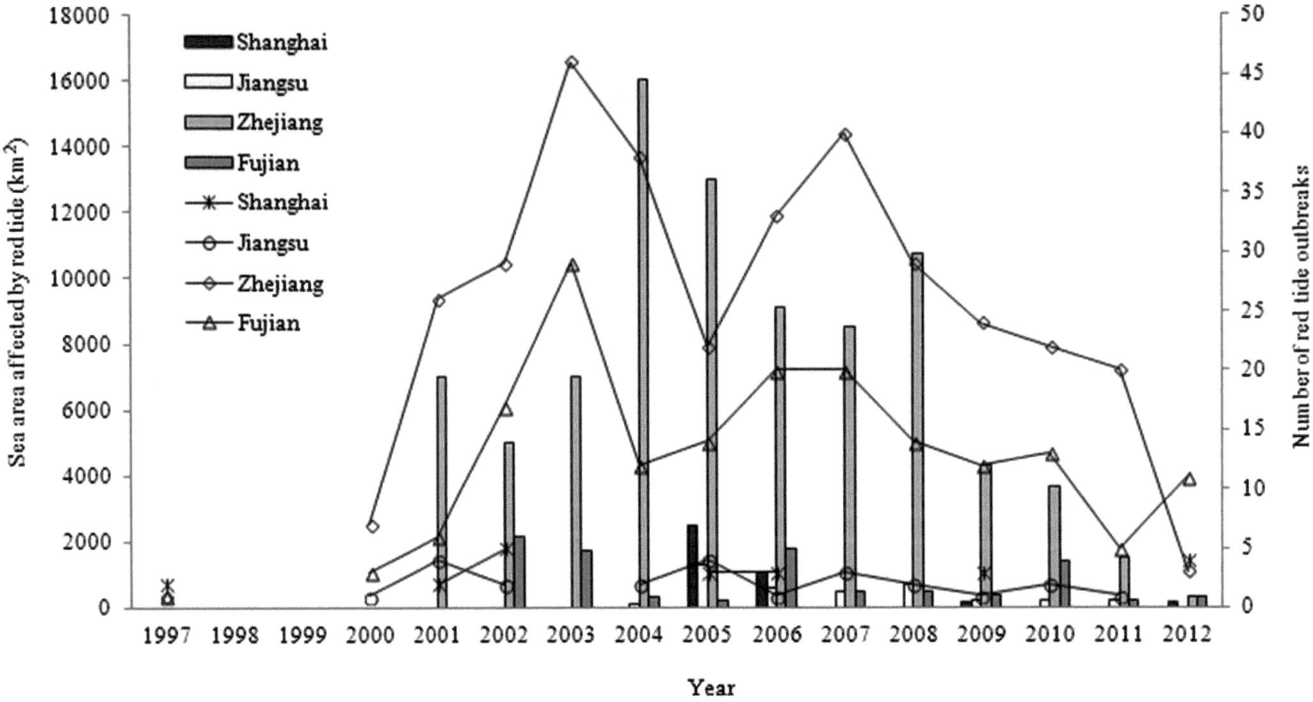

Fig. 9.13 Number of red tide outbreaks and affected areas in each subregion, 1997–2012 (ECSOA 1997–2012a)

Table 9.4 The number of red tide outbreaks per month in 2012 (SOA 2012b)

Month	1	2	3	4	5	6	7	8	9	10	11	12
Number of outbreaks	0	0	2	5	31	11	9	8	2	5	0	0

Table 9.5 Riverine pollutant load per basin area delivered to the coast (tons/10^3 km^2) (SOA 2012a; MWR 2011)

	Changjiang	Qiantangjiang	Minjiang
COD_{Cr}	430	1524	1335
Ammonia	8	40	24
Nitrate	83	88	43
Nitrite	0.5	5	2.8
Total phosphorus (TP)	8	22	10
Petroleum	3.1	3.1	1.6
Heavy metals	2	0.8	3
As	0.1	0.1	0.1

show the fluxes of pollutants from coastal rivers into the East China Sea for the period 2007–2012. The data show that although pollutant fluxes were very high, they gradually decreased over this period.

According to Fang (2008), the main source of nutrients and COD in the East China Sea is coastal rivers, sewage outfall, air pollution, mariculture, and agricultural nonpoint sources. The primary sources of petroleum in this coastal area are rivers, followed by sewage outfall and marine pollution.

In 2012, monitoring of pH, redox potential, organic carbon, sulfide, petroleum, and the contents of Hg, Cd, As, Zn, Cu, and polychlorinated biphenyls (PCBs) in sediment showed the sediment quality outside coastal East China Sea was also Grade I; however, in coastal areas around the East China Sea (e.g., Hangzhou Bay, Xiangshan Harbor, and the Changjiang, Oujiang, and Minjiang estuaries), the index levels occasionally exceeded the national allowable level (ECSOA 2012a).

9.4.4 Suspended Solids

The main sources of suspended solids in the Changjiang, Qiantangjiang, Minjiang, and the East China Sea coastal waters are soil erosion, deforestation, and intensive cultivation. Table 9.6 shows annual runoff and sediment load in major rivers that flow to the East China Sea, i.e., the Changjiang, Minjiang, and Qiantangjiang. The Changjiang has the largest sediment load of all three rivers. With the rapid development of coastal cities, and increased discharge of sewage and industrial wastewater, high concentrations of suspended solids are flushed into the East China Sea, and this acts to increase the overall concentration of suspended solids in coastal waters.

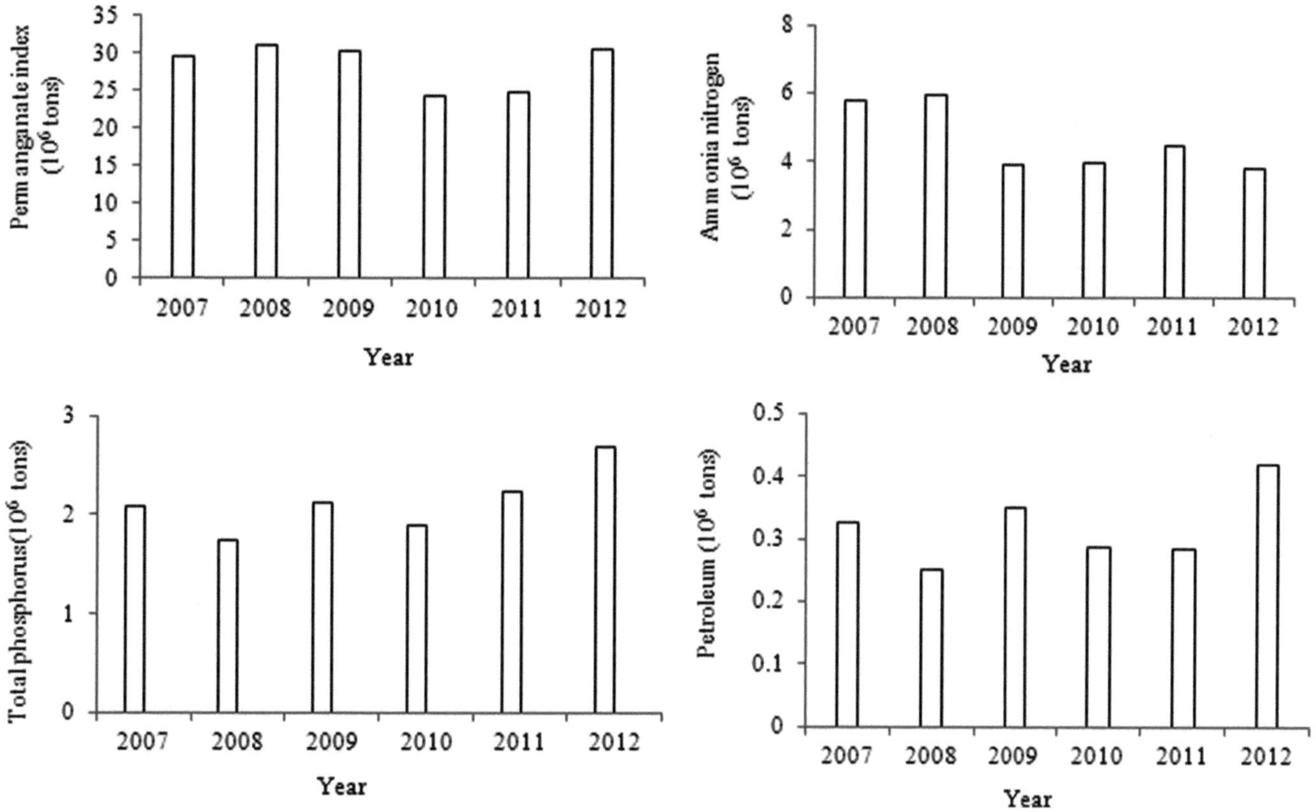

Fig. 9.14 The volume of permanganate index, ammonia nitrogen, total phosphorus, and petroleum carried by coastal rivers into the East China Sea, 2007–2012 (MEP 2007–2012)

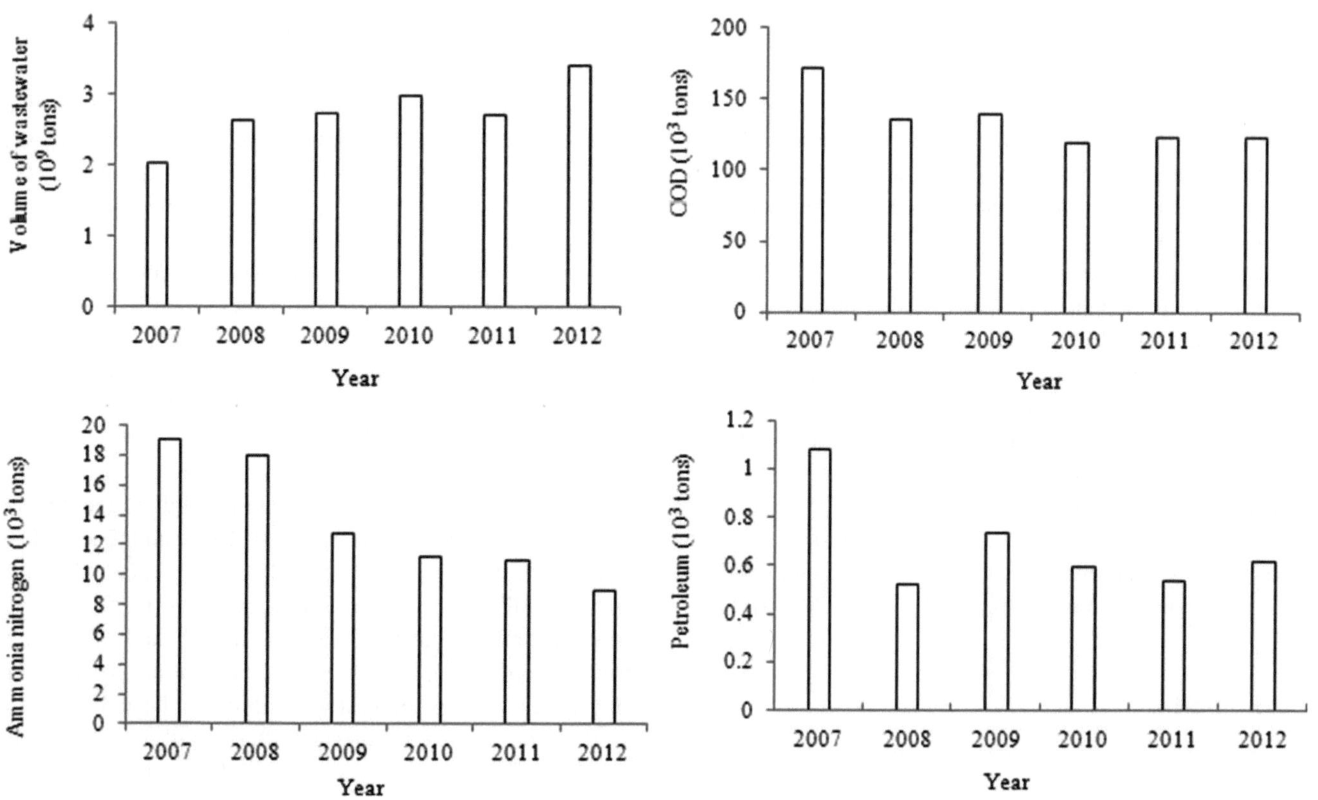

Fig. 9.15 The volume of wastewater, COD, ammonia nitrogen, and petroleum directly discharged into the East China Sea, 2007–2012 (MEP 2007–2012)

Table 9.6 Annual runoff and sediment load for the Changjiang, Minjiang, and Qiantangjiang rivers (MWR 2005–2011)

River	Representative hydrographic stations	Basin area (10^6 km^2)	Year	Annual runoff (10^9 m^3/year)	Annual sediment load (10^6 tons/year)
Changjiang	Datong	17	2005	9015	2160
			2006	6886	848
			2007	7708	1380
			2008	8291	1300
			2009	7819	1110
			2010	10,220	1850
			2011	6671	718
Minjiang	Zhuqi and Yongtai	0.59	2005	683.7	73.7
			2006	715.1	71.7
			2007	432.4	13.8
			2008	442	4.5
			2009	390.9	4.48
			2010	886.8	118
			2011	320.6	4.97
Qiantangjiang	Lanxi, Zhuji, and Huashan	0.23	2005	201.2	17.1
			2006	177.7	14.4
			2007	140	16.9
			2008	153.9	17.8
			2009	160.9	14.3
			2010	316	50.7
			2011	148.3	40.7

9.4.5 Thermal Pollution

In 2012, monitoring of the thermal discharge from the Wushashan, Guohuaninghai, Huanengyuhuan, Zhenengyueqing, and Kemen power plants was showed that the adjacent seawater can be affected by coastal thermal power plants (ECSOA 2012a). The impacts of thermal discharge on the marine environment include the reduced growth of certain temperature-dependent organisms, the miniaturization of zooplankton, low benthos biomass, and decreased biodiversity. The monitoring results also showed that the dominant species near the Zhenengyueqing power plant (in Zhejiang) changed because of changing water temperatures (ECSOA 2011a). The concentration of petroleum, Pb, and Zn in water around the power plants was above Grade II of the Chinese marine water quality standards. The quality of sediment was relatively good, and only Cu levels were above Grade I of the Chinese marine sediment standards (ECSOA 2011a).

9.4.6 Spills

Oil spills, though rare, cause serious pollution to the marine environment. On November 27, 2010, the *Qianhe 12* waste oil tanker and the *Xiamen Port Drag 3* tugboat collided near Xiamen Dongdu Haitian dock (in Fujian), causing the release of five tons of oil into the ocean over an area of 40 km^2 (ECSOA 2010a). The coastal waters around west Xiamen, the channel between Xiamen and Gulangyu, Gulangyu itself, and the beach at Yanwu Bridge were all polluted by this spill (ECSOA 2010a).

On November 28, 2010, a large-scale seawater monitoring program showed that the oil content in surface seawater near west Xiamen was 72 times higher than the Grade IV Chinese marine water quality standard. Residual oil was detected until early December on some sandy beaches in the affected area (ECSOA 2010a).

There has been an increasing number of oil spills in recent decades, with 733 ship-based oil pollution incidents in the oceans around China between 1998 and 2008 (Data source: http://news.sohu.com/20110707/n312681416.shtml). These spills have had adverse impacts on marine ecosystems, can affect marine organism growth and community structure, and disrupt the food chain (Su et al. 2013). Other effects include the suppression of photosynthesis by the film of oil on the surface of the water after a spill, hypoxia of fish and shellfish due to blocked gills, and poisoning of marine organisms by toxic substances.

9.4.7 Nuclear Power Stations

On March 12, 2011, a catastrophic escape of nuclear materials occurred at the Fukushima nuclear power station, on the coast of northeastern Japan, following a high-magnitude earthquake and subsequent tsunami. High concentrations of nuclear elements were found in the coastal waters of the Sea of Japan. To gauge the effect of the accident on the East China Sea, seawater, atmospheric aerosols, and airborne γ radiation were monitored (ECSOA 2011a). In Zhejiang, high concentrations of I-131, Cs-137, and Se-134 were recorded in April 2011, but the amount of radiation from these isotopes was less than one hundred thousandth of background levels (ECSOA 2011a). The monitoring of Cs-137, total U, Sr-90, and total β radiation in seawater at Yangkou Port (in Jiangsu), Yangshan, Shengsi, Wenzhou (in Zhejiang), and Xiamen (in Fujian) showed that the concentrations of these elements were all below the background levels recorded in the East China Sea in 2011 (ECSOA 2011a).

In 2012, the levels of radioactive material in the marine environment around the Qinshan (in Zhejiang) and Tianwan (in Jiangsu) nuclear power stations were all lower than background values (ECSOA 2012a). The monitoring of the nuclear leak from Fukushima nuclear station showed no abnormality in marine environmental quality after the incident (ECSOA 2012a). Wu et al. (2013) measured Cs-137 in seawater samples collected from April to June 2011 from the South China Sea, East China Sea, and Yellow Sea. They found that that negligible Cs-137 was released into the East China Sea from the Fukushima accident.

9.4.8 Engineering Activities

Table 9.7 shows the influence of engineering activities on marine environmental quality in 2011. Engineering activities in marine areas have adverse effects on water and sediment quality, can change currents and the geomorphology in adjacent ocean areas, and can affect biodiversity near construction sites. Pollutants from engineering activities are mainly inorganic nitrogen, active phosphate, and petroleum (ECSOA 2011a). Substandard construction processes and illegal wastewater discharge can cause a significant amount of pollution in adjacent sea areas.

Table 9.7 Marine pollutants identified near engineering activities in 2011 (ECSOA 2011a)

Name of engineering activity	Pollutants
Waterway engineering, Lianyungang (in Jiangsu)	Petroleum
Yangkou Port (in Jiangsu)	Active phosphate
East China Sea Bridge wind power project	Active phosphate, inorganic nitrogen

9.4.9 Ocean Dumping

The East China Sea is the largest marine dumping area in China. The total volume of waste dumped into the East China Sea accounts for 45 % of the Chinese total, and ocean dumping has increased by 2 % annually (Lv and Yang 2011). The largest volume of waste was dumped in waters off Shanghai, followed by Zhejiang, Fujian, and Jiangsu. Table 9.8 shows the dumping frequency and total volume of dumped waste in the East China Sea from 1999 to 2012 and shows that the volume of dumped waste has increased over time. The main source of waste is dredged material from waterways. Ecosystems have been found to be intact after dumping, although the number, density, and biomass of benthic species in some areas have decreased. Illegal dumping is another potential pollution source, and some coastal engineering projects have dumped large amounts of waste into the East China Sea. However, the frequency of illegal dumping has decreased and was 4 times lower in 2012 compared with 2010 (ECSOA 2010–2012a).

9.4.10 Marine Litter

Marine litter from human activities has become an increasingly serious problem in coastal environments, especially

Table 9.8 The number of ocean dumping areas and total volume of waste dumped in the East China Sea over the period 1999–2012 (ECSOA 1999–2012b)

Year	Number of ocean dumping areas	Total volume of dumped waste (10^9 m^3)
1999	25	0.24
2000	22	0.47
2001	24	0.36
2002	22	0.36
2003	24	0.45
2004	31	0.70
2005	31	0.69
2006	38	0.90
2007	36	0.94
2008	28	0.66
2009	26	0.55
2010	32	1.06
2011	30	1.10
2012	27	1.19

after the rapid development of tourism. After entering the sea, 70 % of marine litter sinks to the seafloor, 15 % floats on the surface, and 15 % remains on beaches (Xu 2008). Monitoring of litter in the East China Sea in 2012 showed that plastic and Styrofoam waste was the main component, accounting for 59, 65, and 85 % of beach garbage, floating garbage, and seabed garbage, respectively (ECSOA 2012a). Marine litter not only damages the aesthetics of the ocean, but also pollutes the marine environment.

9.5 Summary

China has experienced rapid urbanization over the past few decades, and Chinese people have seen profound changes to their traditional lifestyles. There has been an influx of rural people into cities, and urbanized areas have expanded rapidly. With increased income and improved living standards, tourism, traffic, and related industries have developed rapidly. The provinces along the coast of the East China Sea are the most developed areas in China and have had an adverse impact on the quality of the marine environment.

Waste discharge
Although industrial wastewater and solid waste output have gradually decreased in all subregions, per capita domestic sewage discharge has increased significantly, especially in cities such as Shanghai. A rapid increase in the numbers of private cars has resulted in air pollution, while wastewater, air pollution, illegal reclamation, and dumping have all affected water quality in the East China Sea.

Changing land use
Urbanization has also resulted in changes in land use. Farmland has been converted into residential or industrial areas. To increase crop yield, fertilizer consumption has increased significantly in all subregions, and agricultural runoff has become a major pollutant and source of eutrophication in the East China Sea.

Unexpected incidents
Spills and nuclear leaks have had adverse effect on marine ecosystems in the East China Sea. Spills have destroyed marine organisms, and though no anomalous radioactivity was found in seawater of the East China Sea, the nuclear leak from the Fukushima nuclear power plant has caused short-term high concentrations of radioactive substances.

Marine environmental quality
The quality of the East China Sea marine environment meets national standards overall, but water quality has followed a deteriorating trend over the last 10 years. Rapid economic growth and urbanization have bought large populations to coastal cities and increased the delivery of human pollutants to the coastal environment. River inputs represent the largest source of water pollution in the East China Sea, but sewage outfalls, mariculture, engineering activities, agriculture, and nonpoint source pollutants are other key contributors. The Changjiang Estuary is one of the most densely populated areas in the study region and has experienced a rapid deterioration in environmental quality. Engineering activities in this region have also polluted adjacent marine areas, and additional measures must be implemented to manage human impacts in the coastal East China Sea.

References

BSC (1949–2012a) China statistical yearbook. National Bureau of Statistics of China. China Statistics Press, Beijing (in Chinese)

BSC (1980–2010b) International statistical yearbook. National Bureau of Statistics of China. China Statistics Press, Beijing (in Chinese)

BSF (1950–2012) Fujian statistical yearbook. Fujian Provincial Bureau of Statistics. China Statistics Press, Beijing (in Chinese)

BSJ (1957–2012) Jiangsu statistical yearbook. Statistics Information Network of Jiangsu. China Statistics Press, Beijing (in Chinese)

BSS (1952–2012) Shanghai statistical yearbook. Bureau of statistic of Shanghai. China Statistics Press, Beijing (in Chinese)

BSZ (1952–2012) Zhejiang statistical yearbook. Zhejiang Provincial Bureau of Statistics. China Statistics Press, Beijing (in Chinese)

CCICED (2011) Ecosystem issues and policy options addressing the sustainable development of China's ocean and coasts. China Environmental Science Press, Beijing, pp 145–188 (in Chinese)

CEPDRC (2001–2010) Taiwan statistical data book. Council for Economic Planning and Development Republic of China. Council for Economic Planning and Development, Taipei (in Chinese)

Chen H, Xu G (2004) The impact of wetland reclamation on environment, Jiangsu. Water Resour Plan Des 18–21 (in Chinese)

ECSOA (1997–2012a) East China Sea marine environment quality bulletin. East China Sea Branch of the State Oceanic Administration. East China Sea Branch of the State Oceanic Administration, Shanghai (in Chinese)

ECSOA (1999–2012b) Bulletin of management of ocean dumping in East China Sea. East China Sea Branch of the State Oceanic Administration. State Oceanic Administration People's Republic of China, Beijing (in Chinese)

Fang Q (2008) The study on sources of main chemical pollutants and fluxes flowing into the East China Sea in recent 30 years. Ocean University of China, Qingdao (in Chinese)

FPDOF (2012) Bulletin of Fujian marine environmental status. Fujian Provincial Department of Ocean and Fisheries, Fujian (in Chinese)

Li J, Yang X, Tong Y (2007) Progress on environmental effects of tidal flat reclamation. Prog Geog 26:43–51 (in Chinese)

Liu Y, Jin YP, Zhou AG (2006) Index system of ecological impact assessment and strategies for sustainable development of tidal flat reclamation on Zhejiang Province. J Mar Sci 26:78–82 (in Chinese)

Liu J, Chen RY, Shen XL, Liu CC, Su YC (2007) Investigation of bacterial contamination status in retail oysters in Shanghai. Chin J Food Hygiene 19:18–20 (in Chinese)

Lv JH, Yang Y (2011) Research on the problems and countermeasures of current management situation of the marine-dumping at East China Sea. Pacific J 19(8):96–104 (in Chinese)

MEP (2007–2012) Bulletin of China environmental status, Ministry of Environmental Protection of the People's Republic of China. State Oceanic Administration, Beijing (in Chinese)
MWR (2005–2011) Bulletin of Chinese river sediment. The Ministry of Water Resources of the People's Republic of China. China Water Power Press, Beijing (in Chinese)
Shen XS, Gu RR, Yu HJ (2005) Microbial survey and its assessment of bacterial contamination of shellfish products in Zhejiang Province. Mar Fisheries 27:64–67 (in Chinese)
SOA (2011–2012a) Bulletin of China marine environment quality, State Oceanic Administration People's Republic of China. State Oceanic Administration People's Republic of China, Beijing (in Chinese)
SOA (2012b) Bulletin of Chinese marine disasters. State Oceanic Administration People's Republic of China. State Oceanic Administration People's Republic of China, Beijing (in Chinese)
Sowana A, Shrestha RP, Parkpian P, Pongquan S (2011) Influence of coastal land use on soil heavy-metal contamination in Pattani Bay. Thailand J Coast Res 27(2):252–262
Su JL, Wang Y, Qin YS et al. (2013) Scientific problems and countermeasures of ecological security in Chinese marine and coastal Engineering. Chin Acad Sci (in Chinese)
Wu JW, Zhou KB, Dai MH (2013) Impacts of the Fukushima nuclear accident on the China Seas: evaluation based on anthropogenic radionuclide Cs-137. Chin Sci Bull 58:552–558
Xia X, Chen X, Liu R, Liu H (2011) Heavy metals in urban soils with various types of land use in Beijing. China. J Hazard Mater 186:2043–2050
Xu LZ (2008) The monitoring and assessment of marine litter. Environ Prot 405:67–68 (in Chinese)

MIX
Papier aus verantwortungsvollen Quellen
Paper from responsible sources
FSC® C105338

If you have any concerns about our products,
you can contact us on
ProductSafety@springernature.com

In case Publisher is established outside the EU,
the EU authorized representative is:
Springer Nature Customer Service Center GmbH
Europaplatz 3, 69115 Heidelberg, Germany

Printed by Libri Plureos GmbH
in Hamburg, Germany